THE FUNCTIONS OF MATHEMATICAL PHYSICS

HARRY HOCHSTADT

Department of Mathematics
Polytechnic Institute of New York

Dover Publications, Inc.
New York

Published in Canada by General Publishing Company, Ltd., 30 Lesmill Road, Don Mills, Toronto, Ontario.
Published in the United Kingdom by Constable and Company, Ltd., 10 Orange Street, London WC2H 7EG.

This Dover edition, first published in 1986, is an unabridged, corrected republication of the work originally published by Wiley–Interscience, New York, 1971, as volume XXIII in the series Pure and Applied Mathematics, edited by R. Courant, L. Bers and J. J. Stoker.

Manufactured in the United States of America
Dover Publications, Inc., 31 East 2nd Street, Mineola, N.Y. 11501

Library of Congress Cataloging-in-Publication Data

Hochstadt, Harry.
The functions of mathematical physics.

"Unabridged, corrected republication . . . originally published by Wiley Interscience, New York, 1971, as volume XXIII in the series Pure and applied mathematics"—CIP t.p. verso.
Bibliography: p.
Includes index.
1. Functions, Special. I. Title.
QA351.H68 1986 530.1′5 86-6383
ISBN 0-486-65214-9

Foreword

The functions of mathematical physics show in a striking manner how a simple mathematical idea, conceived for the solution of a very specific problem, can lead to a far-reaching theory. The oldest example is the trigonometric functions which describe the uniform motion on a circle. Trigonometry deals effectively with some specific problems of astronomy and navigation. But the functions $\sin x$ and $\cos x$ are also the basis for the theory of Fourier series and of the Fourier integral. And this theory, apart from its eminent mathematical interest, has applications to many parts of physics which emerged long after the original problem of describing the motion of the planets had been solved. But even all of these successes of the trigonometric functions do not indicate a surprising number-theoretical property of the function $w = \exp 2i\pi z$. The values of z and w will both be algebraic numbers if and only if z is rational. And the values of w for rational values of z provide us with the means of constructing exactly those finite algebraic extensions of the rational numbers which have an abelian Galois group.

The functions of mathematical physics, which have been studied since the end of the eighteenth century, have an elementary theory that is much more involved than trigonometry. But they, too, are part of, and frequently the motivation for, important general theories, filling the general framework with substance. And they have many aspects, turning up for example as complete sets of orthogonal polynomials and also in approximation and interpolation theory. Their applicability has kept up with the times, and they are as important for the quantum theoretical model of the atom as for the vibrating membrane. Finally, some of them also enter into parts of mathematics far removed from mathematical physics, such as the theory of discontinuous groups.

WILHELM MAGNUS,

New York University

Preface

The topics covered in this book were first studied by the outstanding mathematicians of the eighteenth and nineteenth centuries. Among the many who devoted themselves to these studies are Gauss, Euler, Fourier, Legendre, and Bessel. These men did not recognize the modern and somewhat artificial distinction between pure and applied mathematics. Much of their work was stimulated by physical problems that led to studies of differential equations. Frequently they developed generalizations to obtain results having no immediate or obvious applications. As a consequence mathematics was often ahead of its time in having necessary tools ready before physicists and engineers felt the need for them. This book reflects this historic interplay by presenting topics of obvious interest to applied scientists as well as topics that are, for the present at least, of purely mathematical interest.

In order to cover all the topics of the nine chapters of this book a great deal of selectivity had to be used. Naturally my personal bias played a strong role in making these choices. Entire books have been written on some of the chapters and no claim can be made of having provided an exhaustive treatment. It is hoped, however, that the selection of results and applications, and the methodology used will enable the reader to go on to the more specialized treatises and handbooks with ease, and to derive new results when necessary.

Chapters 1 and 2 are devoted to orthogonal polynomials. In addition to deriving the basic results they cover various applications to numerical integration and approximation problems. Section 10 in Chapter 1 discusses applications of orthogonal polynomials to the construction of conformal mappings. Chapter 3 covers the principal properties of the gamma function. Most of this material is standard, but necessary for later chapters. The final section of this chapter is devoted to the solution of certain algebraic equations. This topic deserves to be better known, and hopefully readers will follow this up on their own.

The results of Chapter 4, on the hypergeometric equation, are needed for the later chapters. Sections 4.6 through 4.9, however, have no obvious applications in mathematical physics and are not needed for later chapters. They were included only because they were mathematically pleasing. They show the relationship between the theory of conformal mapping of curvilinear polygons and the theory of hypergeometric functions. Section 4.10, on nonlinear transformations, is useful in studying the analytic continuation of Legendre functions in the complex domain.

Chapter 5 is concerned with the Legendre functions and some of their applications to solutions of Laplace's equation in spherical coordinates. Chapter 6 continues this theme in an n-dimensional setting. The functions discussed in these chapters can be expressed in terms of hypergeometric functions.

Chapter 7 takes up the confluent hypergeometric functions that arise as limiting cases of the standard hypergeometric functions. Among the most important of the confluent functions are the Bessel functions found in Chapter 8. This is the longest chapter in the book. One might well argue that such a chapter is superfluous, since one can look up Watson's classic treatise on the subject. Surprisingly, however, there are gaps in Watson's book. For example, his proof of the Fourier-Bessel series is valid only for $\nu > -\frac{1}{2}$. A complete proof for $\nu > -1$ is found in Section 8.13.

The ninth and final chapter deals with Hill's equation. In historical terms, this material is the most recent and so far the least adequately covered in the literature. The recent book by Magnus and Winkler is most up to date, but contains no applications to physics. Section 9.8 demonstrates the connection between the study of energy bands in crystals and Hill's equation. Section 9.4 is devoted to the expansion theorems. There are two types, namely, one for the interval $(-\pi, \pi)$ and one for the interval $(-\infty, \infty)$. The former corresponds to the study of a regular differential operator and the latter to that of a singular operator. The first is proved by a standard comparison technique that treats the eigenfunctions as perturbations of trigonometric functions. The singular case is treated as a limiting case of the regular case.

While little claim can be made to originality, it is hoped that there is enough distinction in the selection of material and the type of proof to throw new light on this classical subject. The aim was to present a range of topics such that both mathematicians and applied scientists with a variety of interests will find material that is useful, and mathematically and aesthetically pleasing.

HARRY HOCHSTADT

September 1970
Brooklyn, New York

Contents

CHAPTER 1

Orthogonal Polynomials

1. Linear Spaces

Although this chapter concerns itself with orthogonal polynomials, which can be discussed without reference to the subject of linear spaces, it will prove advantageous to frame much of the following discussion in more abstract language. This will not only allow for greater economy of notation, but also will lead to a richer understanding of many of the notions of abstract spaces which pervade so much of modern mathematics. To make the contents of this chapter self-contained all necessary definitions and basic proofs will be provided.

Definition. A linear space over a field F (in our work F will invariably be the field of real or complex numbers) is a collection X of elements with two defined operations. The first of these is addition of elements in X and the second multiplication of elements in X by scalars in F. In addition we stipulate the following conditions:

1. X forms a commutative group under the additive operation in X. That is if $f, g, h, \ldots \in X$ then

   (a) the operation is closed so that $f + g \in X$,

   (b) the operation is associative:
   $$(f + g) + h = f + (g + h),$$

   (c) there exists an identity 0 for which
   $$f + 0 = f \qquad \text{for} \quad \text{all } f \in X,$$

   (d) for every f there exists an inverse element denoted by $(-f)$ such that
   $$f + (-f) = 0,$$

   (e) the operation is commutative
   $$f + g = g + f \qquad \text{for} \quad \text{all } f, g \in X.$$

2. Multiplication by scalars is closed. That is,

   (a) $1 \cdot f = f$ for all $f \in X$,

   (b) $\alpha f \in X$ for $\alpha \in F$ and $f \in X$,

   (c) for all α, $\beta \in F$, and $f \in X$ $\alpha(\beta f) = (\alpha\beta) f$.

3. The following distributive laws hold.

   (a) $\alpha(f + g) = \alpha f + \alpha g$ for all $\alpha \in F, f, g \in X$.

   (b) $(\alpha + \beta) f = \alpha f + \beta f$.

In all our applications we shall be concerned with inner product spaces. In such spaces an inner product is defined.

Definition. An inner product is a function that assigns to every pair of elements f, $g \in X$ a complex number. We denote this function by (f, g). By definition it has the following properties:

1. $\overline{(f, g)} = (g, f)$.
2. $(\alpha f + \beta g, h) = \alpha(f, h) + \beta(g, h)$.
3. $(f, f) \geq 0$ and $(f, f) = 0$ if and only if $f \neq 0$.

The norm of a vector $\|f\|$ is defined by

$$\|f\| = (f, f)^{1/2}.$$

It has the following properties:

1. $\|f\| \geq 0$ and $\|f\| = 0$ if and only if $f = 0$.
2. $\|\alpha f\| = |\alpha| \, \|f\|$.
3. $\|f + g\| \leq \|f\| + \|g\|$.

(This inequality is known as the triangle inequality.) Property 1 is an immediate consequence of the definition. To verify property 2 we first note that

$$(f, \alpha g) = \overline{(\alpha g, f)} = \bar{\alpha}\overline{(g, f)} = \bar{\alpha}(f, g).$$

Then

$$\|\alpha f\| = (\alpha f, \alpha f)^{1/2} = (\alpha\bar{\alpha})^{1/2}\|f\| = |\alpha|^{1/2}\|f\|.$$

To verify property 3 we first prove the following lemma.

LEMMA For all f, $g \in X$ we have

$$|(f, g)| \leq \|f\| \, \|g\|.$$

Equality is achieved if and only if f and g are linearly dependent; that is, for suitable scalars α and β, $\alpha f + \beta g = 0$. The above is known as the Cauchy-Schwarz inequality. To prove it let

$$(f, g) = |(f, g)|e^{i\theta}, \qquad \alpha = \frac{|(f, g)|}{\|f\|^2} e^{-i\theta}$$

We assume that $f \neq 0$; in that eventuality the inequality is trivially true. Then

$$0 \leq (\alpha f - g, \alpha f - g)\|f\|^2 = \|f\|^2\|g\|^2 - |(f, g)|^2,$$

and the result follows immediately from the above.

To complete the proof we note that if f and g are linearly independent

$$0 < \|\alpha f - g\|$$

for all α and the inequality is strict. Otherwise if, for example,

$$f = \beta g, \qquad \text{where} \quad \beta = |\beta|e^{i\theta}$$

$$(f, g) = \beta\|g\|^2 = |\beta|\ \|g\|^2 e^{i\theta}$$

$$\alpha = \frac{|\beta|\ \|g\|^2}{|\beta|^2\|g\|^2} e^{-i\theta} = \beta^{-1}$$

so that

$$0 = \|f\|^2\|g\|^2 - |(f, g)|^2$$

and equality is achieved. ∎

To prove the triangle inequality we note that, using the lemma, we have

$$\begin{aligned} \|f + g\| &= (f + g, f + g)^{1/2} \\ &= [\|f\|^2 + (f, g) + \overline{(f, g)} + \|g\|^2]^{1/2} \\ &\leq [\|f\|^2 + 2|(f, g)| + \|g\|^2]^{1/2} \leq \|f\| + \|g\|. \end{aligned}$$

Definition. If for two nonzero elements f, g we have $(f, g) = 0$ we say that f and g are orthogonal.

Definition. A set of vectors $f_1, f_2, \ldots, f_n$ is said to be linearly independent if no nontrivial linear relationship between them can exist. In other words the relationship

$$\alpha_1 f_1 + \alpha_2 f_2 + \cdots + \alpha_n f_n = 0$$

can only hold if $\alpha_1 = \alpha_2 = \cdots = \alpha_n = 0$.

A linearly independent set of vectors $f_1, f_2,, \ldots, f_n$ will be said to be orthogonal if

$$(f_i, f_j) = 0 \qquad \text{for} \quad \text{all } i, j \text{ such that } i \neq j.$$

If in addition to the above we have

$$(f_i, f_i) = 1$$

we say that the set is orthonormal.

The set of all vectors of the form

$$f = \sum_{i=1}^{n} \alpha_i f_i, \tag{1}$$

where the $\{f_i\}$ represents a fixed set of vectors, clearly forms a linear space. If the $\{f_i\}$ are linearly independent they form a so-called basis for this space, and the dimension of the space is said to be n. Any element f can be expressed in the form (1). To find the coefficients we merely take the inner product with f_j and thus obtain a system of n equations in n unknowns for the α_i.

$$\sum_{i=1}^{n} \alpha_i (f_i, f_j) = (f, f_j), \qquad j = 1, 2, \ldots, n. \tag{2}$$

That the above system has a unique solution is a consequence of the fact that the determinant

$$D = |(f_i, f_j)| \neq 0.$$

To see this we note that for $f = 0$, the homogeneous system (2) can have only the solution $\alpha_1 = a_2 = \cdots = \alpha_n = 0$, since the $\{f_i\}$ are linearly independent. As a result $D \neq 0$ necessarily.

If the system $\{f_i\}$ were orthonormal, the system (2) could be solved by inspection, since in that case it reduces to

$$\alpha_i = (f, f_i)$$

It is evident that orthonormal systems are more convenient to work with than general systems. Given any linearly independent system $\{f_i\}$ we can replace it by an equivalent orthonormal system. One way of constructing such a system is by the Gram-Schmidt process.

We now let

$$e_1 = \frac{f_1}{\|f_1\|}$$

$$e_2 = \frac{f_2 - (f_2, e_1)e_1}{\|f_2 - (f_2, e_1)e_1\|}$$

and in general

$$e_k = \frac{f_k - \sum_{i=1}^{k-1}(f_k, e_i)e_i}{\|f_k - \sum_{i=1}^{k-1}(f_k, e_i)e_i\|}, \qquad k = 1, 2, \ldots, n. \tag{3}$$

Clearly

$$(e_k, e_k) = 1$$

and

$$(e_2, e_1) = \frac{(f_2, e_1) - (f_2, e_1)(e_1, e_1)}{\|f_2 - (f_2, e_1)e_1\|} = 0.$$

By induction we can easily verify that for $k \neq j$

$$(e_k, e_j) = 0.$$

It is also evident that

$$\begin{aligned} f_k &= \sum_{i=1}^{k} \alpha_{i,k} e_i \\ e_k &= \sum_{i=1}^{k} \beta_{i,k} f_i \end{aligned} \tag{4}$$

and $(e_k, f_j) = 0$ for $j = 1, 2, \ldots, k-1$.

One disadvantage of the Gram-Schmidt process is that it is recursive. However, an explicit expression for the e_k can be obtained. Consider the determinant

$$d_k = \begin{vmatrix} (f_1, f_1) & (f_1, f_2) & \cdots & (f_1, f_k) \\ (f_2, f_1) & (f_2, f_2) & \cdots & (f_2, f_k) \\ \vdots & & & \\ (f_{k-1}, f_1) & (f_{k-1}, f_2) & \cdots & (f_{k-1}, f_k) \\ f_1 & f_2 & \cdots & f_k \end{vmatrix}$$

Using (4), for $k > j$, we see that

$$(e_j, d_k) = \sum_{i=1}^{j} \beta_{i,j}(f_i, d_k) = 0$$

since in (f_i, d_k) the kth and ith rows are identical. It follows therefore that

$$e_k = \frac{d_k}{\|d_k\|}.$$

2. Orthogonal Polynomials

The results and concepts of the preceding section can now be applied to a study of orthogonal polynomials. For our field F we shall select the field of real numbers, and for our space X the set of all polynomials in the variable x with real coefficients. Using the standard operations of addition and multiplication we obtain a linear space. It is an elementary exercise to verify all axioms of the preceding section.

By selecting an arbitrary interval (a, b) and restricting the values of x to that interval we obtain another linear space. We do not preclude the possibility that $b = \infty$, or $a = -\infty$, or both. As yet this space is not an inner product space. To define such a product we select a positive function $w(x)$ defined on (a, b) for which the integral

$$\int_a^b w(x)p(x)\,dx$$

exists for all polynomials $p(x)$.

More generally if the above integral is taken in the sense of Lebesgue $w(x)$ need be positive everywhere except on a set of measure zero. We can now introduce an inner product as follows. If $f(x)$ and $g(x)$ are polynomials

$$(f, g) = \int_a^b w(x)f(x)g(x)\,dx. \tag{1}$$

The above can be shown to satisfy all properties of inner products, as long as $f(x)$ and $g(x)$ are real. More generally we could use

$$(f, g) = \int_a^b w(x)f(x)\overline{g(x)}\,dx,$$

but we will restrict ourselves to real spaces. From (1) it is clear that

$$(f, g) = \overline{(g, f)} = (g, f)$$

in view of the fact that we are dealing with real quantities. Also

$$(\alpha f + \beta g, h) = \alpha(f, h) + \beta(g, h).$$

Lastly we note that

$$(f, f) = \int_a^b w(x)f^2(x)\,dx \geq 0$$

and $(f, f) = 0$ implies that

$$w(x)f^2(x) = 0, \text{ almost everywhere.}$$

Since $f(x)$ is a polynomial, the latter implies that $f(x) \equiv 0$.

To construct a family of orthonormal polynomials we can proceed as follows. First we select as a linearly independent set the polynomials

$$1, x, x^2, \ldots, x^n.$$

That these are indeed linearly independent is a consequence of the fact that

$$\alpha_0 + \alpha_1 x + \cdots + \alpha_n x^n \equiv 0, \qquad x \in (a, b)$$

if and only if $\alpha_0 = \alpha_1 = \alpha_2 = \cdots = \alpha_n = 0$.

Using either the Gram-Schmidt process I.1.3 or the direct formula I.1.5 we can construct an orthonormal set $\phi_k(x)$, where $\phi_k(x)$ is a polynomial of precise degree k.

EXAMPLE Let $(a, b) = (-1, 1)$ and $w(x) = 1$. Then

$$\phi_0(x) = \tfrac{1}{2}^{1/2}$$

$$d_1(x) = \begin{vmatrix} (1, 1) & (1, x) \\ 1 & x \end{vmatrix} = \begin{vmatrix} 2 & 0 \\ 1 & x \end{vmatrix} = 2x$$

$$\phi_1(x) = \frac{2x}{\|2x\|} = \tfrac{3}{2}^{1/2} x$$

$$d_2(x) = \begin{vmatrix} (1, 1) & (1, x) & (1, x^2) \\ (x, 1) & (x, x) & (x, x^2) \\ 1 & x & x^2 \end{vmatrix}$$

$$= \begin{vmatrix} 2 & 0 & 2/3 \\ 0 & 2/3 & 0 \\ 1 & x & x^2 \end{vmatrix} = 4/3x^2 - 4/9$$

$$\phi_2(x) = \frac{4/3x^2 - 4/9}{\|4/3x^2 - 4/9\|} = \tfrac{5}{8}^{1/2}(3x^2 - 1)$$

More generally, as will be shown in the next chapter, the classical Legendre polynomials $P_n(x)$ are related to the above by

$$P_n(x) = \sqrt{\frac{2n + 1}{2}}\, \phi_n(x).$$

Note that the orthonormality requirement does not define these polynomials uniquely, but only up to the choice of a sign. But by also stipulating that the coefficient of the highest power of x be positive a unique polynomial is defined.

We also note for future reference that, using Eq. 1.4,

$$(\phi_n, x^k) = \int_a^b w(x)\phi_n(x)x^k\, dx = 0, \qquad k < n. \tag{2}$$

3. The Recurrence Formula

We let k_n denote the coefficients of x^n in $\phi_n(x)$, so that

$$\phi_n(x) = k_n x^n + \cdots. \tag{1}$$

THEOREM The orthonormal polynomials $\{\phi_n\}$ satisfy the recurrence formula

$$\phi_{n+1} - (A_n x + B_n)\phi_n + C_n \phi_{n-1} = 0 \qquad n = 0, 1, 2, \ldots, \tag{2}$$

where

$$A_n = \frac{k_{n+1}}{k_n}, \qquad C_n = \frac{A_n}{A_{n-1}}, \qquad C_0 = 0. \tag{3}$$

Proof. From (1) and (3) it is evident that

$$\phi_{n+1} - A_n x\phi_n = \sum_{k=0}^{n} \alpha_k x^k = \sum_{k=0}^{n} \beta_k \phi_k(x)$$

for suitable α_k and β_k, since the x^{n+1} term on the left has been cancelled out. Using the orthonormality we obtain from the above

$$-A_n(x\phi_n, \phi_j) = \beta_j \qquad j \leq n.$$

Reference to I.2.2. shows that

$$(x\phi_n, \phi_j) = \int_a^b w(x)x\phi_n(x)\phi_j(x)\,dx = (\phi_n, x\phi_j) = 0 \qquad j \leq n-2$$

since $x\phi_j(x)$ is at most of degree $n-1$. It follows that

$$\beta_0 = \beta_1 = \cdots = \beta_{n-2} = 0,$$

and we let $\beta_n = B_n$, $\beta_{n-1} = -C_n$, which proves the validity of (2). To obtain the explicit value of C_n we note that

$$C_n = A_n(x\phi_n, \phi_{n-1}) = A_n(\phi_n, x\phi_{n-1})$$

but

$$x\phi_{n-1} = k_{n-1}x^n + \cdots = \frac{k_{n-1}}{k_n}[k_n x^n + \cdots]$$

and the bracketed expression must be of the form

$$x\phi_{n-1} = \frac{1}{A_{n-1}}\left[\phi_n(x) + \sum_{j=0}^{n-1} \gamma_j \phi_j(x)\right].$$

It follows that

$$C_n = \frac{A_n}{A_{n-1}}\left(\phi_n, \phi_n + \sum_{j=0}^{n-1} \gamma_j \phi_j\right) = \frac{A_n}{A_{n-1}}. \qquad \blacksquare$$

4. The Christoffel-Darboux Formula

A function that plays a significant role in many aspects of this subject is

$$K_n(x, y) = \sum_{k=0}^{n} \phi_k(x)\phi_k(y).$$

It is possible to find a closed form for the above sum.

LEMMA

$$K_n(x, y) = \sum_{k=0}^{n} \phi_k(x)\phi_k(y) = \frac{k_n}{k_{n+1}}\left[\frac{\phi_n(y)\phi_{n+1}(x) - \phi_n(x)\phi_{n+1}(y)}{x - y}\right] \qquad (1)$$

Formula (1) is known as the Christoffel-Darboux formula.

Proof. Since

$$\phi_0(x) = k_0$$

$$\phi_1(x) = k_1 x - k_1 k_0^2 \int_a^b x w(x)\, dx$$

we have trivially

$$K_0(x, y) = \phi_0^2(x) = k_0^2 = \frac{k_0}{k_1}\left[\frac{\phi_0(y)\phi_1(x) - \phi_0(x)\phi_1(y)}{x - y}\right].$$

To prove the general case, we use the recurrence formula and note that

$$\frac{k_n}{k_{n+1}}\left[\frac{\phi_n(y)\phi_{n+1}(x) - \phi_n(x)\phi_{n+1}(y)}{x - y}\right]$$
$$= \frac{k_n}{k_{n+1}} A_n \phi_n(x)\phi_n(y) + \frac{k_n}{k_{n+1}} C_n\left[\frac{\phi_{n-1}(y)\phi_n(x) - \phi_{n-1}(x)\phi_n(y)}{x - y}\right]$$
$$= \phi_n(x)\phi_n(y) + K_{n-1}(x, y)$$

so that

$$K_n(x, y) = \phi_n(x)\phi_n(y) + K_{n-1}(x, y).$$

From the latter, by iteration we find

$$K_n(x, y) = \sum_{k=1}^{n} \phi_k(x)\phi_k(y) + K_0(x, y) = \sum_{k=0}^{n} \phi_k(x)\phi_k(y). \qquad \blacksquare$$

By letting y approach x in the limit we find

$$K_n(x, x) = \lim_{y \to x} \frac{k_n}{k_{n+1}} \left[\frac{\phi_n(y)\phi_{n+1}(x) - \phi_n(x)\phi_{n+1}(y)}{x - y} \right]$$

$$= \frac{k_n}{k_{n+1}} [\phi_n(x)\phi'_{n+1}(x) - \phi'_n(x)\phi_{n+1}(x)]$$

$$= \sum_{k=0}^{n} \phi_k^2(x) \geq 0 \tag{2}$$

The function $K_n(x, y)$ can be characterized in terms of an extremal property independently of the polynomials $\phi_n(x)$.

THEOREM Among all polynomials $\rho(x)$ of degree n and unit norm, the ones that maximize $|\rho(y)|$, where y is a prescribed point in (a, b), are given by

$$\rho(x) = \frac{\pm K_n(x, y)}{K_n^{1/2}(y, y)}. \tag{3}$$

Proof. Since $\rho(x)$ is of degree n it can be represented in the form

$$\rho(x) = \sum_{k=0}^{n} \alpha_k \phi_k(x)$$

and since $\|\rho\| = 1$ we have

$$\sum_{0}^{n} \alpha_k^2 = 1.$$

The quantity $\rho(y)$ can be construed as an inner product of the two vectors in $n + 1$ dimensional Euclidean space with components $\{\alpha_k\}$ and $\{\phi_k(y)\}$. By means of the Cauchy-Schwarz inequality we see that

$$\rho^2(y) \leq \sum_{k=0}^{n} \alpha_k^2 \sum_{k=0}^{n} \phi_k^2(y) = \sum_{k=0}^{n} \phi_k^2(y).$$

$\rho^2(y)$ will be maximized when equality is achieved. In that case the vectors $\{\alpha_k\}$ and $\{\phi_k(y)\}$ must be linearly dependent, that is, for some scalar λ

$$\alpha_k = \lambda\phi_k(y) \qquad k = 0, 1, 2, \ldots n.$$

We see therefore that

$$\rho(x) = \lambda \sum_{k=0}^{n} \phi_k(y)\phi_k(x) = \lambda K_n(x, y).$$

To determine λ we still require that $\|\rho\| = 1$ and

$$\lambda^2 \sum_{k=0}^{n} \phi_k^2(y) = \lambda^2 K_n(y, y) = 1,$$

which shows that

$$\rho(x) = \pm \frac{K_n(x, y)}{K_n^{1/2}(y, y)}. \qquad \blacksquare$$

N.B. By (2) the denominator is real.

If $p(x)$ is a polynomial of degree n or less we have

$$p(x) = \sum_{k=0}^{n} (p, \phi_k)\phi_k(x).$$

It is also evident that

$$(p(x), K_n(x, y)) = \sum_{k=0}^{n} (p, \phi_k)\phi_k(y)) = p(y).$$

$K_n(x, y)$ is sometimes known as a reproducing kernel, as a result of the above property. In particular we note that for $p(x) = 1$ we obtain

$$(1, K_n(x, y)) = \int_a^b w(x)K_n(x, y)\, dx = 1, \qquad n \geq 0. \tag{4}$$

5. The Weierstrass Approximation Theorem

A result that will prove to be of fundamental importance in the sequel is the Weierstrass Approximation Theorem.

THEOREM Let $f(x)$ be a continuous function defined on the closed and bounded interval $[a, b]$. Given any $\varepsilon > 0$, it is possible to find a polynomial $p(x)$ such that

$$|f(x) - p(x)| < \varepsilon \qquad \text{for} \quad \text{all } x \in [a, b].$$

Proof. The proof to be given will be a constructive one. By means of a sequence of so-called Bernstein polynomials we shall construct an explicit $p(x)$ satisfying the theorem.

By means of a linear transformation the finite interval $[a, b]$ into the interval $[0, 1]$. Now let

$$B_n(x; f) = \sum_{k=0}^{n} \binom{n}{k} f\left(\frac{k}{n}\right) x^k(1 - x)^{n-k}$$

where $\binom{n}{k}$ is the binomial coefficient

$$\binom{n}{k} = \frac{n!}{k!(n-k)!}$$

We shall show that for every $\varepsilon > 0$ we can find $N(\varepsilon)$ so that

$$|f(x) - B_n(x; f)| < \varepsilon \qquad \text{for} \quad n > N(\varepsilon).$$

We shall first prove the theorem for three special cases required for the general proof.

1. By the binomial theorem we have

$$(x+y)^n = \sum_{k=0}^{n} \binom{n}{k} x^k y^{n-k} \tag{1}$$

By differentiating the above with respect to x and multiplying by x/n we obtain

$$x(x+y)^{n-1} = \sum_{k=0}^{n} \binom{n}{k} \frac{k}{n} x^k y^{n-k}. \tag{2}$$

Repeating the above operation we have

$$x^2(x+y)^{n-2} + \frac{xy}{n}(x+y)^{n-2} = \sum_{k=0}^{n} \binom{n}{k} \left(\frac{k}{n}\right)^2 x^k y^{n-k}. \tag{3}$$

For $y = 1 - x$ we have from (1)

$$1 = \sum_{k=0}^{n} \binom{n}{k} x^k (1-x)^{n-k}. \tag{4}$$

(4) shows the theorem to be correct for $f(x) = 1$, that is,

$$B_n(x; 1) = 1 \qquad \text{for} \quad \text{all } n.$$

2. For $f(x) = x$ we have, letting $y = 1 - x$ in (2)

$$B_n(x; x) = x \qquad \text{for} \quad \text{all } n. \tag{5}$$

3. For $f(x) = x^2$ we have, letting $y = 1 - x$ in (3)

$$|x^2 - B_n(x; x^2)| \le \frac{x(1-x)}{n} \le \frac{1}{4n} \qquad \text{for} \quad \text{all } n.$$

In 1., 2., 3., we proved the theorem for the cases $f(x) = 1$, $f(x) = x$, $f(x) = x^2$.

To prove the general case we note that if $|f(x)| < M$ then

$$\left|f(x) - f\left(\frac{k}{n}\right)\right| < 2M, \qquad x \in [0, 1].$$

Also by the uniform continuity of $f(x)$, given ε we can find $\delta(\varepsilon)$ so that

$$\left|f(x) - f\left(\frac{k}{n}\right)\right| < \frac{\varepsilon}{2} \quad \text{for} \quad \left|x - \frac{k}{n}\right| < \delta(\varepsilon).$$

We now see that by (4)

$$\begin{aligned}
|f(x) - B_n(x;f)| &= \left|f(x) - \sum_{k=0}^{n} \binom{n}{k} f\left(\frac{k}{n}\right) x^k(1-x)^{n-k}\right| \\
&= \left|\sum_{k=0}^{n} \binom{n}{k}\left[f(x) - f\left(\frac{k}{n}\right)\right] x^k(1-x)^{n-k}\right| \\
&\le \sum_{|x-k/n|<\delta(\varepsilon)} \binom{n}{k} \left|f(x) - f\left(\frac{k}{n}\right)\right| x^k(1-x)^{n-k} \\
&\quad + \sum_{|x-k/n|\ge\delta(\varepsilon)} \binom{n}{k} \left|f(x) - f\left(\frac{k}{n}\right)\right| x^k(1-x)^{n-k} \\
&= S_1 + S_2.
\end{aligned}$$

For the first sum on the right we have evidently

$$S_1 < \frac{\varepsilon}{2} \sum_{|x-k/n|<\delta(\varepsilon)} \binom{n}{k} x^k(1-x)^{n-k} \le \frac{\varepsilon}{2} \sum_{k=0}^{n} \binom{n}{k} x^k(1-x)^{n-k} = \frac{\varepsilon}{2}.$$

To estimate S_2 we have, using the fact that $|x - k/n| \ge \delta(\varepsilon)$,

$$\begin{aligned}
S_2 &< 2M \sum_{|x-k/n|\ge\delta(\varepsilon)} \binom{n}{k} x^k(1-x)^{n-k} \\
&\le \frac{2M}{\delta^2(\varepsilon)} \sum_{k=0}^{n} \binom{n}{k} \left(x - \frac{k}{n}\right)^2 x^k(1-x)^{n-k} \\
&= \frac{2M}{\delta^2(\varepsilon)} [x^2 B_n(x;1) - 2xB_n(x;x) + B_n(x;x^2)] \\
&= \frac{2M}{\delta^2(\varepsilon)} \left[x^2 - 2x^2 + x^2 + \frac{x(1-x)}{n}\right] \le \frac{2M}{4\delta^2(\varepsilon)n}.
\end{aligned}$$

We summarize these results in the statement

$$|f(x) - B_n(x;f)| < \frac{\varepsilon}{2} + \frac{2M}{4\delta^2(\varepsilon)n} < \varepsilon \qquad \text{for} \quad n > \frac{M}{\delta^2(\varepsilon)\varepsilon},$$

thus proving the theorem. ■

This proof can be given a probabilistic interpretation. We conceive the following game. n numbers are to be selected at random from the interval $[0, 1]$. Let k denote the number falling into $[0, x]$ and $n - k$ the number falling into $(x, 1]$, where x is a preassigned quantity. The payoff is to be $f(k/n)$. The probability of k coming from the interval $[0, x]$ is clearly

$$\binom{n}{k} x^k (1 - x)^{n-k},$$

since x is the probability of a single selection falling into $[0, x]$. The expectation of this game is clearly

$$B_n(x; f) = \sum_{k=0}^{n} \binom{n}{k} f\left(\frac{k}{n}\right) x^k (1 - x)^{n-k}.$$

For n large the expected value of k will be nx and by the law of large numbers we expect that

$$B_n(x; f) \approx f\left(\frac{nx}{n}\right) = f(x).$$

In other words the Weierstrass approximation theorem is nothing but the law of large numbers for a binomial distribution.

6. The Zeros of the Orthogonal Polynomials

In view of the fact that $\phi_n(x)$ is a polynomial of degree n, we know from the fundamental theorem of algebra that it must have precisely n zeros in the field of complex numbers. But we can say even more about it, as a result of its special properties.

THEOREM $\phi_n(x)$ has n real, simple zeros, all in the interval (a, b).

Proof. We suppose that $\phi_n(x)$ has k zeros in the interval (a, b), where it changes sign. The latter is equivalent to saying that we are examining zeros of odd multiplicity. If we can succeed in showing that $k = n$, then $\phi_n(x)$ must have n zeros in (a, b), where it changes sign. But since $\phi_n(x)$ has precisely n zeros they must all be simple and in (a, b).

We now suppose that $k < n$. Then

$$p_k(x) = (x - x_1)(x - x_2)\cdots(x - x_k) = \sum_{j=0}^{k} \alpha_j \phi_j(x)$$

is a polynomial of degree k so that

$$(\phi_n, p_k) = 0.$$

But clearly $\phi_n(x)p_k(x)$ cannot change sign, since it has only zeros of even multiplicity. Therefore if, without loss of generality,

$$\phi_n(x)p_k(x) \geq 0$$

we have

$$(\phi_n, p_k) = \int_a^b w(x)\phi_n(x)p_k(x)\,dx > 0.$$

The contradiction can only be resolved by having $k = n$. ∎

We have another result showing how zeros of successive orthonormal polynomials are related.

THEOREM The zeros of $\phi_n(x)$ and $\phi_{n+1}(x)$ alternate on the interval (a, b), and $\phi_n(x)$ and $\phi_{n+1}(x)$ do not vanish simultaneously.

Proof. Let x_r and x_{r+1} denote two successive zeros of $\phi_n(x)$. Then by 1.4.2

$$\sum_{k=0}^{n} \phi_k^2(x_r) = \frac{k_n}{k_{n+1}} [-\phi_n'(x_r)\phi_{n+1}(x_r)] \geq 0$$

so that

$$\phi_n'(x_r)\phi_{n+1}(x_r) \leq 0$$

and also

$$\phi_n'(x_{r+1})\phi_{n+1}(x_{r+1}) \leq 0.$$

In view of the fact that x_r and x_{r+1} are successive zeros of $\phi_n(x)$ it follows that $\phi_n'(x_r)$ and $\phi_n'(x_{r+1})$ are of opposite sign. Therefore $\phi_{n+1}(x_r)$ and $\phi_{n+1}(x_{r+1})$ also have opposite sign, and being a continuous function of x, $\phi_{n+1}(x)$ must vanish at least once in $[x_r, x_{r+1}]$. Similarly between two successive zeros of $\phi_{n+1}(x)$ there must be at least one zero of $\phi_n(x)$ and the statement of the theorem follows.

To show that $\phi_n(x)$ and $\phi_{n+1}(x)$ do not vanish simultaneously, we see from the recurrence formula that if

$$\phi_{n+1}(x_r) = \phi_n(x_r) = 0$$

then also

$$\phi_{n-1}(x_r) = 0.$$

By repeating this we have finally

$$\phi_{n-2}(x_r) = \cdots = \phi_1(x_r) = \phi_0(x_r) = 0.$$

But $\phi_0 > 0$ so that ϕ_n and ϕ_{n+1} do not vanish simultaneously. ∎

7. Approximation Theory

In many aspects of pure as well as applied mathematics it is often important to investigate how closely one member of a set of functions can be approximated by a member of another class of functions. Suppose, for example, we are given a function $f(x)$ for which

$$\int_a^b w(x)f^2(x)\,dx < \infty.$$

Is it possible to approximate $f(x)$ by a finite linear combination of orthonormal polynomials (over the interval (a, b) with weight function $w(x)$) in some optimum fashion? To decide the answer we must select some criterion by which we can measure the quality of the approximation. One such criterion is given by the norm defined in Section I.1.

We shall say that a polynomial $p_n(x)$ of degree n yields a best approximation to some $f(x)$ over the linear, inner product space X, which consists of functions $g(x)$ for which

$$\|g(x)\| = \left[\int_a^b w(x)g^2(x)\,dx\right]^{1/2}$$

if

$$\|f(x) - p_n(x)\| = \min_{\text{all } q_n(x)} \|f(x) - q_n(x)\|.$$

In the above $q_n(x)$ is an arbitrary polynomial of degree n.

THEOREM Among all polynomials of degree n, there is precisely one for which

$$\|f(x) - p_n(x)\|$$

is minimized. It is given by

$$p_n(x) = \sum_{k=0}^{n} \alpha_k \phi_k(x), \qquad \alpha_k = (f, \phi_k) = \int_a^b w(x)f(x)\phi_k(x)\,dx.$$

Proof. Let $p_n(x)$ be as defined in the theorem and $q_n(x)$ any other polynomial of degree n, say

$$q_n(x) = \sum_{k=0}^{n} \beta_k \phi_k(x).$$

Let

$$g(x) = f(x) - p_n(x)$$

and clearly, by construction

$$(g, \phi_k) = 0, \qquad k \le n,$$

so that

$$(g, q_n) = 0.$$

Also

$$\|g - q_n\|^2 = (g - q_n, g - q_n) = \|g\|^2 + \|q_n\|^2$$

so that

$$\min_{\text{all } q_n(x)} \|g - q_n\| = \|g\|$$

and the minimum is attained if and only if $q_n(x) \equiv 0$. ■

If $f(x)$ is a continuous function we can say even more.

THEOREM Let $f(x)$ be continuous and $p_n(x)$ as in the preceding theorem. Then $f(x) - p_n(x)$ changes sign at least $n + 1$ times in (a, b), or else vanishes identically.

Proof. Denote the points where $f(x) - p_n(x)$ changes sign by $x_1, x_2, \ldots x_k$, where we suppose that $k \le n$. Then

$$q(x) = (x - x_1) \cdots (x - x_k) = \sum_{j=0}^{k} \beta_j \phi_j(x)$$

and since

$$(f - p_n, \phi_j) = 0, \qquad j \le n$$

by construction

$$(f - p_n, q) = 0.$$

The function $(f - p_n)q$ must be of constant sign so that either

$$(f - p_n, q) = \int_a^b w(x)(f(x) - p_n(x))q(x)\, dx$$

does not vanish, or else $f(x) - p_n(x) \equiv 0$. The former contradicts the statement $k \le n$, proving the theorem. ■

It is often of value to find a polynomial that agrees with a given continuous function at n prescribed points. To construct such a polynomial we can make use of the Lagrange Interpolation Formula. Let $f(x)$ be some continuous function on (a, b) and $\phi_n(x)$ an orthonormal polynomial. Consider

$$F(x) = \sum_{k=1}^{n} f(x_k) \frac{\phi_n(x)}{(x - x_k)\phi_n'(x_k)}. \tag{1}$$

Clearly$F(x)$ is a polynomial of degree $n-1$ and also

$$F(x_k) = \lim_{x \to x_k} F(x) = f(x_k), \qquad k \leq n.$$

N.B. For this construction $\phi_n(x)$ did not have to be an orthonormal polynomial. Any polynomial of degree n, with simple zeros would have served.

If $f(x)$ were a polynomial of degree $n-1$, then clearly $f(x) = F(x)$ in (1).

We shall now derive the Gauss Quadrature Formula. To do so we shall assume that $f(x)$ is a polynomial of degree $2n-1$. Then $F(x) - f(x)$ is a polynomial of degree $2n-1$ and it vanishes at $x_1, x_2, \ldots x_n$. Therefore $r(x)$, given by

$$r(x) = \frac{F(x) - f(x)}{\phi_n(x)}$$

must be a polynomial of degree $n-1$. We can now write

$$f(x) = \sum_{k=1}^{n} f(x_k) \frac{\phi_n(x)}{(x - x_k)\phi_n'(x_k)} - r(x)\phi_n(x).$$

By integration we find

$$\int_a^b w(x)f(x)\,dx = \sum_{k=1}^{n} f(x_k) \int_a^b w(x) \frac{\phi_n(x)}{(x - x_k)\phi_n'(x_k)} - \int_a^b w(x)r(x)\phi_n(x)\,dx.$$

We let

$$\int_a^b w(x) \frac{\phi_n(x)}{(x - x_k)\phi_n'(x_k)}\,dx = \lambda_{k,n}$$

and clearly

$$\int_a^b w(x)r(x)\phi_n(x)\,dx = 0$$

since $r(x)$ is of degree $n-1$. The numbers $\lambda_{k,n}$ are known as the Christoffel numbers, and they are independent of $f(x)$.

From the above we have

$$\int_a^b w(x)f(x)\,dx = \sum_{k=1}^{n} f(x_k)\lambda_{k,n}. \tag{2}$$

(2) is known as the Gauss Quadrature Formula. Evidently to find the value of the integral on the left, if $f(x)$ is a polynomial of $2n-1$, we merely need to know the value of $f(x)$ at the n points $x_1, x_2, \ldots, x_n$. If $f(x)$ is not such a polynomial (2) may still furnish a useful approximation for the integral.

For $f(x) = x^j$ where $j \leq 2n - 1$ we have the result

$$\int_a^b w(x)x^j \, dx = \sum_{k=1}^{n} x_k{}^j \lambda_{k,n} \qquad j \leq 2n-1. \tag{3}$$

EXAMPLE Consider

$$\int_{-1}^{1} \frac{dx}{3+x} = \log 2.$$

To obtain an approximate value for the integral we take $w(x) = 1$ and $n = 2$. Then

$$\phi_2(x) = \tfrac{5}{8}^{1/2}(3x^2 - 1)$$

and

$$x_1 = -\tfrac{1}{3}^{1/2}, \; x_2 = \tfrac{1}{3}^{1/2}.$$

$$\lambda_{1,2} = \int_{-1}^{1} \frac{3x^2 - 1}{(x + \frac{1}{3}^{1/2})(-\frac{6}{3^{1/2}})} \, dx = 1$$

$$\lambda_{2,2} = \int_{-1}^{1} \frac{3x^2 - 1}{(x - \frac{1}{3}^{1/2})(\frac{6}{3^{1/2}})} \, dx = 1$$

From (2) we have

$$\int_{-1}^{1} \frac{dx}{3+x} \approx \frac{1}{3 - \frac{1}{3}^{1/2}} + \frac{1}{3 + \frac{1}{3}^{1/2}} = \frac{9}{13} = 0.6923.$$

From tabulated values we have

$$\log 2 = 0.6930 \cdots$$

and our error is less than 0.15%.

There are several different, but equivalent expressions for the Christoffel numbers, as given in the following theorem.

THEOREM

$$\begin{aligned} \lambda_{k,n} &= \int_a^b w(x) \frac{\phi_n(x)}{(x - x_k)\phi_n'(x_k)} \, dx \\ &= -\frac{k_{n+1}}{k_n} \frac{1}{\phi_n'(x_k)\phi_{n+1}(x_k)} \\ &= \frac{1}{\sum_{j=1}^{n-1} \phi_j{}^2(x_k)} \\ &= \int_a^b w(x) \left[\frac{\phi_n(x)}{(x - x_k)\phi_n'(x_k)}\right]^2 dx > 0. \end{aligned}$$

Proof. From I.4.1 we have, since $\phi_n(x_k) = 0$,

$$K_n(x, x_k) = \sum_{j=0}^{n-1} \phi_j(x)\phi_j(x_k) = \frac{k_n}{k_{n+1}} \left[\frac{-\phi_n(x)\phi_{n+1}(x_k)}{x - x_k}\right]. \tag{4}$$

From the above

$$\frac{\phi_n(x)}{(x - x_k)} = \frac{-K_n(x, x_k)k_{n+1}}{k_n\phi_{n+1}(x_k)}$$

so that

$$\lambda_{k,n} = \int_a^b w(x) \frac{\phi_n(x)}{(x - x_k)\phi_n'(x_k)}$$

$$= \frac{-k_{n+1}}{k_n\phi_n'(x_k)\phi_{n+1}(x_k)} \int_a^b w(x)K_n(x, x_k)\, dx$$

and by I.4.4 the latter integral is unity so that

$$\lambda_{k,n} = -\frac{k_{n+1}}{k_n\phi_n'(x_k)\phi_{n+1}(x_k)}$$

By letting $x \to x_k$ in (4) we have

$$K_n(x_k, x_k) = \sum_{j=0}^{n-1} \phi_j^2(x_k) = \frac{-k_n}{k_{n+1}} \phi_n'(x_k)\phi_{n+1}(x_k)$$

so that we also have

$$\lambda_{k,n} = \frac{1}{\sum_{j=0}^{n-1} \phi_j^2(x_k)},$$

thus verifying the first three parts of the theorem.

To prove the last part we use the Gauss quadrature formula (2).

$$\int_a^b w(x)\left[\frac{\phi_n(x)}{(x - x_k)\phi_n'(x_k)}\right]^2 dx = \sum_{j=1}^{n} \lambda_{j,n} \lim_{x \to x_j} \frac{\phi_n(x)}{(x - x_k)\phi_n'(x_k)} = \lambda_{k,n}. \qquad \blacksquare$$

It is possible to obtain error estimates for the Lagrange interpolation formula (1) and the Gauss quadrature formula (2). We suppose that $f(x)$ has n continuous derivatives, then the function

$$g(x) = f(x) - \sum_{k=1}^{n} f(x_k)\rho_k(x),\ \rho_k(x) = \frac{\phi_n(x)}{(x - x_k)\phi_n'(x_k)}$$

has n zeros at $x_1, x_2, \ldots x_n$. We can select a constant K so that $g(x) - K\phi_n(x)$ also vanishes at some prescribed point, say x_{n+1}. Then

$$g(x_i) - K\phi_n(x_i) = 0, \qquad i = 1, 2, \ldots, n + 1.$$

By repeated application of Rolle's theorem the nth derivative of the above function must vanish at least once, say at $x = \eta$. Then

$$K = \frac{g^{(n)}(\eta)}{\phi_n{}^{(n)}(\eta)} = \frac{1}{n!k_n} g^{(n)}(\eta) = \frac{1}{n!k_n} f^{(n)}(\eta).$$

We finally have the Lagrange formula with remainder

$$\begin{aligned} f(x) &= \sum_{k=1}^{n} f(x_k)\rho_k(x) + \frac{f^{(n)}(\eta)}{n!k_n} \phi_n(x) \\ \rho_k(x) &= \frac{\phi_n(x)}{(x - x_k)\phi_n'(x_k)} \end{aligned} \tag{5}$$

if x is allowed to play the role of x_{n+1}.

Another, but related, interpolation formula is the Hermite Interpolation Formula. Here we seek a polynomial of degree $2n - 1$ such that the values of the polynomial and its first derivatives coincide with the corresponding values of $f(x)$ at $x_1, x_2, \ldots, x_n$. To do so we define the functions

$$\begin{aligned} \sigma_k(x) &= \left[1 - (x - x_k)\frac{\phi_n''(x_k)}{\phi_n'(x_k)}\right]\left[\frac{\phi_n(x)}{(x - x_k)\phi_n'(x_k)}\right]^2 \\ \tau_k(x) &= (x - x_k)\left[\frac{\phi_n(x)}{(x - x_k)\phi_n'(x_k)}\right]^2. \end{aligned} \tag{6}$$

A simple computation shows that

$$\sigma_k(x_j) = \sigma_k'(x_j) = \tau_k(x_j) = \tau_k'(x_j) = 0, \qquad j \neq k,$$

and

$$\begin{aligned} \sigma_k(x_k) &= 1, \qquad & \tau_k(x_k) &= 0 \\ \sigma_k'(x_k) &= 0, \qquad & \tau_k'(x_k) &= 1. \end{aligned}$$

Evidently, if $f(x)$ is a given differentiable function and

$$F(x) = \sum_{k=1}^{n} f(x_k)\sigma_k(x) + f'(x_k)\tau_k(x)$$

then

$$\begin{aligned} F(x_k) &= f(x_k) \\ F'(x_k) &= f'(x_k) \end{aligned}$$

and $F(x)$ is a polynomial of degree $2n - 1$.

The function

$$g(x) = f(x) - F(x) - K\phi_n{}^2(x)$$

has double zeros at $x_1, x_2, \ldots, x_n$ so that

$$g(x_k) = g'(x_k) = 0.$$

By a suitable choice of K it can be made to vanish at some other point, say x_{n+1}. As before we find that the $2n$th derivative has to vanish at some point η. Then

$$g^{(2n)}(\eta) = f^{(2n)}(\eta) - Kk_n^2(2n)! = 0,$$

provided of course that $f(x)$ has $2n$ continuous derivatives. Finally we have

$$f(x) = \sum_{k=1}^{n} f(x_k)\sigma_k(x) + f'(x_k)\tau_k(x) + \frac{f^{(2n)}(\eta)\phi_n^2(x)}{k_n^2(2n)!}. \tag{7}$$

From (6), and the last theorem we have

$$\int_a^b w(x)\tau_k(x)\,dx = \left(\phi_n, \frac{\phi_n}{(x - x_k)[\phi_n'(x_k)]^2}\right) = 0$$

since $\phi_n(x)/(x - x_k)$ is of degree $n - 1$. Also

$$\int_a^b w(x)\sigma_k(x)\,dx = \int_a^b w(x)\left[\frac{\phi_n(x)}{(x - x_k)\phi_n'(x_k)}\right]^2 dx = \lambda_{k,n}$$

so that

$$\int_a^b w(x)f(x)\,dx = \sum_1^n f(x_k)\lambda_{k,n} + \int_a^b w(x)\frac{f^{(2n)}(\eta)\phi_n^2(x)}{k_n^2(2n!)}\,dx \tag{8}$$

(8) is again the Gauss Quadrature formula, but with a remainder term, that yields an error estimate.

EXAMPLE In the preceding example we evaluated

$$\int_{-1}^{1} \frac{dx}{3 + x}.$$

For the case $n = 2$ we find

$$\frac{f^{(4)}(x)}{k_2^2 4!} = \frac{24}{(3 + x)^5 \times 45/8 \times 24} \le \frac{1}{180}$$

so that the error E is given by

$$E = \left|\int_{-1}^{1} \frac{f^{(4)}(x)\phi_2^2(x)}{k_2^2 4!}\,dx\right| \le \frac{\max f^{(4)}(x)}{k_2^2 4!} \le \frac{1}{180}$$

which is slightly less than $\frac{1}{2}\%$.

8. More about the Zeros of the Orthonormal Polynomials

It was shown earlier that the zeros of $\phi_n(x)$ and $\phi_{n+1}(x)$ alternate. By a generalization of this result we have the following theorem.

THEOREM If $m > n$, then between any two zeros of $\phi_n(x)$, $\phi_m(x)$ has to vanish at least once.

Proof. Let $x_1, x_2, \ldots, x_m$ denote the zeros of $\phi_m(x)$. We now consider the sequence

$$\phi_n(x_1), \phi_n(x_2), \ldots, \phi_n(x_m).$$

There can be at most n alternations in sign in the above since $\phi_n(x)$ has precisely n zeros. If we succeed in showing that there are at least n sign changes we will have shown that there are precisely n.

We now suppose that there are at most $n - 1$ sign changes in the sequence. In this case it must be possible to construct a polynomial $p(x)$ of degree $n - 1$ such that

$$p(x_k)\phi_n(x_k) \geq 0, \qquad k = 1, 2, \ldots, m.$$

and for at least some values of k the above will be a strict inequality. By using I.7.8 and the fact that $p(x)$ is of degree $n - 1$ we have

$$\int_a^b w(x)\phi_n(x)p(x)\,dx = \sum_{k=1}^{m} \lambda_{k,m}\phi_n(x_k)p(x_k) = 0.$$

But since all $\lambda_{k,m} > 0$ the above is a contradiction, so that the aforementioned sequence has precisely n sign changes. Accordingly we can select a subsequence of zeros of $\phi_m(x)$, say $x_{k_1}, x_{k_2}, \ldots, x_{k_n}$, so that

$$\phi_n(x_{k_1}), \phi_n(x_{k_2}), \ldots, \phi_n(x_{k_n})$$

represents a sequence of alternately positive and negative numbers. Therefore between any two successive x_{k_i}, $\phi_n(x)$ vanishes precisely once, and between every two zeros of $\phi_n(x)$ lies exactly one x_{k_i}. ■

A theorem that provides some information regarding the distribution of zeros for large values of the index n is the following.

THEOREM Let (α, β) be a subinterval of the finite and bounded interval (a, b). Then for sufficiently large n $\phi_n(x)$ vanishes at least once in (α, β).

Note that this theorem does not apply to orthogonal polynomials in infinite intervals.

Proof. First we construct the function

$$f(x) = (x-\alpha)(\beta - x) - \varepsilon, \qquad x \in (\alpha, \beta)$$
$$= -\varepsilon \qquad , \qquad \text{elsewhere.}$$

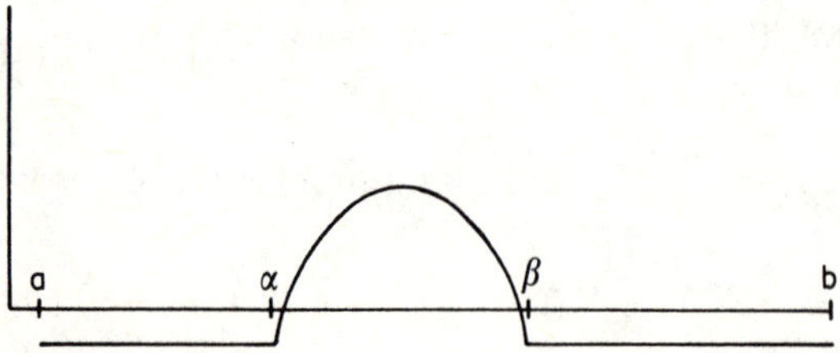

By the Weierstrass theorem we can find a polynomial $p(x)$ such that

$$|f(x) - p(x)| < \varepsilon/2 \qquad \text{in } (a, b).$$

For ε sufficiently small we certainly have

$$\int_a^b w(x)f(x)\,dx > 0$$

and also

$$\int_a^b w(x)p(x)\,dx > 0.$$

By construction $p(x) < 0$ outside the interval (α, β).

We now consider the polynomial $\phi_n(x)$, assume that n is so large that

$$\deg p(x) \le 2n - 1$$

and that $\phi_n(x)$ does not vanish in (α, β). Without loss of generality, $\phi_n(x) > 0$ in (α, β). Then by the Gauss quadrature formula

$$0 < \int_a^b w(x)p(x)\,dx = \sum_{k=1}^{n} p(x_k)\lambda_{k,n}.$$

But if all x_k are outside (α, β), and $\lambda_{k,n} > 0$ we necessarily have

$$\sum_{k=1}^{n} p(x_k)\lambda_{k,n} < 0.$$

The contradiction can be resolved only by having $\phi_n(x)$ vanish at least once in (α, β). ■

There is another theorem that yields more precise information on the distribution of zeros of the orthogonal polynomials. We define a sequence

$$a = y_0 < y_1 < y_2 < \cdots < y_{n-1} < y_n = b$$

as follows. The function

$$I(y) = \int_a^y w(x)\, dx$$

is a continuous and increasing function of y. We define y_k to be the unique value for which

$$I(y_k) = \sum_{j=1}^{k} \lambda_{j,n}.$$

SEPARATION THEOREM The zeros of $\phi_n(x)$, $x_1, x_2, \ldots x_n$ interlace with the above sequence. That is

$$a = y_0 < x_1 < y_1 < x_2 < \cdots < y_{n-1} < x_n < y_n = b.$$

Proof. We first construct a polynomial $\rho(x)$, of degree $2n - 2$, that simulates a stepfunction. It will be defined by the $2n - 1$ requirements

$$\begin{aligned} \rho(x_i) &= 1 & 1 &\le i \le k \\ &= 0 & k+1 &\le i \le n \\ \rho'(x_i) &= 0 & i &\ne k. \end{aligned}$$

For the case $n = 5$, $k = 3$, $\rho(x)$ has the following structure.

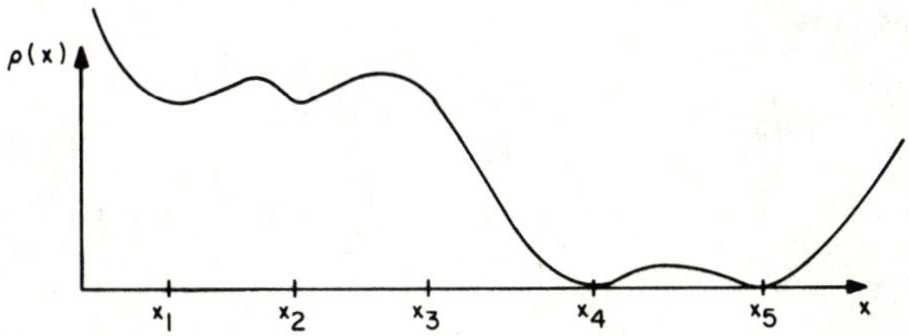

We wish to show that

$$\begin{aligned} \rho(x) &\ge 0 & &\text{for all } x \\ \rho(x) &\ge 1 & &\text{for } x \le x_k. \end{aligned}$$

It is clear that the above conditions define a unique polynomial of degree $2n - 2$. Its derivative is of degree $2n - 3$. By definition it vanishes at all x_i $(i \ne k)$, and these points are $n - 1$ in number. Also between any two successive points where $\rho(x) = 1$, $\rho'(x)$ vanishes at least once, and similarly for points where $\rho(x) = 0$. Therefore $\rho'(x)$ has to vanish at at least $n - 2$ additional points, yielding at least $2n - 3$ points where $\rho'(x)$ vanishes. But since $\rho'(x)$ is of degree $2n - 3$ this accounts for all zeros. If $\rho(x) \le 1$ for $x \le x_k$ then $\rho'(x_k) > 0$ and $\rho'(x)$ would have to vanish in (x_k, x_{k+1}), which is impossible, since all zeros were already accounted for.

The following equality is obviously correct

$$\int_a^b w(x)\rho(x)\,dx = \sum_{j=1}^{n} \lambda_{j,n}\rho(x_j) = \sum_{j=1}^{k} \lambda_{j,n} = \int_a^{y_k} w(x)\,dx.$$

We also note that

$$\int_a^b w(x)\rho(x)\,dx > \int_a^{x_k} w(x)\rho(x)\,dx > \int_a^{x_k} w(x)\,dx$$

and combining the last two statements we have

$$\int_a^{y_k} w(x)\,dx > \int_a^{x_k} w(x)\,dx$$

so that $y_k > x_k$, thus proving part of the theorem.

To complete the proof we define a new polynomial of degree $2n - 2$ such that

$$\begin{aligned} \tilde{\rho}(x_i) &= 0 \qquad 0 \le i \le k \\ &= 1 \qquad k+1 \le i \le n \\ \tilde{\rho}'(x_i) &= 0 \qquad i \ne k \end{aligned}$$

It is such that

$$\begin{aligned} \tilde{\rho}(x) &\ge 0 \qquad \text{all } x \\ \tilde{\rho}(x) &\ge 1 \qquad x \ge x_k. \end{aligned}$$

As before we have

$$\int_a^b w(x)\tilde{\rho}(x)\,dx = \sum_{j=k+1}^{n} \lambda_{j,n} = \int_a^b w(x)\,dx - \int_a^{y_k} w(x)\,dx$$

$$> \int_{x_{k+1}}^b w(x)\tilde{\rho}(x)\,dx > \int_{x_{k+1}}^b w(x)\,dx$$

from which

$$\int_a^{y_k} w(x)\,dx < \int_a^{x_{k+1}} w(x)\,dx.$$

From the above we have

$$x_{k+1} > y_k$$

completing the proof. ■

9. The Completeness of the Orthonormal Polynomials in the Space of Square-Integrable Functions

We now consider the space $L_2(w)$, which consists of all real functions for which

$$\int_a^b w(x) f^2(x)\,dx < \infty. \tag{1}$$

That the above under the standard operations of addition and multiplication by real scalars forms a linear space is a standard result in the theory of Lebesgue Integration. By introducing the inner product

$$(f, g) = \int_a^b w(x)f(x)g(x)\,dx \tag{2}$$

the space becomes an inner product space. With the above we can form the norm

$$\|f\| = \left[\int_a^b w(x)f^2(x)\,dx\right]^{1/2} = (f, f)^{1/2}. \tag{3}$$

Definition. A sequence of functions in $L_2(w)$ denoted by $\{f_n(x)\}$ is called a Cauchy sequence if for every $\varepsilon > 0$ we can find $N(\varepsilon)$ such that

$$\|f_n - f_m\| < \varepsilon \qquad n, m > N(\varepsilon).$$

Definition. A normed linear space is said to be complete if every Cauchy sequence has a limit in the space. That is, if $\{f_n(x)\}$ is a Cauchy sequence then there exists a function $f(x)$ in the space such that $\text{Lim}_{n\to\infty}\|f - f_n\| = 0$.

We are now in a position to define the important concept of a Hilbert Space.

Definition. A linear, inner product space, that is complete under the norm induced by the inner product is known as a Hilbert Space.

With these results we can now state one of fundamental theorems in the theory of Lebesgue Integration.

THEOREM The space $L_2(w)$ consisting of real function $f(x)$ for which (1) holds, with the inner product (2) and norm (3) is a Hilbert Space.

The proof of this theorem is outside the scope of the present work. Proofs can be found in numerous books: see for example the work by Riesz and Nagy on functional analysis.

Although the above theorem is true for all intervals (a, b) regardless of whether these are finite or not we shall from now on assume that (a, b) is finite, and without loss of generality the interval $(0, 1)$.

We had already shown in Section 1.7 that if we let

$$\alpha_k = (f, \phi_k)$$

then the polynomial $p(x)$ of degree n for which

$$\|f - p\|$$

is minimized is given by

$$p(x) = \sum_{k=0}^{n} \alpha_k \phi_k(x).$$

Also, since the ϕ_n are orthonormal

$$0 \le \left\| f - \sum_0^n \alpha_k \phi_k \right\|^2 = \|f\|^2 - \sum_0^n \alpha_k^2$$

so that

$$\sum_0^n \alpha_k^2 \le \|f\|^2$$

for all n. The above is known as Bessel's inequality. Letting n approach infinity we see that

$$\sum_0^\infty \alpha_k^2 \le \|f\|^2$$

and accordingly the series on the left converges. Should the above reduce to an equality then necessarily

$$\lim_{n\to\infty} \left\| f - \sum_0^n \alpha_k \phi_k \right\|^2 = \|f\|^2 - \sum_0^\infty \alpha_k^2 = 0. \tag{4}$$

The latter would imply that the quantity $\|f - p\|$ can be made as small as desirable, by a suitable choice of the polynomial $p(x)$. We shall show that, as a consequence of the completeness of $L_2(w)$, (4) is indeed correct.

Definition. An orthonormal set $\{\phi_n\}$ will be said to be complete in $L_2(w)$ if for every $f(x)$ in $L_2(w)$ we have

$$\|f\|^2 = \sum_0^\infty \alpha_k^2, \qquad \alpha_k = (f, \phi_k).$$

The latter is known as Parseval's equality.

A concept intimately related to completeness is closedness.

Definition. A set of elements $\{\phi_n\}$ in a Hilbert space is said to be closed if from

$$(f, \phi_n) = 0, \qquad n \ge 0$$

we can conclude that $f = 0$.

We shall prove later that in a Hilbert Space complete sets are closed and vice versa. In dealing with the orthonormal polynomials we have observed that the set is countably infinite. We can make the following stronger assertion.

THEOREM An orthonormal set $\{\phi\}$ in $L_2(w)$ is countably infinite.

Proof. We are now operating on the finite interval [0, 1]. The rationals $\{r_n\}$ form a countable set on (0, 1). As a result we can define a countable class of step functions $\{S_{r_i}\}$ by

$$\begin{aligned} S_{r_i}(x) &= 1 \qquad 0 \le x \le r_i \\ &= 0 \qquad r_i < x \le 1. \end{aligned}$$

Suppose $g(x)$ is a function such that

$$\int_0^1 w(x)S_{r_i}(x)g(x)\,dx = \int_0^{r_i} w(x)g(x)\,dx = 0, \qquad \text{all } i.$$

Then necessarily $g(x)$ is a null function; that is $g(x)$ vanishes, except on a set of measure zero. For

$$G(x) = \int_0^x w(t)g(t)\,dt$$

is a continuous function of x, vanishing at all rationals. Hence $G(x)$ vanishes identically. Then

$$\frac{d}{dx}\int_0^x w(t)g(t)\,dt = 0$$

and $g(x)$ vanishes almost everywhere.

Next we consider a subset $\{\phi\}_i$ of the orthonormal set $\{\phi\}$ consisting of those elements for which

$$\int_0^1 w(x)S_{r_i}(x)\phi(x)\,dx \neq 0.$$

In this fashion we can construct a countable infinity of subsets $\{\phi\}_i$ of $\{\phi\}$. At least one of these must be uncountably infinite, or else $\{\phi\}$ would be countable. Let $\{\phi\}_n$ be such a one.

We now subdivide the positive reals into intervals

$$[1, \infty), [\tfrac{1}{2}, 1), [\tfrac{1}{3}, \tfrac{1}{2}), \ldots, \left[\frac{1}{k+1}, \frac{1}{k}\right), \ldots$$

and denote these by A_k. From $\{\phi\}_n$ we select those ϕ for which the number

$$C^2 = \left[\int_0^1 w(x)S_{r_n}(x)\phi^2(x)\,dx\right]^2$$

lies in A_k. We can in this fashion subdivide the set $\{\phi\}_n$ into new sets $\{\phi\}_{n_i}$, at least one of which is uncountably infinite, say $\{\phi\}_{n_k}$.

To prove the theorem we try to approximate S_{r_n} using elements in $\{\phi\}_{n_k}$. But if

$$\alpha_j = (S_{r_n}, \phi_j), \qquad \phi_j \in \{\phi\}_{n_k}$$

then $\sum \alpha_j^2$ cannot converge since for every j, $\alpha_j^2 \geq (1/k + 1)$, thus violating Bessel's inequality. Accordingly each set $\{\phi\}_{n_i}$ is at most finite and $\{\phi\}_n$ at most countable. ■

Our next theorem is the following

THEOREM In a Hilbert Space an orthonormal set $\{\phi_n\}$ is complete if and only if it is also closed.

Proof. We shall first suppose that $\{\phi_n\}$ is closed. Let $f(x)$ be any function in $L_2(w)$ and let

$$g_n(x) = f(x) - \sum_{0}^{n} \alpha_k \phi_k(x), \qquad \alpha_k = (f, \phi_k).$$

The sequence $\{g_n(x)\}$ is a Cauchy sequence, since if $n > m$

$$\|g_n - g_m\| = \sum_{m+1}^{n} \alpha_k^2$$

and by Bessel's inequality the above becomes arbitrarily small for large n and m. Therefore there exists $g(x)$ in $L_2(w)$

$$\lim_{n\to\infty} \|g_n - g\| = 0.$$

For fixed k and $n > k$, $(g_n, \phi_k) = 0$ so that

$$|(g, \phi_k)| = |(g_n - g, \phi_k)| \leq \|g_n - g\|$$

by the Cauchy inequality. Letting n become infinite we have

$$(g, \phi_k) = 0, \qquad k \geq 0.$$

Since $\{\phi_n\}$ is closed g is a null function. Then

$$\lim_{n\to\infty} \left[\|f\|^2 - \sum_{1}^{n} \alpha_k^2\right] = \lim_{n\to\infty} \|g_n\|^2 = 0$$

so that $\{\phi_n\}$ is complete.

Now we suppose that $\{\phi_n\}$ is complete. Suppose $f(x)$ is such that

$$\alpha_k = (f, \phi_k) = 0, \qquad k \geq 0.$$

Then

$$\lim_{n\to\infty}\left[\|f\|^2 - \sum_0^n \alpha_k^2\right] = \lim_{n\to\infty} \|f\|^2 = \|f\|^2 = 0$$

so that f is a null function and $\{\phi_n\}$ is closed. ■

With this result we have concluded the preliminaries and we are now in a position to prove the key result of this section.

THEOREM The set of orthonormal polynomials $\{\phi_n\}$ is complete in the space $L_2(w)$.

Proof. To accomplish the proof we shall first show that if

$$\int_0^1 w(x)f(x)x^n\,dx = 0, \qquad n \geq 0 \tag{5}$$

and if $f(x)$ is continuous then $f(x) \equiv 0$. For every $\varepsilon > 0$ we can construct a polynomial $p(x)$ such that

$$f(x) = p(x) + r(x) \qquad \text{where } |r(x)| < \varepsilon.$$

Clearly,

$$0 = \int_0^1 w(x)f(x)p(x)\,dx = \int_0^1 w(x)f^2(x)\,dx - \int_0^1 w(x)f(x)r(x)\,dx,$$

so that

$$0 \leq \int_0^1 w(x)f^2(x)\,dx = \left|\int_0^1 w(x)f(x)r(x)\,dx\right| \leq \varepsilon \int_0^1 w(x)\,|f(x)|\,dx.$$

The above must hold for all $\varepsilon > 0$, so that the right side can be made arbitrarily small. It follows that

$$\int_0^1 w(x)f^2(x)\,dx = 0$$

and since $f(x)$ is continuous $f(x) \equiv 0$.

To complete the proof we now assume (5) to be true for some element $f(x)$ in $L_2(w)$. Let

$$F(x) = \int_0^x w(t)f(t)\,dt$$

so that $F(0) = F(1) = 0$ (by (5) for $n = 0$). Equivalently, we can write

$$\int_0^1 x^n F'(x)\,dx = 0, \qquad n \geq 0$$

and by an integration by parts

$$x^n F(x)\Big|_0^1 - n\int_0^1 x^{n-1}F(x)\,dx = -n\int_0^1 x^{n-1}F(x)\,dx = 0, \qquad n \geq 0.$$

Since $F(x)$ is continuous we can conclude from the above that

$$F(x) = \int_0^x w(t)f(t)\,dt \equiv 0.$$

It follows that

$$w(x)f(x) = 0 \qquad \text{almost everywhere}$$

and since $w(x) > 0$ almost everywhere $f(x)$ must be a null function. ∎

It should be noted how strongly this proof depends on the finiteness of the interval. The Weierstrass theorem is not valid on infinite intervals. For counter examples on infinite intervals, see, for example, Kaczmarz and Steinhaus. In the chapter on the classical orthogonal polynomials we shall encounter the Laguerre and Hermite polynomials defined on the intervals $(0, \infty)$ and $(-\infty, \infty)$ respectively. Separate proofs for their completeness will have to be furnished.

In the preceding theorem we showed that the orthogonal polynomials are closed in $L_2(w)$ and hence also complete. In other words if $f(x)$ is any element in $L_2(w)$ and if

$$\alpha_k = (f, \phi_k)$$

then

$$\lim_{n\to\infty} \left\| f - \sum_{k=0}^{n} \alpha_k \phi_k \right\| = 0.$$

10. Generalizations and an Application to Conformal Mappings

Suppose C is a non-selfintersecting, rectifiable arc in the complex plane. If s denotes the arclength parameter and $w(s)$ a suitable integrable and non-negative function then we could build up an analogous theory of orthonormal polynomials to the one discussed earlier. There C was an interval on the real axis. One of the chief differences lies in the fact that we are now dealing with complex quantities. As inner product we define

$$(\phi_i, \phi_j) = \int_C w(s)\phi_i(z)\overline{\phi_j(z)}\,ds.$$

In particular we shall now restrict ourselves to the case where C is a finite closed curve, and $w(s) \equiv 1$. In analogy to previous results we define the function

$$K_n(a, z) = \sum_{k=0}^{n} \overline{\phi_k(a)}\phi_k(z).$$

The above is a reproducing kernel on the space of polynomials of degree n. We obtain, exactly as in I.4

$$(p(z), K_n(a, z)) = \int_C p(z)\overline{K_n(a, z)}\, ds = p(a).$$

As in the theorem of I.4 we find that the polynomial

$$G_n(z) = \frac{\varepsilon K_n(a, z)}{K_n^{1/2}(a, a)}, \quad |\varepsilon| = 1$$

is such that

$$\|G_n\|^2 = (G_n, G_n) = \int_C |G_n(z)|^2\, dz = 1$$

and

$$|G_n(a)| \geq |p(a)|$$

where $p(z)$ is any polynomial of unit norm and degree n. What is more $G_n(z)$ uniquely satisfies the above maximum property, except for the choice of ε. We shall show that these functions allow us to construct a broad class of conformal mapping functions.

First we recall the following facts. Consider a domain D enclosed by C which is simply connected and has at least two boundary points. According to the Riemann mapping theorem there exists an analytic function, say $\gamma(z)$, mapping this domain into the unit circle in a one to one manner. Such a mapping is not unique. However if a denoted a point in the domain and we required that $\gamma(a) = 0$ and $\gamma'(a) > 0$, and hence of course real, then $\gamma(z)$ is unique. The proof of this theorem is not constructive in character, but by means of the functions $K_n(a, z)$ we can approximate $\gamma(z)$ as closely as desired.

THEOREM Let C be a simple, closed, oriented, rectifiable curve enclosing a finite domain, and

$$K_n(a, z) = \sum_{k=0}^{n} \overline{\phi_k(a)}\, \phi_k(z)$$

where the $\phi_k(z)$ are orthonormal polynomials on C. The mapping function, mapping the domain into the unit circle, such that

$$\gamma(a) = 0, \qquad \gamma'(a) > 0$$

is given by

$$\gamma(z) = \lim_{n\to\infty} \frac{2\pi}{K_n(a, a)} \int_a^z K_n^2(a, \zeta)\, d\zeta.$$

Proof. To prove the above we first show that the function

$$K(a, z) = \lim_{n\to\infty} K_n(a, z)$$

exists.

We can show that in any closed subdomain of the interior of C $K_n(a, z)$ is uniformly (in n) bounded. From the Cauchy integral formula we have

$$K_n(a, a) = \frac{1}{2\pi i} \int_C \frac{K_n^2(a, z)}{(z - a)K_n(a, a)}\, dz \le \frac{1}{2\pi} \int \frac{|G_n(z)|^2\, ds}{|z - a|}$$

$$\le \frac{1}{2\pi\delta} \int_C |G_n(z)|^2\, ds = \frac{1}{2\pi\delta}.$$

Here we use the fact that $G_n(z)$ has unit norm and δ is the distance from a to C.

If δ denotes the smaller of the distances of a and z from C we have by means of the Cauchy-Schwarz inequality.

$$|K_n(a, z)|^2 \le K_n(a, a)K_n(z, z) \le \frac{1}{4\pi^2\delta^2}.$$

Hence $K_n(a, z)$ is uniformly bounded in every closed subdomain so that $K(a, z)$ exists. Correspondingly, the function $G(z)$ defined by

$$G(z) = \lim_{n\to\infty} \frac{K_n(a, z)}{K_n^{1/2}(a, a)}$$

exists. It also satisfies

$$\int_C |G(z)|^2\, ds = 1.$$

The function $G(z)$ is also a solution of a maximum problem. Let $\gamma(z)$ denote the mapping function of the interior of C into the unit circle, and $g(\gamma)$ its inverse mapping. Then

$$ds = \frac{d\theta}{|\gamma'(z)|} = |g'(\gamma)|\, d\theta$$

where $d\theta$ is the arclength on the unit circle, and by integrating in the γ plane

$$\int_C |G(z)|^2\, ds = \int_0^{2\pi} |G(g(\gamma))|^2\, |g'(\gamma)|\, d\theta$$

so that

$$\int_0^{2\pi} |G(g(\gamma))g'(\gamma)^{1/2}|^2\, d\theta = 1,$$

where $\gamma = e^{i\theta}$.

If we let $F(\gamma) = G(g(\gamma))g'(\gamma)^{1/2} = \sum_{n=0}^{\infty} a_n \gamma^n$ we have, by the above integral

$$2\pi \sum_0^{\infty} |a_n|^2 = 1.$$

To maximize $|G(a)|^2$ we must equivalently maximize $|F(\gamma)g'(\gamma)^{-1/2}|^2$ at $\gamma = 0$. To do so it is necessary that $F(\gamma) = F(0) = a_0$, where $2\pi|a_0|^2 = 1$. Then

$$\max |G(a)|^2 = \frac{1}{2\pi\,|g'(0)|} = \frac{\gamma'(a)}{2\pi}.$$

Recall in the above that $\gamma'(a) > 0$.

Lastly we consider the following integral

$$J_n = \int_C |K_n(a, z) - \frac{1}{2\pi}\{\gamma'(a)\gamma'(z)\}^{1/2}|^2\, ds.$$

We shall show that $\lim_{n\to\infty} J_n = 0$, which in essence is equivalent to the assertion of the theorem. By expansion of the above integral we have

$$J_n = \int_C |K_n(a, z)|^2\, ds + \frac{\gamma'(a)}{(2\pi)^2}\int_C |\gamma'(z)|\, ds - \frac{1}{\pi}\, Re \int_C K_n(a, z)[\gamma'(a)\,\overline{\gamma'(z)}]^{1/2}\, ds.$$

For the first integral we have

$$\int_C |K_n(a, z)|^2\, ds = K_n(a, a)\int_C |G_n(z)|^2\, ds = K_n(a, a).$$

To evaluate the second we perform the integration in the γ plane.

$$\frac{\gamma'(a)}{(2\pi)^2}\int_c |\gamma'(z)|\, ds = \frac{\gamma'(a)}{(2\pi)^2}\int_0^{2\pi} d\theta = \frac{\gamma'(a)}{2\pi}.$$

Lastly we have

$$\begin{aligned}\int_c K_n(a, z)[\gamma'(a)\overline{\gamma'(z)}]^{1/2}\, ds &= \int_0^{2\pi} K_n(a, g(\gamma))\left[\frac{\gamma'(a)}{\overline{g'(\gamma)}}\right]^{1/2} |g'(\gamma)|\, d\theta \\ &= \gamma'(a)^{1/2}\int_0^{2\pi} K_n(a, g(\gamma))g'(\gamma)^{1/2}\, d\theta \\ &= \gamma'(a)^{1/2} 2\pi K_n(a, g(0))g'(0)^{1/2} = 2\pi K_n(a, a)\end{aligned}$$

where we used the mean-value theorem for harmonic functions. Then

$$J_n = K_n(a, a) + \frac{\gamma'(a)}{2\pi} - 2K_n(a, a) = \frac{\gamma'(a)}{2\pi} - K_n(a, a).$$

As we saw before

$$\max |G(a)|^2 = \frac{\gamma'(a)}{2\pi} = \lim_{n\to\infty} K_n(a, a)$$

so that $\lim_{n\to\infty} J_n = 0$,

Now it follows that

$$\gamma'(z) = \lim_{n\to\infty} \frac{(2\pi)}{\gamma'(a)} K_n^2(a, z) = \lim_{n\to\infty} 2\pi \frac{K_n^2(a, z)}{K_n(a, a)}$$

and since $\gamma(a) = 0$

$$\gamma(z) = \lim_{n\to\infty} \frac{2\pi}{K_n(a, a)} \int_a^z K_n^2(a, \zeta)\, d\zeta. \qquad \blacksquare$$

As an application of the above we consider the problem of mapping a unit circle onto itself so that $\gamma(a) = 0$ and $\gamma'(a) > 0$. We let C be the unit circle in the z plane. Clearly

$$\phi_k(z) = \frac{1}{(2\pi)^{1/2}} z^k$$

since

$$(\phi_j, \phi_k) = \int_C \phi_j(z) \overline{\phi_k(z)}\, ds = \frac{1}{2\pi} \int_0^{2\pi} e^{i(j-k)\theta}\, d\theta = \delta_{j,k}.$$

Also

$$K(a, z) = \sum_{k=0}^{\infty} \overline{\phi_k(a)}\, \phi_k(z) = \frac{1}{2\pi} \sum_0^{\infty} (\bar{a}z)^k = \frac{1}{2\pi(1 - \bar{a}z)}.$$

Finally we have

$$\gamma(z) = 4\pi^2(1 - |a|^2) \int_a^z \frac{d\zeta}{4\pi^2(1 - \bar{a}\zeta)^2} = \frac{z - a}{1 - \bar{a}z}$$

which is of course, the familiar bilinear mapping.

Exercises

1. Show that the set of polynomials of degree at most n, with real coefficients, in a real variable x forms a linear space, under the standard operations.

2. Show that if $w(x)$ is a positive function of x, then for the linear space of exercise 1 the expression

$$(p, q) = \int_{-1}^{1} w(x)\, p(x)\, q(x)\, dx$$

yields an inner product.

3. Show that the trigonometric polynomials with real coefficients, that is sums of the type

$$p(x) = \tfrac{1}{2}a_0 + \sum_{k=1}^{n} a_k \cos k\,x + b_k \sin k\,x,$$

form a linear space. Show that

$$(p, q) = \int_{-\pi}^{\pi} p(x)\, q(x)\, dx$$

yield an inner product and that the polynomials

$$p_k(x) = \cos k\,x, \quad q_k(x) = \sin k\,x$$

form an orthogonal set.

4. The dimension of a finite dimensional linear space was defined as the number of elements in a basis. A basis is not unique. Show that the dimension is independent of the basis selected.

5. Show that the space of continuous real functions on the interval (0, 1) forms a linear space under the standard operations of addition and multiplication.

6. Show that

$$(f, g) = \int_{0}^{1} f(x)\, g(x)\, dx$$

forms an inner product on the space of exercise 5.

7. Show that the norm defined by

$$\|f(x)\| = \sup_{x \in (0,1)} |f(x)|$$

has all the properties of a norm on the space of exercise 5. Note that a different norm can be constructed from the inner product of exercise 6.

8. For the following weight function and intervals construct $\phi_0, \phi_1, \phi_2, \phi_3$.

(a) $w(x) = \sin \pi\, x$, $(-1, 1)$

(b) $w(x) = \sqrt{1 - x^2}$, $(-1, 1)$

(c) $w(x) = e^x$ $(0, 1)$

(d) $w(x) = x\, e^{-x}$, $(0, \infty)$

(e) $w(x) = e^{-x^2/2}$, $(-\infty, \infty)$

9. If x_0 is a zero of $\phi_n(x)$ show that

$$x_0 = \frac{\int_a^b w(x)x\,[\{\phi_n(z)\}/(x-x_0)]^2\,dx}{\int_a^b w(x)\,[\{\phi_n(x)\}/(x-x_0)]^2\,dx}.$$

10. Use the Gauss Quadrature Formula to obtain estimates for π from

$$\frac{\pi}{4} = \int_0^1 \frac{dx}{1+x^2}.$$

Estimate the error. Obtain another estimate from

$$\frac{\sqrt{\pi}}{2} = \int_0^\infty e^{-x^2}\,dx$$

11. Show that the results of 1.10 apply to ordinary Fourier series.

12. Prove that the orthogonal set $\{\cos nx, \sin nx\}$ is complete on the space of square integrable functions on the interval $(-\pi, \pi)$. (*Hint:* Consider the properties of the polynomial $R_n(x) = [\delta + \cos x]^n$.)

13. Provide a detailed proof of the results in regard to the function $G_n(z)$ used in Section 1.11. There it is stated that $\|G_n(z)\| = 1$ and among all polynomials of degree n $G_n(z)$ is one for which $|G_n(z)|$ is maximized.

CHAPTER 2

The Classical Orthogonal Polynomials

1. Rodrigues' Formula and the Classical Orthogonal Polynomials

We shall first consider the case of finite intervals, and without loss of generality follow the convention of assuming the interval to be $[-1, 1]$. The functions

$$p_n(x) = \frac{1}{w(x)} \left(\frac{d}{dx}\right)^n w(x)(1 - x^2)^n, \qquad n = 0, 1, 2, \ldots, \tag{1}$$

where $w(x)$ is sufficiently differentiable, form a set with the property that

$$\int_{-1}^{1} w(x)x^k p_n(x)\, dx = 0, \qquad 0 \leq k < n, \tag{2}$$

as will be shown shortly. In other words each $p_n(x)$ is orthogonal to every polynomial of degree less than n. Should $w(x)$ be so constituted that $p_n(x)$ is indeed a polynomial then clearly, by (2), these must be orthogonal polynomials with weight function $w(x)$ over the interval $[-1, 1]$.

To prove (2), we substitute (1) in (2) and integrate by parts repeatedly

$$\int_{-1}^{1} x^k \left(\frac{d}{dx}\right)^n w(x)(1 - x^2)^n\, dx$$

$$= \sum_{l=0}^{k} (-1)^l k(k-1)\cdots(k-l+1)x^{k-l}\left(\frac{d}{dx}\right)^{n-l-1} w(x)(1-x^2)^n \Big|_{-1}^{1} = 0. \tag{3}$$

Of course, we suppose that $w(x)$ is sufficiently regular at the endpoints that each of the above terms vanishes.

To obtain some information about the structure of $w(x)$ we examine

$$p_1(x) = \frac{w'(x)}{w(x)}(1 - x^2) - 2x.$$

For $p_1(x)$ to be a polynomial of the first degree it is necessary that

$$\frac{w'(x)}{w(x)}(1-x^2)-2x = Ax + B.$$

The above differential equation is solved easily and one finds that

$$w(x) = (1-x)^\alpha(1+x)^\beta \tag{4}$$

where α and β can be expressed in terms of A and B. For $\alpha > -1$, $\beta > -1$ the integrals in (2) exists for all k and n such that $k < n$ and also each term in (3) vanishes. Furthermore each $p_n(x)$ in (1) is a polynomial. To see this we apply Leibnitz' rule

$$\left(\frac{d}{dx}\right)^n f(x)\,g(x) = \sum_{k=0}^{n} \frac{n!}{k!(n-k)!}\left[\left(\frac{d}{dx}\right)^k f(x)\right]\left[\left(\frac{d}{dx}\right)^{n-k} g(x)\right]$$

to (1), when $w(x)$ is given by (4).

$$p_n(x) = \sum_{k=0}^{n} \frac{n!}{k!(n-k)!}\left[(1-x)^{-\alpha}\left(\frac{d}{dx}\right)^k(1-x)^{n+\alpha}\right]$$
$$\left[(1+x)^{-\beta}\left(\frac{d}{dx}\right)^{n-k}(1+x)^{n+\beta}\right]$$

It is easy to see that each of the above bracketed terms is a polynomial, and that each term in the above sum is a polynomial of precise degree n.

(1) does not define a unique set of orthogonal polynomials. The terms in (1) may be multiplied by arbitrary constants. The Jacobi polynomials are defined by convention as

$$P_n^{\alpha,\beta}(x) = \frac{(-1)^n}{2^n n!}(1-x)^{-\alpha}(1+x)^{-\beta}\left(\frac{d}{dx}\right)^n[(1-x)^\alpha(1+x)^\beta(1-x^2)^n]. \tag{5}$$

For $\alpha = \beta = 0$ we obtain the Legendre polynomials

$$P_n(x) = \frac{(-1)^n}{2^n n!}\left(\frac{d}{dx}\right)^n(1-x^2)^n. \tag{6}$$

For $\alpha = \beta = -\frac{1}{2}$, and a different multiplicative constant we have the Tchebicheff polynomials

$$T_n(x) = \frac{(-1)^n 2^n}{(2n)!}(1-x^2)^{1/2}\left(\frac{d}{dx}\right)^n(1-x^2)^{n-1/2}. \tag{7}$$

As a generalization of the above polynomials we have the Gegenbauer polynomials. These are special cases of the Jacobi polynomials, given by $\alpha = \beta \equiv \gamma - \frac{1}{2}$, with $\gamma > -\frac{1}{2}$. We then have

$$C_n^\gamma(x) = \frac{(-1)^n(n+2\gamma-1)(n+2\gamma-2)\cdots(2\gamma)}{n!2^n(n+\gamma-\frac{1}{2})(n+\gamma-\frac{3}{2})\cdots(\gamma+\frac{1}{2})} \left[(1-x^2)^{-\gamma+\frac{1}{2}}\left(\frac{d}{dx}\right)^n(1-x^2)^{n+\gamma-\frac{1}{2}}\right] \tag{8}$$

We can treat the case of the semi-infinite interval $(0, \infty)$ in a similar fashion. In analogy to (1) we let

$$p_n(x) = \frac{1}{w(x)}\left(\frac{d}{dx}\right)^n w(x)x^n, \qquad n = 0, 1, 2, \ldots, \tag{9}$$

(2) can be shown to hold for this case. To deduce the nature of $w(x)$ we have

$$p_1(x) = \frac{w'(x)x}{w(x)} + 1 = B - x.$$

In the above the coefficient of x was taken to be -1, since by a linear transformation on the interval $(0, \infty)$ this can always be accomplished. Now we find

$$w(x) = x^\alpha e^{-x} \tag{10}$$

where $\alpha > -1$, and α is related to B. We can verify as before that (9) with the above choice of $w(x)$ leads to polynomials.

(10) can be considered as a limiting case of (4). To see this let

$$x = 1 - 2\rho t \qquad \beta = \frac{1}{\rho}$$

in (4). By introducing a suitable constant

$$w_\rho(t) = \frac{(2\rho)^\alpha 2^\beta t^\alpha(1-\rho t)^{1/\rho}}{(2\rho)^\alpha 2^\beta}$$

$$w(t) = \lim_{\rho\to 0} w_\rho(t) = t^\alpha e^{-t}.$$

The Laguerre polynomials are traditionally defined by

$$L_n^\alpha(x) = \frac{1}{n!}x^{-\alpha}e^x\left(\frac{d}{dx}\right)^n e^{-x}x^{n+\alpha}. \tag{11}$$

In a similar fashion we can arrive at the Hermite polynomials. We consider the set of function defined by

$$p_n(x) = \frac{1}{w(x)}\left(\frac{d}{dx}\right)^n w(x), \qquad n = 0, 1, 2, \ldots, \tag{12}$$

defined over $(-\infty, \infty)$. As before we can show that (2) holds. From

$$p_1(x) = \frac{w'(x)}{w(x)} = -x$$

(where by a suitable linear transformation of $(-\infty, \infty)$ the general polynomial of first degree takes the above form) we have

$$w(x) = e^{-x^2/2}. \tag{13}$$

The Hermite polynomials are given by

$$H_n(x) = (-1)^n e^{x^2/2} \left(\frac{d}{dx}\right)^n e^{-x^2/2}. \tag{14}$$

Equation (14), as is the case with (10), can be considered as a limiting case of (4). If we let $\alpha = \beta = 1/(2\rho^2)$ and $x = \rho t$ we have

$$w_\rho(x) = (1 - \rho^2 t^2)^{1/2\rho^2}.$$

Letting $\rho \to 0$ we have finally

$$\lim_{p\to 0} (1 - \rho^2 t^2)^{1/2\rho^2} = e^{-t^2/2}.$$

From the previous chapter we see that these orthogonal polynomials must satisfy a recurrence relationship. In Table I the coefficients for these are shown. Furthermore, in view of the fact that for the recurrence formula to be usable two initial cases must be known, the first two polynomials for each of these are indicated.

TABLE 1

Polynomials	Weight Function	Interval	A_n
Jacobi $P_n^{\alpha,\beta}(x)$	$(1-x)^\alpha(1+x)^\beta$ $\alpha, \beta > -1$	$(-1, 1)$	$\dfrac{(2n+\alpha+\beta+1)(2n+\alpha+\beta+2)}{(2n+2)(n+\alpha+\beta+1)}$
Gegenbauer: $C_n^\gamma(x)$	$(1-x^2)^{\gamma-\frac{1}{2}}$ $\alpha=\beta=\gamma-\frac{1}{2}, \gamma>-\frac{1}{2}$	$(-1, 1)$	$\dfrac{2(\gamma+n)}{n+1}$
Tchebicheff: $T_n(x)$	$(1-x^2)^{-\frac{1}{2}}$ $\gamma=0$	$(-1, 1)$	2
1 Legendre: $P_n(x)$	1 $\gamma=\frac{1}{2}$	$(-1, 1)$	$\dfrac{2n+1}{n+1}$
2 Laguerre: $L_n^\alpha(x)$	$x^\alpha e^{-x}$ $\alpha>-1$	$(0, \infty)$	$-\dfrac{1}{n+1}$
3 Hermite: $H_n(x)$	$e^{-x^2/2}$	$(-\infty, \infty)$	1

1 ORBITAL ANGULAR MOMENTUM EIGENFNS
2. HYDROGEN ATOM RADIAL EIGENFNS
3. QUANTUM HARMONIC OSCILLATOR ENERGY EIGENFNS
2/24/88 [illegible]

TABLE 1 (*Continued*)

B_n	C_n	P_0	P_1
$\dfrac{(n+\alpha+\beta+1)(\alpha^2-\beta^2)}{(2n+2)(n+\alpha+\beta+1)(2n+\alpha+\beta)}$	$\dfrac{(2n+\alpha)(n+\beta)(2n+\alpha+\beta+2)}{(2n+2)(n+\alpha+\beta+1)(2n+\alpha+\beta)}$	1	$\dfrac{\alpha-\beta}{2}+\left(1+\dfrac{\alpha+\beta}{2}\right)x$
0	$\dfrac{2\gamma+n-1}{n+1}$	1	$2\gamma x$
0	1	1	x
0	$\dfrac{n}{n+1}$	1	x
$\dfrac{2n+\alpha+1}{n+1}$	$\dfrac{n+\alpha}{n+1}$	1	$1+\alpha-x$
0	n	1	x

2. The Differential Equations Satisfied by the Classical Orthogonal Polynomials

In solving the partial differential equations of mathematical physics by the method of separation of variables we encounter different second order differential equations. Many of these are solvable in terms of the classical orthogonal polynomials or related functions to be encountered in subsequent chapters.

To derive the differential equation satisfied by the Jacobi polynomial we consider the expression

$$\frac{d}{dx}\{(1-x^2)w(x)\phi_n'(x)\}$$
$$= w(x)\{(1-x^2)\phi_n''x) + [\beta-\alpha-(\alpha+\beta+2)x]\phi_n'(x)\} \tag{1}$$

where $w(x) = (1-x)^\alpha(1+x)^\beta$, $\phi_n(x) = P_n^{\alpha,\beta}(x)$. The expression in brackets on the right side in the above is clearly a polynomial of degree n, and we can therefore write

$$\frac{d}{dx}\{(1-x^2)w(x)\phi_n'(x)\} = w(x)\sum_{j=0}^{n}\alpha_j\phi_j(x). \tag{2}$$

By the orthogonality property we can evaluate the coefficient.

$$\alpha_k\|\phi_k\|^2 = \int_{-1}^{1}\phi_k(x)\left\{\frac{d}{dx}(1-x^2)w(x)\phi_n'(x)\right\}dx.$$

Integrating by parts twice, and noting that the integrated terms vanish we have

$$\alpha_k \|\phi_k\|^2 = \int_{-1}^{1} \phi_n(x)\left\{\frac{d}{dx}(1-x^2)w(x)\phi_k'(x)\right\}dx. \tag{3}$$

We can rewrite (3) in the form

$$\alpha_k \|\phi_k\|^2 = \int_{-1}^{1} w(x)\phi_n(x)p(x)\,dx.$$

where $p(x)$ is a polynomial of degree k so that

$$\alpha_k = 0 \qquad k < n.$$

(2) can now be rewritten as

$$\frac{d}{dx}\{(1-x^2)w(x)\phi_n'(x)\} = w(x)\alpha_n\,\phi_n(x).$$

To evaluate α_n, the easiest approach is to use II.1.5 and to compare the coefficients of the highest powers of x. In this fashion we are led to

$$\left\{(1-x^2)\left(\frac{d}{dx}\right)^2 + [(\beta-\alpha)-(\alpha+\beta+2)x]\frac{d}{dx} + n(n+\alpha+\beta+1)\right\} P_n^{\alpha,\beta}(x) = 0 \tag{4}$$

and similarly

$$\left\{(1-x^2)\left(\frac{d}{dx}\right)^2 - (2\gamma+1)x\frac{d}{dx} + n(n+2\gamma)\right\}C_n^{\gamma}(x) = 0 \tag{5}$$

$$\left\{(1-x^2)\left(\frac{d}{dx}\right)^2 - x\frac{d}{dx} + n^2\right\}T_n(x) = 0 \tag{6}$$

$$\left\{(1-x^2)\left(\frac{d}{dx}\right)^2 - 2x\frac{d}{dx} + n(n+1)\right\}P_n(x) = 0. \tag{7}$$

A similar analysis applies to the Hermite and Laguerre polynomials. We can show that

$$\left\{x\left(\frac{d}{dx}\right)^2 + (\alpha+1-x)\frac{d}{dx} + n\right\}L_n^{\alpha}(x) = 0 \tag{8}$$

$$\left\{\left(\frac{d}{dx}\right)^2 - x\frac{d}{dx} + n\right\}H_n(x) = 0. \tag{9}$$

3. On the Zeros of the Jacobi Polynomials

We know, of course, that $P_n^{\alpha,\beta}(x)$ has n real zeros in the interval $(-1, 1)$. These have an interesting application in potential theory. Suppose that we place two positive charges of strengths p and q at the points $+1$ and -1 respectively. In addition we shall distribute n unit charges throughout the interval at points $x_1, x_2, \ldots, x_n$. We shall show that when this system is in equilibrium the set $\{x_i\}$ represents the zeros of a certain Jacobi polynomial.

Let

$$T = \prod_{k=1}^{n}(1 - x_k)^p(1 + x_k)^q \prod_{\substack{j,k=1 \\ j<k}}^{n} |x_j - x_k|.$$

Then $\ln T^{-1}$ represents the potential energy of the system. A condition that is necessary at an equilibrium state is that the potential energy take on an extremal value; then

$$\frac{\partial}{\partial x_k} \ln T^{-1} = 0, \qquad k = 1, 2, \ldots, n.$$

Using the expression for T we find for the above condition

$$\frac{p}{x_k - 1} + \frac{q}{x_k + 1} + \sum_{j \neq k} \frac{1}{x_k - x_j} = 0, \qquad k = 1, 2, \ldots n. \tag{1}$$

The latter represents a system of n equations in n unknowns. The system is nonlinear and in general such systems are difficult, if not impossible, to solve. To solve this particular system we define the polynomial

$$f(x) = (x - x_1)(x - x_2) \cdots (x - x_n),$$

and seek the value of $f''(x)/f'(x)|_{x=x_k}$. We write

$$f(x) = (x - x_k)g(x), \qquad g(x) = \frac{f(x)}{x - x_k}.$$

Then

$$f'(x) = (x - x_k)g'(x) + g(x)$$

$$f''(x) = (x - x_k)g''(x) + 2g'(x)$$

so that

$$\frac{1}{2}\frac{f''(x_k)}{f'(x_k)} = \frac{g'(x_k)}{g(x_k)}.$$

But

$$\frac{g'(x)}{g(x)} = \sum_{j \neq k} \frac{1}{x - x_j}$$

so that

$$\frac{1}{2}\frac{f''(x_k)}{f'(x_k)} = \sum_{j \neq k} \frac{1}{x_k - x_j}. \tag{2}$$

By means of (2), (1) can be rewritten as

$$\frac{p}{x_k - 1} + \frac{q}{x_k + 1} + \frac{1}{2}\frac{f''(x_k)}{f'(x_k)} = 0$$

or equivalently

$$(1 - x_k^2)\, f''(x_k) + [2(q - p) - 2(q + p)x_k]\, f'(x_k) = 0 \tag{3}$$

By inspection we note that the polynomial

$$(1 - x^2)f''(x) + [2(q - p) - 2(q + p)x]f'(x)$$

is of degree n and vanishes at all n values x_k by (3). It follows necessarily that

$$(1 - x)^2 f''(x) + [2(q - p) - 2(q + p)x]f'(x) = cf(x)$$

for a suitable constant c. To determine the latter we compare the coefficients of x^n on both sides of the equation. Then we find that

$$c = -n(n + 2q + 2p - 1)$$

so that

$$(1 - x^2)f''(x) + [2(q - p) - 2(q + p)x]f'(x) + n(n + 2q + 2p - 1)f(x) = 0$$

A comparison with Eq. 2.2.4 now shows that

$$f(x) = cP_n^{2p-1,\, 2q-1}(x)$$

thereby proving that the points $\{x_k\}$ coincide with the zeros of a certain Jacobi polynomial.

4. An Alternative Approach to the Tchebicheff Polynomials

We first state and prove a theorem due to Pólya.

THEOREM Let $k(x)$ and $g(x)$ denote two positive, and continuously differentiable functions. Consider the differential equation

$$\frac{d}{dx} k(x)y' + g(x)y = 0. \tag{1}$$

If $k(x)g(x)$ is non-increasing (non-decreasing) then the relative maxima of y^2 are non-decreasing (non-increasing).

Proof. Let

$$f(x) = y^2 + \frac{(ky')^2}{kg}.$$

Whenever y^2 has a relative maximum y' must vanish. Then

$$f(x) = y^2 \qquad \text{when} \quad y' = 0.$$

Also

$$f'(x) = 2yy' + \frac{2ky'}{kg}(ky')' + \left(\frac{1}{kg}\right)'(ky')^2 = \left(\frac{1}{kg}\right)'(ky')^2$$

if we use (1) in the above. If we suppose that kg is non-increasing it follows that $f'(x) \geq 0$ so that $f(x)$ is non-decreasing.

Suppose x_1 and x_2 are two successive zeros of y', such that $x_1 < x_2$. Since $f(x_2) \geq f(x_1)$ we have also $y^2(x_2) \geq y^2(x_1)$. Thus the relative maxima of y^2 form a non-decreasing set. ■

The Tchebicheff polynomials satisfy

$$(1 - x^2)y'' - xy' + n^2y = 0$$

which can be rewritten in the form

$$\frac{d}{dx}(1 - x^2)^{1/2}y' + n^2(1 - x^2)^{-1/2}y = 0.$$

If Pólya's theorem is applied to this equation we find

$$kg = n^2$$

so that kg is constant. It follows that all relative maxima of $T_n^{\,2}(x)$ must assume the same value.

The above property of $T_n(x)$ may be used as the defining property. Suppose we seek a polynomial $f_n(x)$ of degree n, such that

$$f_n^{\,2}(x) = A^2$$

whenever $f_n{}'(x) = 0$. That is, at all x, where $f_n^{\,2}(x)$ has a relative maximum $f_n^{\,2}(x)$ attains the same value. Clearly at these points $f_n^{\,2}(x) - A^2$ and $[f'(x)]^2$ both have double zeros. Then the function

$$\frac{f_n^{\,2}(x) - A^2}{[f_n'(x)^2]}$$

is a rational function and all zeros of the denominator also occur in the numerator. Therefore the above is a quadratic, and without loss of generality we have

$$\frac{f_n^2(x) - A^2}{[f_n'(x)]^2} = k(x^2 - 1). \tag{2}$$

This implies that

$$f_n^2(\pm 1) = A^2.$$

(2) can be considered to be a differential equation for $f_n(x)$. Letting

$$f_n(x) = A \cos \theta, \qquad x = \cos \phi$$

(2) becomes

$$\left(\frac{d\theta}{d\phi}\right)^2 = \frac{1}{k}$$

so that

$$\theta = \pm \frac{\phi}{\sqrt{k}} + c.$$

Finally we have

$$f_n(x) = A \cos\left[\frac{1}{\sqrt{k}} \cos^{-1} x + c\right]$$

but we have not, as yet, made sure that $f_n(x)$ is a polynomial. We return to (2), divide by x^2 and let x approach infinity. Then, inserting for $f_n(x)$ a polynomial of degree n, we obtain

$$\frac{1}{n^2} = k$$

so that

$$f_n(x) = A \cos [n \cos^{-1} x + c].$$

To determine c we note that

$$f_n^2(\pm 1) = A^2 = A^2 \cos [n \cos^{-1} (\pm 1) + c]$$

so that $c = 0$ and

$$f_n(x) = A \cos [n \cos^{-1} x]. \tag{3}$$

We still wish to show that the polynomials (3) are essentially the Tchebicheff polynomials. One can, of course, verify that they are orthogonal on $[-1, 1]$

with weight function $(1 - x^2)^{-1/2}$, by a direct calculation. An easier method, however, is to return to (2) in the form

$$n^2 f_n^2(x) - n^2 A^2 = (x^2 - 1)[f_n'(x)]^2.$$

After a differentiation and a rearrangement of terms we find

$$(1 - x^2) f_n''(x) - x f_n'(x) + n^2 f_n(x) = 0$$

which is the differential equation satisfied by the Tchebicheff polynomials. Note that (3) yields an explicit representation for the Tchebicheff polymomials.

5. An Application of the Hermite Polynomials to Quantum Mechanics

We now consider the Hermite differential equation

$$H_n'' - xH_n' + nH_n = 0 \tag{1}$$

and the related equation

$$U_n'' + [n + \tfrac{1}{2} - \tfrac{1}{4}x^2]U_n = 0. \tag{2}$$

A simple calculation shows that

$$U_n(x) = e^{-x^2/4} H_n(x).$$

From the general theory of orthogonal polynomials we know that $H_n(x)$ has precisely n real zeros, and the same is clearly true for $U_n(x)$. From Rodrigues formula II.1.14 we see that for n even $H_n(x)$ is an even function of x, and for n odd $H_n(x)$ is an odd function of x. Accordingly all zeros of $H_n(x)$ are placed symmetrically with respect to the origin. It follows that between any two consecutive zeros of $U_n(x)$, $U_n'(x)$ vanishes at least once. Since

$$U_n'(x) = e^{-x^2/4}\left[H_n'(x) - \frac{x}{2} H_n(x)\right]$$

$U_n'(x)$ must vanish precisely $n + 1$ times. Denote the zeros of $H_n(x)$ by $x_1 < x_2 < \cdots < x_n$. Then, in view of the fact that $U_n(x)$ vanishes in the limit as x becomes large, $U_n'(x)$ must have precisely two zeros outside the interval (x_1, x_n), and $n - 1$ inside.

Similarly we note that $U_n''(x)$ must have at least one zero between any two consecutive zeros of $U_n'(x)$ and at least two more. Thus $U_n''(x)$ must have at least $n + 2$ zeros. But

$$U_n''(x) = -[n + \tfrac{1}{2} - \tfrac{1}{4}x^2] e^{-x^2/4} H_n(x)$$

has precisely $n + 2$ zeros, namely those of $H_n(x)$ and the two corresponding to the quadratic factor. Thus we see that

$$|x_k| < 2\sqrt{n + \tfrac{1}{2}}, \qquad k = 1, 2, \ldots, n,$$

yielding a bound on the magnitude of the zeros of $H_n(x)$.

If we now apply Polya's theorem to (2) with

$$k(x)g(x) = n + \tfrac{1}{2} - \tfrac{1}{4}x^2$$

which is a decreasing function for $x > 0$, we see that the relative maxima of $U_n{}^2(x)$, all of which lie in the interval $(-2\sqrt{n + \frac{1}{2}}, 2\sqrt{n + \frac{1}{2}})$, form a decreasing set for negative x and an increasing set for positive x. A typical case is furnished by $U_7(x)$.

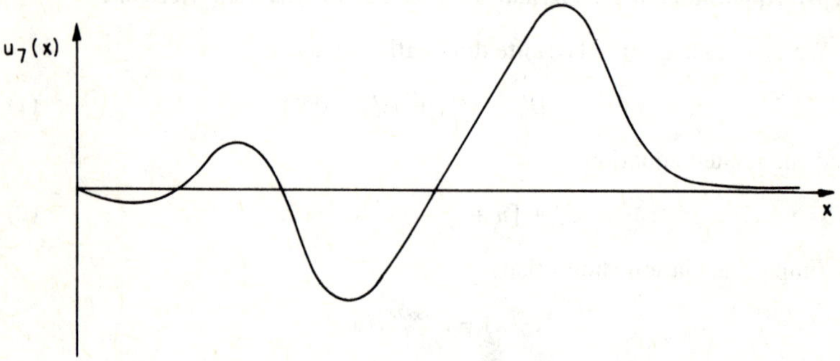

Note that the Hermite equation (2) can be written in the form

$$\frac{d}{dx} e^{-x^2/2} H_n' + n e^{-x^2/2} H_n = 0.$$

Again by Polya's theorem for this case,

$$k(x)g(x) = ne^{-x^2}$$

which is a decreasing function for $x > 0$ we see that the relative maxima of $H_n{}^2(x)$ form an increasing sequence for positive x.

In studying the linear harmonic oscillator in the theory of quantum mechanics the differential equation

$$V'' + [E - \tfrac{1}{4}x^2]V = 0 \tag{3}$$

has to be investigated. In general solutions of the above will not satisfy the condition

$$\int_{-\infty}^{\infty} V^2(x)\, dx < \infty \tag{4}$$

unless E assumes one of the eigenvalues of the system. By comparing (3) with (2) we see that whenever

$$E = E_n = 2n + 1$$

$$V_n(x) = c_n e^{-x^2/2} H_n(\sqrt{2}\,x). \tag{5}$$

The above clearly satisfies (4), and one can show that the above values of E are the only ones for which (3) has solutions satisfying (4).

By a suitable choice of c_n we can have

$$\int_{-\infty}^{\infty} V_n^2(x)\,dx = 1.$$

The function $V_n^2(x)$ has, according to the theory of quantum mechanics, the interpretation of a probability density. The probability that the amplitude of the oscillation lies in some interval (a, b) is given by

$$P[a \le x \le b] = \int_a^b V_n^2(x)\,dx.$$

From the nature of the function $V_n(x)$ we see that

$$\lim_{A\to\infty} P[|x| > A] = 0$$

To study the properties of the above probability function we consider the associated probability density

$$p(x) = V_n^2(x).$$

By repeated differentiation and use of (3) we easily find

$$p^{(''')} + (4E - 2x^2)p' - 2xp = 0 \tag{6}$$

For $|x| \ll \sqrt{2E}$ we approximate the above by

$$(4E - 2x^2)p' - 2x\,p = 0$$

and find

$$p = \frac{c}{\sqrt{2E - x^2}}.$$

Using the above it is easy to verify that the term p''' in (6) is indeed small compared to the other terms in (6) for $|x| \ll \sqrt{2E}$. To evaluate the constant c we assume that, in view of the fact that (5) shows that $p(x)$ becomes exponentially small for $x \ll \sqrt{2E}$,

$$\int_{-\infty}^{\infty} p(x)\,dx \approx \int_{-\sqrt{2E}}^{\sqrt{2E}} \frac{c}{\sqrt{2E - x^2}}\,dx = 1$$

and

$$c = \frac{1}{\pi}$$

so that

$$p(x) \approx \frac{1}{\pi\sqrt{2E - x^2}}, \qquad |x| \ll \sqrt{2E}. \tag{7}$$

The above approximation is valid only for high energy states. One would intuitively expect that for these the quantum mechanical results and the classical results should be comparable. In the classical theory we would expect x to satisfy the differential equation

$$\frac{d^2x}{dt^2} + \omega^2 x = 0,$$

so that

$$x = A \sin(\omega t + \phi).$$

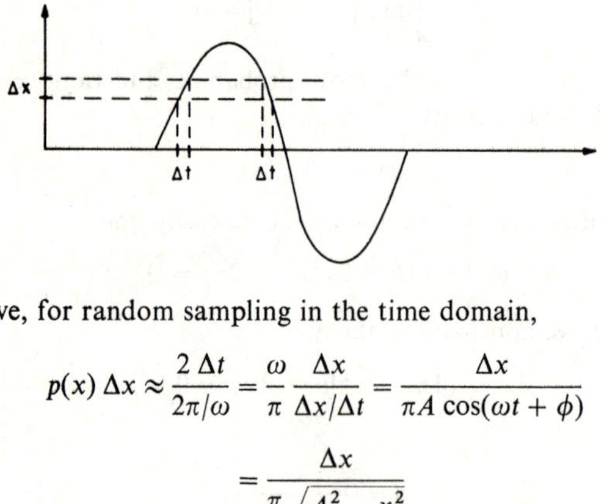

Now we have, for random sampling in the time domain,

$$p(x)\,\Delta x \approx \frac{2\,\Delta t}{2\pi/\omega} = \frac{\omega}{\pi}\,\frac{\Delta x}{\Delta x/\Delta t} = \frac{\Delta x}{\pi A \cos(\omega t + \phi)}$$

$$= \frac{\Delta x}{\pi\sqrt{A^2 - x^2}}$$

But we know that the energy and amplitude are related by

$$E = \frac{1}{2}\left[\frac{1}{\omega^2}x'^2 + x^2\right] = \frac{1}{2}A^2$$

so that

$$p(x) = \frac{1}{\pi\sqrt{2E - x^2}}$$

which agrees with the previous quantum mechanical approximation for high energy states.

6. The Completeness of the Hermite and Laguerre Polynomials

As was noted in the previous chapter, the general proof of the completeness of the orthogonal polynomials was restricted to the case of finite intervals. Accordingly it does not apply to the Hermite and Laguerre polynomials. Nevertheless these polynomials are complete, as will now be demonstrated.

The proof that completeness is a necessary and sufficient condition for closure applies to any separable Hilbert Space. The space of all real functions $f(x)$ for which

$$\int_{-\infty}^{\infty} e^{-x^2/2} f^2(x)\, dx < \infty$$

forms a Hilbert space. We shall show that if

$$\int_{-\infty}^{\infty} e^{-x^2/2} x^n f(x)\, dx = 0, \qquad n = 0, 1, 2, \ldots \tag{1}$$

then $f(x)$ is a null function. Consider the function

$$F(z) = \int_{-\infty}^{\infty} e^{izx} e^{-x^2/2} f(x)\, dx. \tag{2}$$

It is the Fourier transform of $e^{-x^2/2}f(x)$, and represents an analytic function of the complex variable z. The integral converges absolutely and uniformly in every finite domain of the z-plane. Therefore $F(z)$ can have no singularities and must be an entire function of z. In other words we can represent $F(z)$ in the form

$$F(z) = \sum_{n=0}^{\infty} a_n z^n$$

and the radius of convergence of the latter series must be infinite. Then

$$a_n = \frac{1}{n!} F^{(n)}(0) = \frac{1}{n!} i^n \int_{-\infty}^{\infty} x^n e^{-x^2/2} f(x)\, dx = 0$$

by (1). It follows that

$$F(z) \equiv 0.$$

From the uniqueness of the inverse Fourier transforms it now follows from (2) that $f(x)$ is a null function. Accordingly the Hermite polynomials form a closed and therefore also a complete set.

The corresponding result for the Laguerre polynomials follows by use of the preceding discussion. We now consider the set of all functions $f(x)$ for which

$$\int_0^\infty x^\alpha e^{-x} f^2(x)\,dx < \infty.$$

Suppose that

$$\int_0^\infty x^\alpha e^{-x} x^n f(x)\,dx = 0, \qquad n = 0, 1, 2, \ldots$$

Now let $x = t^2/2$, so that the above can be written as

$$\int_0^\infty t^{2n} e^{-t^2/2} g(t)\,dt = 0, \qquad n = 0, 1, 2, \ldots$$

where $g(t) = t^{2\alpha+1} f(t^2/2)$.
We will now continue $g(t)$ for negative values of t by letting

$$g(t) = g(-t), \qquad t < 0$$

so that we have

$$\int_{-\infty}^\infty t^{2n} e^{-t^2/2} g(t)\,dt = 0, \qquad n = 0, 1, 2 \ldots$$

and since $g(t)$ is an even function we can write

$$\int_{-\infty}^\infty t^n e^{-t^2/2} g(t)\,dt = 0, \qquad n = 0, 1, 2, \ldots$$

From this expression we see that $g(t)$ is a null function and $f(x)$ is also. As a result the Laguerre polynomials are closed and complete on the space under discussion.

7. Generating Functions

A generating function, for our purposes, is a function $F(t, x)$ which has a series expansion in the variable t of the form

$$F(t, x) = \sum_{n=0}^\infty a_n \phi_n(x) t^n$$

where $\phi_n(t)$ is one of the classical orthogonal polynomials. Such generating functions play a significant role in many areas of mathematics.

Many of these generating functions can be derived by use of the Rodrigues formulas. From the Cauchy integral theorem, we have by repeated differentiation

$$f(z) = \frac{1}{2\pi i}\int_C \frac{f(\zeta)\,d\zeta}{(\zeta - z)}$$
$$f^{(n)}(z) = \frac{n!}{2\pi i}\int_C \frac{f(\zeta)\,d\zeta}{(\zeta - z)^{n+1}} \tag{1}$$

We shall now try to sum the series

$$\sum_{n=0}^{\infty} \frac{H_n(x)}{n!} t^n$$

by setting

$$H_n(x) = (-1)^n e^{x^2/2}\left(\frac{d}{dx}\right)^n e^{-x^2/2} = (-1)^n e^{x^2/2} \frac{n!}{2\pi i}\int_C \frac{e^{-\zeta^2/2}\,d\zeta}{(\zeta - x)^{n+1}}.$$

Then

$$\sum_{n=0}^{\infty} \frac{H_n(x)t^n}{n!} = \frac{e^{x^2/2}}{2\pi i}\int_C e^{-\zeta^2/2}\,d\zeta \sum_{n=0}^{\infty} \frac{(-1)^n t^n}{(\zeta - x)^{n+1}}$$
$$= \frac{e^{x^2/2}}{2\pi i}\int_C \frac{e^{-\zeta^2/2}\,d\zeta}{\zeta - x + t}$$

if we sum the resultant geometric series. We of course assumed that the point x was inside the contour C, and to insure that the geometric series converged we require that the point $x - t$ should also be inside the contour. Lastly we evaluate the above integral and find

$$e^{tx - t^2/2} = \sum_{n=0}^{\infty} \frac{H_n(x)t^n}{n!}. \tag{1}$$

Similarly one can show that

$$\frac{e^{-(xt/1-t)}}{(1-t)^{\alpha+1}} = \sum_{n=0}^{\infty} L_n^{\alpha}(x)t^n. \tag{2}$$

Such generating functions are not unique. We shall now find another generating function for the Laguerre polynomials. To do so, however, we shall presuppose the knowledge of a generating function for the Bessel functions, namely

$$\exp\frac{x}{2}\left(t - \frac{1}{t}\right) = \sum_{n=-\infty}^{\infty} J_n(x)t^n, \tag{3}$$

where $J_n(x)$ denotes the Bessel function of order n. Bessel functions and their connection with (3) will be discussed in a future chapter. Using the Rodrigues formula

$$L_n^0(x) = \frac{1}{n!} e^x \left(\frac{d}{dx}\right)^n x^n e^{-x}$$

we have

$$\sum_{n=0}^{\infty} \frac{1}{n!} L_n^0(x) t^n = e^x \frac{1}{2\pi i} \int_C \frac{e^{-\zeta}\, d\zeta}{\zeta - x} \sum_{n=0}^{\infty} \frac{(t\zeta/(\zeta - x))^n}{n!}$$

$$= e^x \frac{1}{2\pi i} \int_C \frac{e^{-[\zeta - (t\zeta/(\zeta - x))]}}{\zeta - x}\, d\zeta.$$

The change of variable

$$\zeta = x - \sqrt{tx}\, u$$

coupled with the introduction of (3) leads to

$$\sum_{n=0}^{\infty} \frac{1}{n!} L_n^0(x) t^n = \frac{e^t}{2\pi i} \int_C \frac{e^{(2\sqrt{xt}/2)(u - 1/u)}}{u}\, du = e^t J_0(2\sqrt{xt}).$$

More generally one can show

$$e^t (xt)^{-\alpha/2} J_\alpha(2\sqrt{xt}) = \sum_{n=0}^{\infty} \frac{L_n^\alpha(x) t^n}{(n + \alpha)!}. \tag{4}$$

Next we shall derive a generating function for the Gegenbauer polynomials. These, as can be seen from Table I, satisfy the following recursion formula

$$(n + 1)\phi_{n+1}(x) - 2(\gamma + n)x\phi_n(x) + (2\gamma + n - 1)\phi_{n-1}(x) = 0$$

with $\phi_0(x) = 1$, $\phi_1(x) = 2\gamma x$, and for convenience we set $\phi_{-1}(x) = 0$. The correctness of this formula may be verified by the use of the Rodrigues formula. We shall seek an expression in closed form for

$$G(x, t) = \sum_{n=0}^{\infty} \phi_n(x) t^n.$$

Rather than use the Rodrigues formula we shall use the recursion formula to sum the above. This technique is at times more convenient. First we note that

$$\frac{\partial G}{\partial t} = \sum_{n=0}^{\infty} n\phi_n(x) t^{n-1} = \sum_{n=0}^{\infty} (n + 1)\phi_{n+1}(x) t^n$$

$$= \sum_{n=0}^{\infty} [2(\gamma + n)x\phi_n(x) - (2\gamma + n - 1)\phi_{n-1}(x)] t^n.$$

By straightforward rearrangements we find that

$$\frac{\partial G}{\partial t} = 2\gamma x G + 2xt\frac{\partial G}{\partial t} - 2\gamma t G - t^2\frac{\partial G}{\partial t}$$

which leads to the partial differential equation

$$\frac{\partial G/\partial t}{G} = \frac{2\gamma(x-t)}{1-2xt+t^2}.$$

Coupled with the initial condition $G(x, 0) = 1$ we finally have

$$\frac{1}{(1-2xt+t^2)^\gamma} = \sum_{n=0}^{\infty} \phi_n(x)t^n. \tag{5}$$

One could use (5) to define the Gegenbauer polynomials and all their crucial properties could be derived from the above.

A case of particular significance in mathematical physics relates to the Legendre polynomials. For $\gamma = 1/2$ we find

$$\frac{1}{\sqrt{1-2xt+t^2}} = \sum_{n=0}^{\infty} P_n(x)t^n.$$

To find the normalization constant for these polynomials we proceed as follows. Squaring the last expression, and then integrating over $(-1, 1)$ and using the orthogonality property we have

$$\sum_{0}^{\infty} \|P_n(x)\|^2 t^{2n} = \int_{-1}^{1} \frac{dx}{1-2xt+t^2} = \frac{1}{t}\log\frac{1-t}{1-t} = \sum_{n=0}^{\infty} \frac{2}{2n+1}\, t^{2n}.$$

Thus we find

$$\int_{-1}^{1} P_n^2(x)\, dx = \frac{2}{2n+1}.$$

The Legendre polynomials have an interesting historical origin, in potential theory. Consider an electric field with a point charge e, one unit away from the origin.

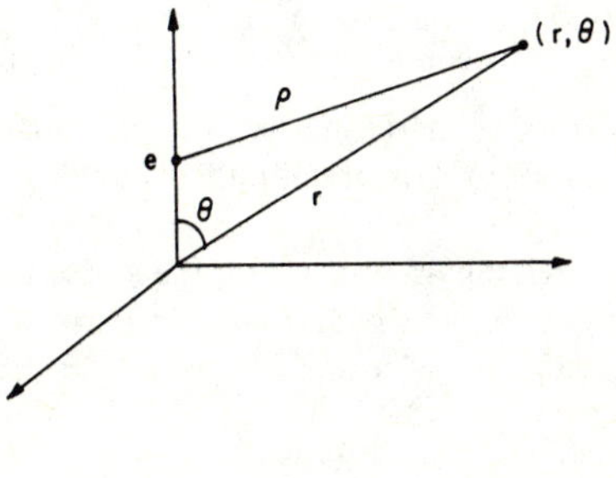

The potential at the point (r, θ) in a spherical coordinate system is given by

$$u(r, \theta) = \frac{1}{\rho} = \frac{1}{\sqrt{1 - 2r\cos\theta + r^2}}.$$

We can also seek to construct the potential $u(r, \theta)$ as a solution of the Laplace equation. In spherical coordinates this becomes

$$\frac{\partial^2 u}{\partial r^2} + \frac{2}{r}\frac{\partial u}{\partial r} + \frac{1}{r^2 \sin\theta}\frac{\partial}{\partial\theta}\sin\theta\frac{\partial u}{\partial\theta} + \frac{1}{r^2\sin^2\theta}\frac{\partial^2 u}{\partial\phi^2} = 0$$

From the symmetry of the problem we see that u will be independent of ϕ. Furthermore, if we apply the method of separation of variables we let

$$u(r, \theta) = R(r)\Theta(\theta)$$

and find that the partial differential equation separates into two ordinary differential equations,

$$r^2 R'' + 2rR' - \lambda^2 R = 0$$

$$\frac{1}{\sin\theta}\frac{d}{d\theta}\sin\theta\Theta' + \lambda^2\Theta = 0$$

where λ^2 is the separation constant. The solution of the first of these is

$$R = c_1 r^n + \frac{c_2}{r^{n+1}}, \qquad \text{if } \lambda^2 = n(n+1).$$

The second equation is transformed by the substitution $\cos\theta = x$ into

$$(1 - x^2)\Theta'' - 2x\Theta' + n(n+1)\Theta = 0,$$

whose solution is given by $P_n(x)$ if n is an integer. For $r < 1$, i.e. near the origin we can now set $c_2 = 0$ and a typical solution is

$$u_n(r, \theta) = c_n P_n(\cos\theta)\, r^n.$$

From the superposition principle for linear homogeneous equations we finally have

$$\frac{1}{\sqrt{1 - 2r\cos\theta + r^2}} = \sum_{n=0}^{\infty} c_n P_n(\cos\theta) r^n.$$

The appropriate coefficients c_n in the above will of course depend on the normalization selected for $P_n(x)$. Historically the normalization of $P_n(x)$ was so defined that $c_n = 1$.

The more general expression (5) arises in the corresponding potential theoretic problem for higher dimensional spaces. In particular for $\gamma = \frac{1}{2}p - 1$ we obtain the potential function of a unit charge in a p dimensional space.

Exercises

1. Using the definitions of the various classical orthogonal polynomials, verify the coefficients listed in Table 1 for their recurrence formulas.

2. Derive the differential equations for the Laguerre and Hermite polynomials stated in Section 2.2.

3. Show that the Legendre polynomials may also be normalized by the requirement that $P_n(1) = 1$.

4. Show that

$$H_n(x) = \frac{1}{\sqrt{2\pi}} \int_{-\infty}^{\infty} (x + it)^n \, e^{-t^2/2} \, dt$$

5. Derive the following generating function

$$\sum_{n=0}^{\infty} H_n(x) \, H_n(y) \frac{t^n}{n!} = \frac{1}{\sqrt{1-t^2}} \exp \frac{2xyt - t^2(x^2 + y^2)}{2(1-t^2)}$$

6. Derive the generating function

$$\sum_{n=0}^{\infty} P_n(x) \, t^n = \frac{1}{\sqrt{1 - 2xt + t^2}}$$

using Rodrigues formula and Cauchy's integral formula.

7. Show that

$$\sum_{m=0}^{n} \binom{n}{m} H_{n-m}(\sqrt{2}t) \, H_m(\sqrt{2}\tau) = 2^{n/2} \, H_n(t + \tau)$$

8. Show that

$$\frac{d}{dx} L^{\alpha}_{n+1}(x) = - L_n^{\alpha+1}(x).$$

9. Show that

$$P_n(x) = \sqrt{\frac{2}{\pi}} \frac{1}{n!} \int_0^{\infty} e^{-t^2/2} \, t^n \, H_n(xt) \, dt.$$

10. Show that

$$H_{2n}(t) = \frac{(-1) \, 2^n \, n!}{(2n)!} L_n^{-1/2} \, (t^2/2)$$

$$H_{2n+1}(t) = \frac{(-1)^{n+1} \, 2^n n!}{(2n+1)!} L_n^{1/2}(t^2/2), \qquad n = 0, 1, 2, \ldots.$$

11. In order to prove the completeness of the Laguerre polynomials, use the Laplace Transform

$$F(s) = \int_0^\infty e^{-sx} x^\alpha e^{-x} f(x)\, dx$$

Supply a discussion similar to the one given for the Hermite polynomials which involve Fourier transforms.

12. Show that the set of orthogonal functions $\{\phi_n(x)\}$ is complete in the space of square integrable functions on the finite interval $[a, b]$ if and only if

$$\sum_{n=1}^{\infty} \left(\int_0^x \phi_n(x)\, dx \right)^2 = x - a.$$

13. Same as Exercise 12 for the condition

$$\sum \int_a^b \left[\int_a^x \phi_n(x)\, dx \right]^2 dx = \frac{(b-a)^2}{2}.$$

14. Show that if $\{\phi_n(x)\}$ is a complete set of orthogonal polynomials corresponding to a weight function $w_1(x)$, then $w_1(x)$ is uniquely determined by $\{\phi_n(x)\}$. (The word unique here is to be interpreted in the sense of almost everywhere.)

CHAPTER 3

The Gamma Function

1. Definitions and Basic Properties

In some of the formulas introduced in the previous chapter we arrived at the expressions of the type $\alpha!$. For α a positive integer this function is well defined, but in many problems we will be interested in generalizing our results to non-integral indices. For such reasons we now define and study the Gamma function, which is the appropriate generalization of the factorial function.

We shall present three different definitions, due to different mathematicians. Each of these displays and emphasizes different features and properties of the Gamma function. It will, of course, be necessary subsequently to prove that these definitions are indeed equivalent. The symbol $\Gamma(z)$ will be used to denote the function, and z will be taken to be a complex variable.

Definition 1. (*Euler*)

$$\Gamma(z) = \int_0^\infty e^{-t} t^{z-1}\, dt, \qquad \operatorname{Re} z > 0. \tag{1}$$

Clearly the above makes sense if and only if the integral converges. Hence the restriction $\operatorname{Re} z > 0$.

Definition 2. (*Gauss*).

$$\Gamma(z) = \lim_{n\to\infty} \frac{n!\, n^z}{z(z+1)\cdots(z+n)}, \qquad z \neq 0, -1, -2, \ldots. \tag{2}$$

The above shows that if the limit exists $\Gamma(z)$ is defined for all z, except possibly for the nonpositive integers. Lastly we have

Definition 3 (*Weierstrass*).

$$\frac{1}{\Gamma(z)} = z e^{\gamma z} \prod_{n=1}^{\infty} \left(1 + \frac{z}{n}\right) e^{-z/n}, \tag{3}$$

where γ is the Euler-Mascheroni constant, defined by

$$\gamma = \lim_{n\to\infty} \sum_{k=1}^{n} \frac{1}{k} - \ln(n+1) \approx 0.5772157.$$

It is easy to show that the above limit exists. Consider

$$u_k = \int_0^1 \frac{t}{k(t+k)}\, dt < \frac{1}{k^2}\int_0^1 t\, dt = \frac{1}{2k^2}.$$

But we also have

$$u_k = \int_0^1 \left(\frac{1}{k} - \frac{1}{t+k}\right) dt = \frac{1}{k} \ln \frac{1}{k+1} + \ln \frac{1}{k}$$

so that

$$\sum_{k=1}^{n} u_k = \sum_{k=1}^{n} \frac{1}{k} - \ln(n+1).$$

$$\gamma = \lim_{n\to\infty} \sum_{k=1}^{n} u_k \le \frac{1}{2} \sum_{k=1}^{\infty} \frac{1}{k^2} = \frac{\pi^2}{12}.$$

From Definitions 2 and 3 we can note immediately that, as a function of the complex variable z, $\Gamma(z)$ has simple poles at $z = 0, -1, -2, \ldots$. This fact may also be observed from Definition 1 if we use Prym's Decomposition. We write

$$\Gamma(z) = \int_0^\infty e^{-t} t^{z-1}\, dt = \int_0^1 e^{-t} t^{z-1}\, dt + \int_1^\infty e^{-t} t^{z-1}\, dt.$$

The last term on the right is defined for all values of z, and therefore represents an entire function. We also have

$$\int_0^1 e^{-t} t^{z-1}\, dt = \sum_{n=0}^{\infty} \frac{(-1)^n}{n!} \int_0^1 t^{n+z-1}\, dt = \sum_{n=0}^{\infty} \frac{(-1)^n}{n!(n+z)}.$$

Since the above integral converges uniformly the performed operations are legitimate. Thus we have

$$\Gamma(z) = \sum_{n=0}^{\infty} \frac{(-1)^n}{n!(n+z)} + \int_1^\infty e^{-t} t^{z-1}\, dt. \tag{4}$$

From the latter we note again that $\Gamma(z)$ has simple poles at all nonpositive integers. To obtain the residues at these poles we compute

$$\lim_{z\to -n} (z+n)\Gamma(z) = \frac{(-1)^n}{n!}.$$

(4), of course, is the Mittag-Leffler expansion of $\Gamma(z)$.

Using (1) or (3) we can derive the recurrence formula for $\Gamma(z)$. By integration by parts we have

$$\Gamma(z+1) = \int_0^\infty e^{-t} t^z \, dt = z \int_0^\infty e^{-t} t^{z-1} \, dt = z\Gamma(z).$$

Since $\Gamma(z+1)$ is defined for all z, such that $\operatorname{Re} z > -1$ we see that with

$$\Gamma(z) = \frac{\Gamma(z+1)}{z}$$

we have obtained the analytic continuation of $\Gamma(z)$ into the strip $-1 < \operatorname{Re} z < 0$.

To verify the equivalence of the three definitions we need the following lemma.

LEMMA

$$\lim_{n\to\infty} \int_0^n \left[e^{-t} - \left(1 - \frac{t}{n}\right)^n\right] t^{z-1} \, dt = 0, \qquad \text{if} \quad \operatorname{Re} z > 0.$$

Proof. Consider the function

$$f_n(t) = 1 - e^t\left(1 - \frac{t}{n}\right)^n, \qquad 0 \le t \le n.$$

$$f_n'(t) = e^t\left(1 - \frac{t}{n}\right)^{n-1} - e^t\left(1 - \frac{t}{n}\right)^n$$

$$= e^t\left(1 - \frac{t}{n}\right)^{n-1} \frac{t}{n} \ge 0.$$

It follows that

$$f_n(t) = \int_0^t f_n'(\xi) \, d\xi \ge 0$$

and also

$$f_n(t) = \int_0^t f_n'(\xi) \, d\xi = \int_0^t e^\xi\left(1 - \frac{\xi}{n}\right)^{n-1} \frac{\xi}{n} \, d\xi \le \frac{e^t}{n} \int_0^t \xi \, d\xi = \frac{e^t t^2}{2n}.$$

Using these estimates we have

$$0 \le e^{-t} - \left(1 - \frac{t}{n}\right)^n \le \frac{t^2}{2n}, \qquad 0 \le t \le n.$$

Then we obtain, with the above

$$\left(1 - \frac{t}{n}\right)^n \le e^{-t}$$

so that, for every fixed a such that $0 \le a \le n$

$$\left|\int_a^n \left[e^{-t} - \left(1 - \frac{t}{n}\right)^n\right] t^{z-1}\, dt\right| \le 2\int_a^n e^{-t} t^{x-1}\, dt,$$

where $x = \operatorname{Re} z > 0$. By the absolute and uniform convergence of the last integral, we can, for every positive ε, select $N(\varepsilon)$ so that

$$2\int_a^n e^{-t} t^{x-1}\, dt < \varepsilon \qquad \text{if} \quad a, n > N(\varepsilon).$$

Furthermore

$$\left|\int_0^a \left[e^{-t} - \left(1 - \frac{t}{n}\right)^n\right] t^{z-1}\, dt\right| \le \frac{1}{2n}\int_0^a t^{x+1}\, dt = \frac{a^{x+2}}{2n(x+2)}$$

so that

$$|I_n| = \left|\int_0^n \left[e^{-t} - \left(1 - \frac{t}{n}\right)^n\right] t^{z-1}\, dt\right| \le \frac{a^{x+2}}{2n(x+2)} + \varepsilon.$$

Thus we have

$$\lim_{n\to\infty} |I_n| < \varepsilon$$

and since ε is arbitrary the proof is completed. ■

THEOREM. Definitions 1, 2, and 3 are equivalent.

Proof. Using Definition 1 and the lemma we see that

$$\Gamma(z) = \lim_{n\to\infty} \int_0^n e^{-t} t^{z-1}\, dt = \lim_{n\to\infty} \int_0^n \left(1 - \frac{t}{n}\right)^n t^{z-1}\, dt.$$

Next we let $t = n\tau$ in the last integral and integrate by parts n times. Then

$$\Gamma(z) = \lim_{n\to\infty} n^z \int_0^1 (1-\tau)^n \tau^{z-1}\, d\tau = \lim_{n\to\infty} \frac{n!\, n^z}{z(z+1)\cdots(z+n)}$$

Showing that the limit on the right exists and thus Definition 1 implies Definition 2.

To show that Definition 2 implies Definition 3 we write

$$\frac{1}{\Gamma(z)} = \lim_{n\to\infty} n^{-z} \frac{z(z+1)\cdots(z+n)}{n!}$$

$$= \lim_{n\to\infty} ze^{-z\ln n} \prod_{k=1}^{n} e^{z/k} \prod_{k=1}^{n} \left(1+\frac{z}{k}\right)e^{-z/k}$$

Since

$$\lim_{n\to\infty} \sum_{k=1}^{n} \frac{1}{k} - \ln n = \gamma$$

we have

$$\lim_{n\to\infty} e^{-z\ln n} \prod_{k=1}^{n} e^{z/k} = e^{\gamma z}.$$

The product

$$\lim_{n\to\infty} \prod_{k=1}^{n} \left(1+\frac{z}{k}\right)e^{-z/k} = \prod_{1}^{\infty} \left(1+\frac{z}{k}\right)e^{-z/k}$$

exists since for z fixed and k large

$$\left(1+\frac{z}{k}\right)e^{-z/k} = 1 - \frac{z^2}{k^2} + 0\left(\frac{1}{k^3}\right).$$

Accordingly the product converges like $\sum_{k=1}^{\infty} 1/k^2$. Then

$$\frac{1}{\Gamma(z)} = ze^{\gamma z} \prod_{k=1}^{\infty} \left(1+\frac{z}{k}\right)e^{-z/k}.$$

Since the steps are retracable all three definitions are equivalent. ■

2. Analytic Continuations and Integral Representations

A result that yields an immediate analytic continuation from the right to the left half plane is the following

LEMMA.

$$\Gamma(z)\Gamma(1-z) = \frac{\pi}{\sin \pi z}, \qquad z \neq 0, \pm 1, \pm 2, \ldots \tag{1}$$

Proof. Using Definition 3 we have immediately

$$\phi(z) = \frac{1}{\Gamma(z)\Gamma(1-z)} = \frac{1}{-z\Gamma(z)\Gamma(-z)} = z \prod_{k=1}^{\infty} \left(1-\frac{z^2}{k^2}\right)$$

Now let

$$\psi(z) = -\frac{\phi'(z)}{\phi(z)} = \sum_{k=-\infty}^{\infty} \frac{-z}{z^2 - k^2}$$

so that

$$\psi'(z) = \sum_{k=-\infty}^{\infty} \frac{1}{(z-k)^2}$$

Inspection shows that

$$\psi'(z+1) = \psi'(z)$$

so that $\psi'(z)$ is periodic and we need concern ourselves only with the strip $|\text{Re } z| \leq \frac{1}{2}$. The function has a double pole at $z = 0$, and is regular elsewhere in the strip. Then

$$\left| \psi'(z) - \frac{1}{z^2} \right| \leq \left(\sum_{k=-\infty}^{-1} + \sum_{k=1}^{\infty} \right) \frac{1}{|x - k + iy|^2} \leq 2 \sum_{k=1}^{\infty} \frac{1}{(k - \frac{1}{2})^2}.$$

By the uniform convergence of the series we are allowed to interchange limits and sums so that

$$\lim_{Imz \to \infty} \psi'(z) = 0.$$

The function $\pi^2 \operatorname{cosec}^2 \pi z$ has the same period as $\psi'(z)$ and the same poles as well. Hence

$$\psi'(z) - \pi^2 \operatorname{cosec}^2 \pi z$$

has no singularities, and therefore must be an entire function. But

$$\lim_{Imz \to \infty} \pi^2 \operatorname{cosec}^2 \pi z = 0$$

so that, by Liouville's theorem,

$$\psi'(z) = \pi^2 \operatorname{cosec}^2 \pi z.$$

By an integration

$$\psi(z) = -\pi \cot z \pi + c.$$

To evaluate the constant c we note that $\psi(z)$ and $\pi \cot \pi z$ are odd function with identical poles. Hence $c = 0$. By a second integration

$$\ln \phi(z) = \ln \sin \pi z + c$$

so that

$$\phi(z) = c \sin \pi z.$$

To evaluate c we have

$$\lim_{z\to 0}\frac{\phi(z)}{z} = 1, \qquad \lim_{z\to 0}\frac{\sin \pi z}{z} = \pi$$

so that $c = 1/\pi$, and combining these we have

$$\Gamma(z)\Gamma(1-z) = \frac{\pi}{\sin \pi z}. \qquad \blacksquare$$

For $z = \frac{1}{2}$ we find immediately that

$$\Gamma\left(\frac{1}{2}\right) = \int_0^\infty e^{-t}t^{-1/2}\,dt = 2\int_0^\infty e^{-t^2}\,dt = \sqrt{\pi}.$$

THEOREM. When z is not an integer we have

$$\Gamma(z) = \frac{1}{e^{2\pi i z} - 1}\int_C e^{-\zeta}\zeta^{z-1}\,d\zeta$$

where C is the following contour.

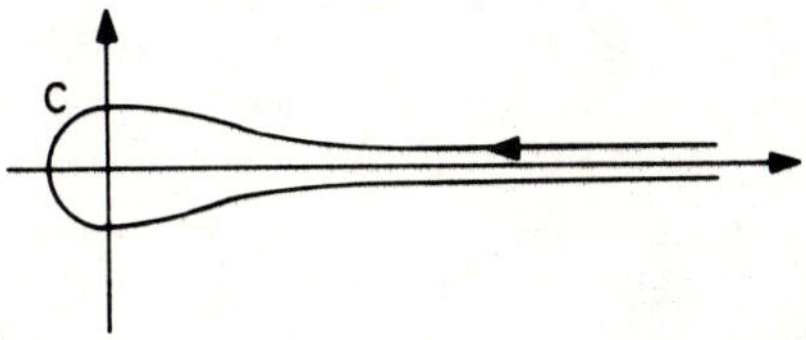

Proof. We deform C into a contour consisting of two straight lines and a circle.

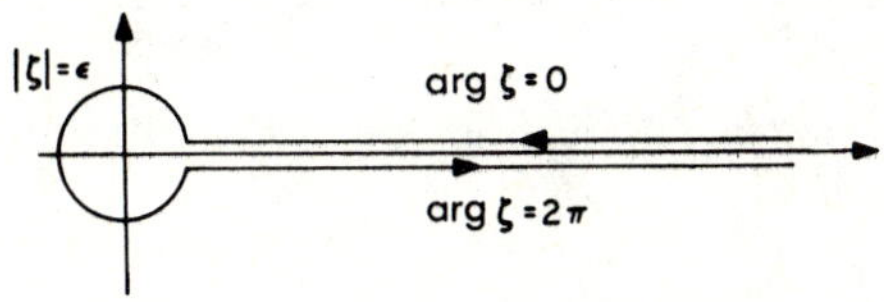

Then

$$I(z) = \int_C e^{-\zeta}\zeta^{z-1}\,d\zeta = \int_\infty^\varepsilon e^{-t}t^{z-1}\,dt + \int_{|\zeta|=\varepsilon} e^{-\zeta}\zeta^{z-1}\,d\zeta + \int_\varepsilon^\infty e^{-t}(e^{2\pi i}t)^{z-1}\,dt,$$

and we easily calculate

$$\lim_{\varepsilon\to 0}\int_{\infty}^{\varepsilon} e^{-t}t^{z-1}\,dt = -\Gamma(z)$$

$$\lim_{\varepsilon\to 0}\int_{\varepsilon}^{\infty} e^{-t}(e^{2\pi i}t)^{z-1}\,dt = e^{2\pi iz}\Gamma(z),$$

if Re $z > 0$. For the middle integral we find

$$\left|\int_{|\zeta|=\varepsilon} e^{-\zeta}\zeta^{z-1}\,d\zeta\right| = \left|\int_0^{2\pi} e^{-\varepsilon e^{i\theta}}(\varepsilon e^{i\theta})^{z-1} i\varepsilon e^{i\theta}\,d\theta\right|$$
$$\le \varepsilon^x \int_0^{2\pi} e^{-\varepsilon\cos\theta}\,d\theta$$

and for Re $z = x > 0$

$$\lim_{\varepsilon\to 0}\int_{|\zeta|=\varepsilon} e^{-\zeta}\zeta^{z-1}\,d\zeta = 0.$$

Finally

$$I(z) = (e^{2\pi iz} - 1)\,\Gamma(z)$$

which is equivalent to the theorem. The proof assumed that Re $z > 0$, but the integral converges exponentially at infinity, and therefore exists for all z. By analytic continuation the result is valid in the entire z plane, except at integral points, where the denominator vanishes. ■

We can obtain from the above representation another one, one that is often referred to as Hankel's representation

$$\Gamma(z) = \frac{e^{-i\pi z}}{e^{i\pi z} - e^{-i\pi z}}\int_C e^{-\zeta}\zeta^{z-1}\,d\zeta$$
$$= \frac{e^{-i\pi}}{2i\sin\pi z}\int_C e^{-\zeta}(e^{-i\pi}\zeta)^{z-1}\,d\zeta = \frac{-1}{2i\sin\pi z}\int_C e^{-\zeta}(-\zeta)^{z-1}\,d\zeta.$$

Another representation for the reciprocal of $\Gamma(z)$ is the following

THEOREM.

$$\frac{1}{\Gamma(z)} = -\frac{1}{2\pi i}\int_C e^{-\zeta}(-\zeta)^{-z}\,d\zeta$$

This representation is valid for all z and C is the same contour as in the preceding theorem.

Proof. By Hankel's representation and the former lemma

$$\Gamma(1-z) = \frac{-1}{2i \sin \pi z} \int_C e^{-\zeta}(-\zeta)^{-z}\, d\zeta$$

$$= \frac{-1}{2\pi i} \Gamma(z)\Gamma(1-z) \int_C e^{-\zeta}(-\zeta)^{-z}\, d\zeta$$

so that

$$\frac{1}{\Gamma(z)} = -\frac{1}{2\pi i} \int_C e^{-\zeta}(-\zeta)^{-z}\, d\zeta. \qquad \blacksquare$$

3. Asymptotic Expansions

One of the most powerful tools in the study of transcendental functions comes under the heading of asymptotic expansions. In dealing with complicated functions it is often convenient to seek representations in terms of simpler functions. As an example of such we have Stirling's formula for the Gamma function

$$\Gamma(z) = \sqrt{\frac{2\pi}{z}} \left(\frac{z}{e}\right)^z \left[1 + \frac{1}{12z} + \cdots\right].$$

The expression in the brackets is a series in powers of $1/z$. It does not converge, but it is asymptotic. For large values of z it furnishes useful approximations. For example

$$\Gamma(10) = 9! = 362{,}880$$

$$\sqrt{\frac{2\pi}{10}} \left(\frac{10}{e}\right)^{10} = 360{,}778.\ldots$$

We see that the error is 0.55%. We shall now make the ideas underlying asymptotic series more precise and prove a lemma, that we shall use frequently to develop such series.

Definition. A series $\sum_{k=1}^{\infty} a_k/z^k$ will be called an asymptotic expansion of $f(z)$ if

$$\lim_{|z|\to\infty} z^n \left\{ f(z) - \sum_0^n \frac{a_k}{z^k} \right\} = 0 \qquad \text{for} \quad \text{all } n = 0, 1, \ldots.$$

If the above is the case we write

$$f(z) \approx \sum_{k=0}^{\infty} \frac{a_k}{z^k}.$$

Note that such representations need not be unique. For example if

$$f(z) \approx \sum_{k=0}^{\infty} \frac{a_k}{z^k}.$$

then also

$$f(z) + e^{-z} \approx \sum_{k=0}^{\infty} \frac{a_k}{z^k} \qquad \text{for} \quad \operatorname{Re} z \gg 0.$$

WATSON'S LEMMA Suppose that $F(\tau)$ is a possibly multivalued function of τ, which has the following expansion near $\tau = 0$.

$$F(\tau) = \sum_{n=1}^{\infty} a_n \tau^{n/r-1}, \qquad |\tau| \le a,\ r > 0.$$

Furthermore suppose that $F(\tau)$ is of, at most, exponential growth for large τ, so that for suitable positive constants K, b

$$|\mathrm{F}(\tau)| < \mathrm{K}\, e^{b|\tau|}, \qquad |\tau| \ge a.$$

Then for $|v|$ large

$$\mathscr{F}(v) \equiv \int_0^{\infty} e^{-v\tau} F(\tau)\, d\tau \approx \sum_{n=1} a_n \Gamma\left(\frac{n}{r}\right) v^{-n/r} \qquad \text{for} \quad |\arg v| < \pi/2.$$

Proof. We have already

$$S_n(v) \equiv \int_0^{\infty} e^{-v\tau} \sum_{k=1}^{n} a_k \tau^{(k/r)-1}\, d\tau = \sum_{k=1}^{n} a_k \Gamma\left(\frac{k}{r}\right) v^{-k/r}$$

since for a finite sum the interchange of summation and integration is permissible.

To establish the lemma it suffices to show that

$$v^{n/r}\{\mathscr{F}(v) - S_n(v)\} \to 0 \quad \text{as} \quad v \to \infty \quad \text{and} \quad |\arg v| < \frac{\pi}{2}.$$

By virtue of the representation of $F(\tau)$ near $\tau = 0$ and its growth estimate for large τ, there must exist a constant C_n such that

$$\left| F(\tau) - \sum_{k=1}^{n} a_k \tau^{k/r-1} \right| \le C_n e^{b|\tau|} |\tau|^{(n+1/r)-1}.$$

It follows that

$$|v|^{n/r} |\mathscr{F}(v) - S_n(v)| \le |v|^{n/r} C_n \int_0^{\infty} e^{-|v|(\cos\phi + b)\tau} \tau^{(n+1/r)-1}\, d\tau$$

$$= \frac{|v|^{n/r} C_n}{(|v|(\cos\phi + b))^{(n+1)/r}} \Gamma\left(\frac{n+1}{r}\right).$$

For $|\arg v| < \pi/2$, $\cos \phi > 0$, and an inspection of the above shows that the lemma is correct. ■

THEOREM. For $\Gamma(z)$ we have the representation

$$\Gamma(z) \approx \sqrt{\frac{2\pi}{z}} \left(\frac{z}{e}\right)^z \sum_{k=0}^{\infty} \frac{a_k}{z^k}, \qquad \operatorname{Re} z > 0 \tag{1}$$

where the above series is asymptotic. In particular we have

$$a_0 = 1, \qquad a_1 = \frac{1}{12}, \ldots.$$

The above asymptotic representation is known as Stirling's formula. Before moving on to a formal proof we shall present a somewhat looser argument, but one that contains many of the fundamental ingredients of the proof. We begin with

$$\Gamma(z+1) = \int_0^\infty e^{-t} t^z \, dt = \int_0^\infty e^{-(t - z \ln t)} \, dt \tag{2}$$

and let

$$w(t) = (t - z \ln t) - (z - z \ln z).$$

Then

$$w'(t) = 1 - \frac{z}{t}$$

$$w''(t) = \frac{z}{t^2}$$

so that

$$w(z) = w'(z) = 0, \qquad w''(z) = \frac{1}{z}$$

and expressing $w(t)$ as a Taylor series about the point z we can write

$$w(t) = \frac{1}{2z}(t - z)^2 + \cdots. \tag{3}$$

For z real and positive the following figures show some of the chief characteristics of $w(t)$.

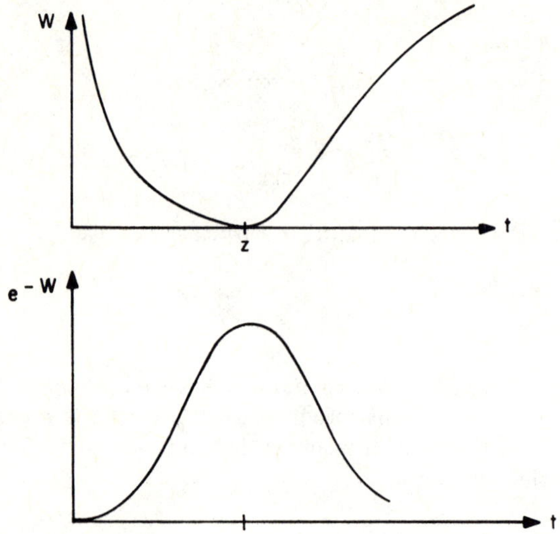

If we neglect all terms above the quadratic in (3) we can rewrite (2) in the form

$$\Gamma(z+1) \approx e^{-(z - z \ln z)} \int_0^{\infty} \exp\left\{-\frac{1}{2z}(t-z)^2\right\} dt$$

$$\approx e^{-(z - z \ln z)} \int_{-\infty}^{\infty} \exp\left\{-\frac{1}{2z}(t-z)^2\right\} dt.$$

For large z the contribution to the integral over $(-\infty, 0)$ is small compared to that of $(0, \infty)$, so that its addition will not matter terribly. Finally by evaluating the last integral we have

$$\Gamma(z+1) \approx e^{-(z - z \ln z)}\sqrt{2\pi z} = \sqrt{2\pi z}\left(\frac{z}{e}\right)^z.$$

But using

$$\Gamma(z+1) = z\Gamma(z)$$

we have

$$\Gamma(z) \approx \sqrt{\frac{2\pi}{z}}\left(\frac{z}{e}\right)^z.$$

Proof of Stirling's Formula. To prove the result we will have to show that all terms neglected in the foregoing heuristic argument are indeed negligible. We return to the function

$$w(t) = (t - z \ln t) - (z - z \ln z)$$

and let $t = z(\tau + 1)$. Then

$$\frac{w(t)}{z} = \tau - \ln(\tau + 1) = \frac{\tau^2}{2} - \frac{\tau^3}{3} + \frac{\tau^4}{4} - \cdots$$

and we try to invert the above so that we express τ as a function of w. We see that $\tau(w)$ will have a branch point near the origin, so that we can find two branches

$$\tau_1 = \sum_{k=1}^{\infty} (-1)^k a_k \left(\frac{w}{z}\right)^{k/2} \qquad a_1 = \sqrt{2}, \quad a_2 = \frac{2}{3}, \quad a_3 = \frac{\sqrt{2}}{18}, \cdots$$

$$\tau_2 = \sum_{k=1}^{\infty} a_k \left(\frac{w}{z}\right)^{k/2}.$$

and correspondingly

$$t_1 = z(1 + \tau_1)$$

$$t_2 = z(1 + \tau_2)$$

Using these functions we can rewrite (2) in the form

$$\frac{\Gamma(z+1)}{(z/e)^z} = \int_0^z e^{-w} dt + \int_z^\infty e^{-w}\, dt = -\int_0^\infty e^{-w} \frac{dt_1}{dw}\, dw + \int_0^\infty e^{-w} \frac{dt_2}{dw}.$$

By Watson's Lemma we find, using the series representations for dt_1/dw and dt_2/dw,

$$\frac{\Gamma(z+1)}{(z/e)^z} \approx \sum_{k=0}^{\infty} \frac{(2k+1)a_{2k+1}\Gamma(k+\frac{1}{2})}{z^{k+1/2}} = \sqrt{\frac{2\pi}{z}}\left(1 + \frac{1}{12z} + \cdots\right)$$

which is in essence Stirling's formula. ■

A result that can be proved by means of Stirling's formula is the following.

MULTIPLICATION THEOREM.

$$\Gamma(z)\Gamma\left(z+\frac{1}{m}\right)\Gamma\left(z+\frac{2}{m}\right)\cdots\Gamma\left(z+\frac{m-1}{m}\right) = (2\pi)^{1/2(m-1)} m^{1/2-mz}\Gamma(mz). \quad (4)$$

Proof. We denote the left side of (4) by $G(z)$. Then by the use of Gauss' definition we have

$$\frac{1}{G(z)} = \lim_{n\to\infty} \prod_{k=0}^{m-1} \frac{(z+k/m)(z+k/m+1)\cdots(z+k/m+n)}{n!n^{z+k/m}}$$

$$= \lim_{n\to\infty} \frac{\prod_{k=0}^{nm+m-1}(z+k/m)}{n!^m n^{(mz+\sum_1^{m-1} k/m)}} = \lim_{n\to\infty} \frac{\prod_{k=0}^{nm+m-1}(mz+k)}{n!^m n^{mz+\frac{1}{2}(m-1)} m^{m(n+1)}}$$

By a slight modification of Gauss' definition we also have

$$\Gamma(mz) = \lim_{n\to\infty} \frac{(mn)!(mn)^{mz}}{\prod_{k=0}^{mn}(mz+k)}$$

so that

$$\frac{\Gamma(mz)}{G(z)} = \lim_{n\to\infty} \frac{(mn)!(mn)^{mz} n^{m-1} \prod_{k=nm+1}^{nm+m-1}[(mz+k)/n]}{n!^m n^{m-z+\frac{1}{2}(m-1)} m^{m(n+1)}}$$

$$= m^{mz} \lim_{n\to\infty} \frac{(mn)!n^{\frac{1}{2}(m-1)}}{n!^m m^{mn+1}}, \tag{5}$$

since

$$\lim_{n\to\infty} \prod_{k=nm+1}^{nm+m-1} \left[\frac{mz+k}{n}\right] = m^{m-1}.$$

Using Stirling's formula we have

$$\frac{(mn)!n^{\frac{1}{2}(m-1)}}{n!^m m^{mn+1}} \sim \frac{\sqrt{2\pi mn}\,(mn/e)^{mn} n^{\frac{1}{2}(m-1)}}{(2\pi n)^{m/2}(n/e)^{mn} m^{mn+1}} = (2\pi)^{\frac{1}{2}(m-1)} m^{-\frac{1}{2}}$$

so that finally

$$G(z) = (2\pi)^{\frac{1}{2}(m-1)} m^{\frac{1}{2}-mz}\,\Gamma(mz). \qquad \blacksquare$$

A well-known special case of the above is the following immediate result.

DUPLICATION FORMULA.

$$\Gamma(z)\Gamma(z+\tfrac{1}{2}) = 2^{1-2z}\sqrt{\pi}\,\Gamma(2z).$$

Proof. Let $m = 2$ in the multiplication theorem. ■

Another method of proof, that is worth mentioning, since it does not make use of Stirlings formula is to return to (5). We note that

$$\frac{m^{-mz}\Gamma(mz)}{G(z)}$$

is independent of z. Rather than take the limit in (5), we evaluate the above for a particular value of z, say $z = 1/m$. Then

$$\frac{m^{-mz}\Gamma(mz)}{G(z)} = \frac{m^{-1}\Gamma(1)}{G(1/m)} = \frac{1}{m\Gamma(1/m)\Gamma(2/m)\cdots\Gamma((m-1)/m)}. \tag{7}$$

To evaluate the latter we can write

$$\left[\frac{1}{G(1/m)}\right]^2 = \prod_{k=1}^{m-1} \frac{1}{\Gamma(k/m)\Gamma\{1-(k/m)\}} = \prod_{k=1}^{m-1} \frac{\sin \pi k/m}{\pi}$$

by use of III, 2.1. If we let

$$\varepsilon = e^{i\pi/m}$$

so that

$$\sin k\pi/m = \frac{\varepsilon^k - \varepsilon^{-k}}{2i}$$

and note that

$$\varepsilon^{1+2+\cdots+m-1} = \varepsilon^{\frac{1}{2}m(m-1)} = e^{i\pi\frac{1}{2}(m-1)} = i^{m-1}$$

we have

$$\left[\frac{1}{G(1/m)}\right]^2 = \frac{\prod_{k=1}^{m-1}[\varepsilon^k - \varepsilon^{-k}]}{(2\pi i)^{m-1}} = \frac{\prod_{k=1}^{m-1}(\varepsilon^{2k}-1)}{(2\pi i)^{m-1}\varepsilon^{1+2+\cdots+m-1}}$$

$$= \frac{\prod_{k=1}^{m-1}(1-\varepsilon^{2k})}{(2\pi)^{m-1}}.$$

Since the ε^k are certain roots of unity we can write

$$x^{2m} - 1 = \prod_{k=0}^{m-1}(x^2 - \varepsilon^{2k})$$

so that

$$\left[\frac{1}{G(1/m)}\right]^2 = \frac{1}{(2\pi)^{m-1}} \lim_{x\to 1} \frac{x^{2m}-1}{x^2-1} = \frac{m}{(2\pi)^{m-1}}.$$

This result combined with (7) again yields the multiplication theorem.

4. Beta Functions

A function closely related to the gamma function is the beta function, defined by

$$B(x, y) = \int_0^1 t^{x-1}(1-t)^{y-1}\,dt, \qquad \operatorname{Re} x, y > 0.$$

THEOREM.

$$B(x, y) = \frac{\Gamma(x)\Gamma(y)}{\Gamma(x + y)}.$$

Proof.

$$\Gamma(x)\Gamma(y) = \int_0^\infty e^{-t} t^{x-1}\, dt \int_0^\infty e^{-\tau} \tau^{y-1}\, d\tau$$
$$= \int_0^\infty \int_0^\infty e^{-(t+\tau)} t^{x-1} \tau^{y-1}\, dt\, d\tau.$$

We now introduce the new coordinates

$$t = r\cos^2\theta, \qquad \tau = r\sin^2\theta$$

so that

$$\Gamma(x)\Gamma(y) = 2\int_0^\infty \int_0^{\pi/2} e^{-r} r^{x+y-1} \cos^{2x-1}\theta \sin^{2y-1}\theta\, d\theta\, dr$$
$$= 2\Gamma(x + y) \int_0^{\pi/2} \cos^{2x-1}\theta \sin^{2y-1}\theta\, d\theta.$$

To evaluate the last integral we let $\cos\theta = t^{1/2}$ so that

$$\Gamma(x)\Gamma(y) = \Gamma(x + y) \int_0^1 t^{x-1}(1 - t)^{y-1}\, dt = \Gamma(x + y)B(x, y). \qquad \blacksquare$$

The Beta function was defined only for Re $x, y > 0$. It is of course desirable to find the analytic continuation into other regions. The preceding theorem furnishes such in terms of the Gamma function, which is known in the whole plane. We shall now derive an integral representation for the beta function that is valid for more general values of x and y.

THEOREM.

$$B(x, y) = (1 - e^{2\pi i x})^{-1}(1 - e^{2\pi i y})^{-1} \int^{(1+,\,0+,\,1-,\,0-)} \zeta^{x-1}(1 - \zeta)^{y-1}\, d\zeta.$$

In the above the integral is to be taken over a path that circles the points 1 and 0 first in a positive sense and then in a negative sense, as shown in the following diagram.

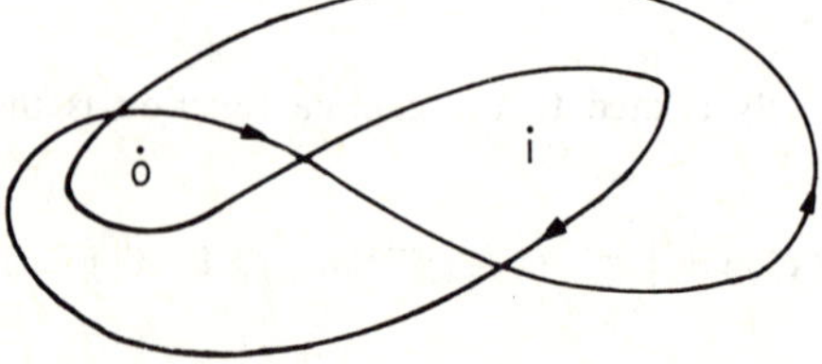

Since $\zeta^{x-1}(1-\zeta)^{y-1}$ has branch points at $\zeta = 0$ and $\zeta = 1$, the function is single valued on the ζ plane cut on (0, 1). In view of the fact that the above path crosses the cut an even number of times, the path may be viewed as a closed path on the appropriate Riemann surface.

Proof. For purposes of proof it will be convenient to replace the path of integration by an equivalent one consisting entirely of circular loops and straight segments along the real axis, as shown in the following.

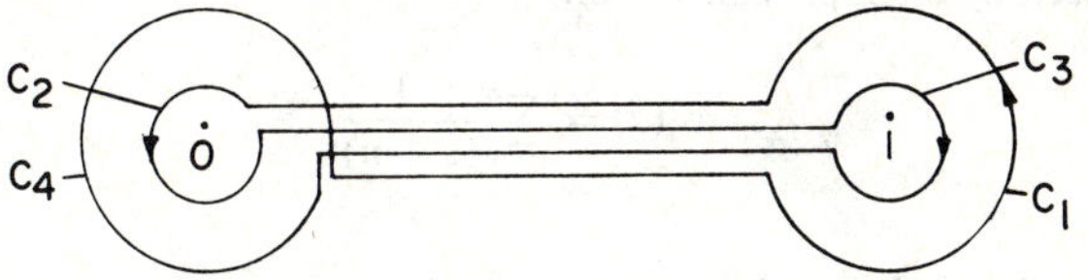

The contribution from C_1 will be

$$\lim_{\varepsilon\to 0}\int_{-\pi}^{\pi}(1+\varepsilon e^{i\theta})^{x-1}(-\varepsilon e^{i\theta})^{y-1}\varepsilon^{i\theta} i\,d\theta = 0 \qquad \text{if } \operatorname{Re} y > 0.$$

By a similar argument the contributions from C_2, C_3, C_4 can be shown to vanish. Then we have, omitting the integrands for the sake of brevity,

$$\int^{(1+,0+,1-,0-)}$$

$$= \lim_{\varepsilon\to 0}\left\{\int_{\varepsilon}^{1-\varepsilon} + e^{2\pi i(y-1)}\int_{1-\varepsilon}^{\varepsilon} + e^{2\pi i(y-1)}e^{2\pi i(x-1)}\int_{\varepsilon}^{1-\varepsilon} + e^{2\pi i(x-1)}\int_{1-\varepsilon}^{\varepsilon}\right\}$$

$$= (1 - e^{2\pi iy} + e^{2\pi ix}e^{2\pi iy} - e^{2\pi ix})\int_0^1 t^{x-1}(1-t)^{y-1}\,dt$$

$$= (1 - e^{2\pi ix})(1 - e^{2\pi iy})B(x, y).$$

We used the fact that Re x, Re $y > 0$, but by analytic continuation the results must be correct for all non-integral x and y. ∎

5. The Logarithmic Derivative of the Gamma Function

A function that arises frequently in application is

$$\psi(z) = \frac{d}{dz}\ln\Gamma(z) = \frac{\Gamma'(z)}{\Gamma(z)}.$$

Using Weierstrass' definition we readily obtain

$$\psi(z) = -\gamma + \sum_{n=0}^{\infty}\frac{1}{n+1} - \frac{1}{z+n}.$$

From the above it is evident that $\psi(z)$ has simple poles at all non-positive integers. At the positive integers

$$\psi(1) = -\gamma$$

$$\psi(k+1) = -\gamma + 1 + \frac{1}{2} + \cdots + \frac{1}{k}, \qquad k = 1, 2, \ldots.$$

The derivative of $\psi(z)$ is of course also meromorphic and has double poles, as can be seen by a simple calculation.

$$\psi'(z) = \sum_{n=0}^{\infty} \frac{1}{(z+n)^2}.$$

6. Mellin-Barnes Integrals

We consider here the following integral.

$$I = \int_{-i\infty}^{i\infty} \Gamma(\alpha + s)\Gamma(\beta + s)\Gamma(\gamma - s)\Gamma(\delta - s)\, ds. \tag{1}$$

The path of integration is such that all the poles of $\Gamma(\alpha + s)\,\Gamma(\beta + s)$ lie on the left of the path and all the poles of $\Gamma(\gamma - s)\,\Gamma(\delta - s)$ lie on the right. The parameters α, β, γ, δ must be such that no pole in one of these sets occurs in the other.

In order to be able to use residue integration we need to show that we can close the contour on the right. This will be permissible if

$$\operatorname{Re}(\alpha + \beta + \gamma + \delta - 1) < 0. \tag{2}$$

Denote the integrand by $J(s)$. Then we can write

$$J(s) = \frac{\pi^2\Gamma(\alpha + s)\Gamma(\beta + s)}{\Gamma(1 - \gamma + s)\Gamma(1 - \delta + s)\sin \pi(\gamma - s)\sin \pi(\delta - s)}$$

and we can examine the asymptotic behavior of the above for large s by means of Stirling's formula. Letting

$$\Gamma(\alpha + s) \approx \sqrt{\frac{2\pi}{\alpha + s}} \left(\frac{\alpha + s}{e}\right)^{\alpha + s}$$

and similarly for the other terms we have

$$J(s) \approx \frac{K(\alpha + s)^{\alpha + s}(\beta + s)^{\beta + s}}{(1 - \gamma + s)^{1 - \gamma + s}(1 - \delta + s)^{1 - \delta + s}\sin \pi(\gamma - s)\sin \pi(\delta - s)}$$

where K is a suitable constant, independent of s. Now

$$\begin{aligned}\ln(\alpha+s)^{\alpha+s} &= (\alpha+s)\ln(\alpha+s)\\ &= (\alpha+s)\,[\ln s+\ln(1+\alpha/s)]\\ &\approx (\alpha+s)\ln s\end{aligned}$$

so that

$$\frac{(\alpha+s)^{\alpha+s}(\beta+s)^{\beta+s}}{(1-\gamma+s)^{1-\gamma+s}(1-\delta+s)^{1-\delta+s}} \approx s^{(\alpha+\beta+\gamma+\delta-2)}.$$

Then

$$J(s) \approx \frac{Ks^{(\alpha+\beta+\gamma+s-2)}}{\sin\pi(\gamma-s)\sin\pi(\delta-s)}.$$

For $0 < |\arg s| < \pi/2$ $J(s)$ will decrease exponentially at infinity. For $|\arg s| = \pi/2$, under the hypothesis (2) the integral (1) will exist, and can be evaluated by residue integration. The residue of $\Gamma(\gamma - s)$ at $s = n + \gamma$ is $(-1)^n/n!$. We then find

$$\begin{aligned}I = -2\pi i \sum_{n=0}^{\infty} \frac{(-1)^n}{n!}\{&\Gamma(\alpha+\gamma+n)\Gamma(\beta+\gamma+n)\Gamma(\delta-\gamma-n)\\ &+\Gamma(\alpha+\delta+n)\Gamma(\beta+\delta+n)\Gamma(\gamma-\delta-n)\}\end{aligned}$$

To simplify the above we consider a typical term

$$\begin{aligned}&\Gamma(\alpha+\gamma+n)\Gamma(\beta+\gamma+n)\Gamma(\delta-\gamma-n)\\ &\quad= \frac{\Gamma(\alpha+\gamma+n)\Gamma(\beta+\gamma+n)}{\Gamma(1-\delta+\gamma+n)} \times \frac{\pi}{\sin\pi(\delta-\gamma-n)}\\ &\quad= \frac{(-1)^n\pi\Gamma(\beta+\gamma+n)}{\sin\pi(\delta-\gamma)\Gamma(1-\alpha-\delta)} \times \frac{\Gamma(\alpha+\gamma+n)\Gamma(1-\alpha-\delta)}{\Gamma(1-\delta+\gamma+n)}\\ &\quad= \frac{(-1)^n\pi\Gamma(\beta+\gamma+n)}{\sin\pi(\delta-\gamma)\Gamma(1-\alpha-\delta)} \times B(\alpha+\gamma+n,\,1-\alpha-\delta)\\ &\quad= \frac{(-1)^n\pi\Gamma(\beta+\gamma+n)}{\sin\pi(\delta-\gamma)\Gamma(1-\alpha-\delta)} \int_0^1 t^{\alpha+\gamma+n-1}(1-t)^{-\alpha-\delta}\,dt.\end{aligned}$$

We also note that the binomial theorem may be written as

$$(1-t)^{-\rho} = \sum_{n=0}^{\infty} \frac{\Gamma(\rho+n)t^n}{\Gamma(\rho)n!}.$$

By using these results we can rewrite I as

$$I = -2\pi i\left\{\frac{\pi\Gamma(\beta+\gamma)}{\sin\pi(\delta-\gamma)\Gamma(1-\alpha-\delta)}\int_0^1 (1-t)^{-\alpha-\delta}t^{\alpha+\gamma-1}\right.$$

$$\times\sum_{n=0}^{\infty}\frac{\Gamma(\beta+\gamma+n)}{n!\Gamma(\beta+\gamma)}t^n\,dt - \frac{\pi\Gamma(\beta+\delta)}{\sin\pi(\delta-\gamma)\Gamma(1-\alpha-\gamma)}$$

$$\left.\times\int_0^1 (1-t)^{-\alpha-\gamma}t^{\alpha+\delta-1}\sum_{n=0}^{\infty}\frac{\Gamma(\beta+\delta+n)}{n!\Gamma(\beta+\delta)}t^n\,dt\right\}$$

$$= -2\pi i\left\{\frac{\pi\Gamma(\beta+\gamma)B(\alpha+\gamma, 1-\alpha-\beta-\gamma-\delta)}{\sin\pi(\delta-\gamma)\Gamma(1-\alpha-\delta)}\right.$$

$$\left.-\frac{\pi\Gamma(\beta+\delta)B(\alpha+\delta, 1-\alpha-\beta-\gamma-\delta)}{\sin\pi(\delta-\gamma)\Gamma(1-\alpha-\gamma)}\right\}.$$

Note that

$$\frac{\pi\Gamma(\beta+\gamma)B(\alpha+\gamma, 1-\alpha-\beta-\gamma-\delta)}{\sin\pi(\delta-\gamma)\Gamma(1-\alpha-\delta)}$$

$$= \frac{\pi\Gamma(\beta+\gamma)\Gamma(\alpha+\gamma)\Gamma(1-\alpha-\beta-\gamma-\delta)}{\sin\pi(\delta-\gamma)\Gamma(1-\alpha-\delta)\Gamma(1-\beta-\delta)}$$

$$= \frac{\Gamma(\beta+\gamma)\Gamma(\alpha+\gamma)\Gamma(\alpha+\delta)\Gamma(\beta+\delta)\sin\pi(\alpha+\delta)\sin\pi(\beta+\delta)}{\sin\pi(\delta-\gamma)\Gamma(\alpha+\beta+\gamma+\delta)\sin\pi(\alpha+\beta+\gamma+\delta)}$$

and also

$$\sin\pi(\alpha+\delta)\sin\pi(\beta+\delta) - \sin\pi(\alpha+\gamma)\sin\pi(\beta+\gamma)$$

$$= \tfrac{1}{2}[\cos(\alpha+\beta+2\gamma) - \cos(\alpha+\beta+2\delta)]$$

$$= -\sin\pi(\delta-\gamma)\sin\pi(\alpha+\beta+\gamma+\delta).$$

Collecting all these results we obtain finally

$$I = \frac{2\pi i\Gamma(\alpha+\gamma)\Gamma(\beta+\gamma)\Gamma(\alpha+\delta)\Gamma(\beta+\delta)}{\Gamma(\alpha+\beta+\gamma+\delta)}.$$

7. Mellin Transforms

In the next section we shall need some results that pertain to Mellin transforms. The function

$$F(z) = \int_0^{\infty} f(x)x^{z-1}\,dx \tag{1}$$

is known as the Mellin Transform of $f(x)$. The above can be inverted in terms of the integral

$$f(x) = \frac{1}{2\pi i} \int_L F(z) x^{-z}\, dz \tag{2}$$

where L is a path extending from $c - i\infty$ to $c + i\infty$, for a suitable constant c. Although a detailed proof would carry us too far afield, we shall sketch the main ingredients of the proof.

We consider the integral

$$\begin{aligned} I &= \frac{1}{2\pi i} \int_L \left[\int_0^\infty f(y) y^{z-1}\, dy\, x^{-z}\, dz \right] \\ &= \frac{1}{2\pi i} \int_0^\infty \frac{f(y)}{y} \left[\int_L \left(\frac{y}{x}\right)^z dz \right] dy. \end{aligned}$$

To evaluate the inner integral we first consider

$$\int_{-iT}^{iT} \left(\frac{y}{x}\right)^z dz = \int_{-iT}^{iT} e^{z \ln y/x}\, dz = \frac{2i \sin (T \ln y/x)}{\ln y/x}$$

so that

$$I = \frac{1}{\pi} \lim_{T\to\infty} \int_0^\infty \frac{f(y)}{y} \frac{\sin [T \ln y/x]}{\ln y/x}\, dy.$$

With the substitution

$$u = T \ln y/z$$

$$\begin{aligned} I &= \frac{1}{\pi} \lim_{T\to\infty} \int_{-\infty}^\infty f(xe^{u/T}) \frac{\sin u}{u}\, du \\ &= f(x) \frac{1}{\pi} \int_{-\infty}^\infty \frac{\sin u}{u}\, du = f(x). \end{aligned}$$

In general this last result is only valid for continuous functions, but we will be concerned only with such.

8. Applications to Algebraic Equations

Mellin first investigated the integrals named after him in connection with his attempts to solve the trinomial equation

$$y^n + xy^p - 1 = 0 \qquad n > p. \tag{1}$$

In general every solution of the above will be a function of the parameter x. In particular for $x = 0$ we have

$$y^n(0) - 1 = 0$$

and if we let

$$\varepsilon = e^{2\pi i/n},$$

then

$$y_k(0) = \varepsilon^k, \qquad k = 0, 1, 2, \ldots, n-1.$$

If $y(x)$ is a solution of (1), so is $\varepsilon^k y(\varepsilon^{pk} x)$. To see this we write

$$\varepsilon^{nk} y^n(\varepsilon^{pk} x) + \varepsilon^{pk} x y^p(\varepsilon^{pk} x) - 1 = y^n(\tau) + \tau y^p(\tau) - 1 = 0$$

for $\tau = \varepsilon^{pk} x$. Thus if we know one solution of (1) we can generate all by this process.

For x sufficiently small (1) must have n distinct and analytic solutions, as can be seen by a direct application of Rouché's theorem.
We write

$$y^n + xy^p - 1 = y^n - 1 + xy^p.$$

On the circle $y = 1 + \rho e^{i\theta}$ we have, since

$$\frac{y^n - 1}{y - 1} = \frac{(1 + \rho e^{i\theta})^n - 1}{\rho e^{i\theta}} = n + \varepsilon(\rho e^{i\theta})$$

where $|\varepsilon(\rho e^{i\theta})|$ can be as small as we please, if we let ρ be sufficiently small,

$$\rho(n - |\varepsilon|) \leq |y^n - 1|.$$

But we also have

$$xy^p = x(1 + \rho e^{i\theta})^p = x + xp\,\rho e^{i\theta} + \rho e^{i\theta}\,\delta(\rho e^{i\theta})$$

where $|\delta(\rho e^{i\theta})|$ is small, so that

$$|xy^p| \leq |x| + p|x|\rho + \rho|\delta|.$$

It follows that

$$|xy^p| < |y^n - 1|$$

on $y = 1 + \rho e^{i\theta}$, for ρ small and x sufficiently small. Since $y^n - 1$ will have precisely one zero in the circle, so will (1).

In order for (1) to have multiple roots for some value of x it will be necessary that

$$y^n + xy^p - 1 = 0$$

and simultaneously its derivative with respect to y

$$ny^{n-1} + xpy^{p-2} = 0.$$

For both of these conditions to hold simultaneously we require that

$$x = -\frac{n}{p}\, y^{n-p}$$

and that, by eliminating x

$$y^n = -\frac{p}{n-p}.$$

These two conditions finally yield

$$x^n = \frac{(-1)^p}{(p/n)^p((n-p)/n)^{n-p}}.$$

Hence for x such that

$$|x| < \frac{1}{(p/n)^{p/n}((n-p)/n)^{(n-p)/n}} \le 2^*,$$

(1) will have n analytic, distinct solutions.

In order to study the structure of $y(x)$ it is convenient to investigate its Mellin transform. More generally we shall consider $y^\mu(x)$ and we let

$$Y(z) = \int_0^\infty y^\mu(x) x^{z-1}\, dx, \qquad \frac{\mu}{p} > \operatorname{Re} z > 0. \tag{2}$$

We will select $y(x)$ so that $y(0) = 1$. For x small $y^\mu(x) \approx 1$ and the above integral certainly converges near the origin for $\operatorname{Re} z > 0$.

From (1), differentiation, we obtain easily

$$y' = \frac{-y^p}{ny^{n-1} + xpy^{p-1}} < 0.$$

We thus see that $y(0) = 1$ and y is a decreasing function. Since $y = 0$ does not satisfy (1) it follows that $0 < y(x) \le 1$. Thus for large x we expect y to be small, so that the middle term in (1) predominates.

$$y^p = \frac{1 - y^n}{x} \approx \frac{1}{x}$$

and

$$y \approx x^{-1/p}$$

(2) will converge at infinity if $\operatorname{Re} z < \mu/p$.

* This follows from the fact that $t^t(1-t)^{1-t} \ge \frac{1}{2}$ for $0 \le t \le 1$.

To evaluate (2) we go from an x integration to a y integration. Then

$$x = y^{-p} - y^{n-p}$$

$$dx = -[py^{-p-1} + (n-p)y^{n-p-1}]\,dy$$

and

$$Y(z) = p\int_0^1 y^{\mu-pz-1}(1-y^n)^{z-1}\,dy + (n-p)\int_0^1 y^{\mu-pz+n-1}(1-y^n)^{z-1}\,dy.$$

By letting $y^n = t$, the latter can be integrated in terms of beta functions.

$$\begin{aligned} Y(z) &= \frac{p}{n}B\left(\frac{\mu-pz}{n}, z\right) + \frac{n-p}{n}B\left(\frac{\mu-pz}{n}+1, z\right) \\ &= \frac{\Gamma(z)}{n}\left\{\frac{p\Gamma((\mu-pz)/n)}{\Gamma((\mu-pz)/n)+z)} + \frac{(n-p)\Gamma((\mu-pz)/n)+1)}{\Gamma((\mu-pz)/n)+1+z)}\right\} \\ &= \frac{\mu\Gamma(z)\Gamma((\mu-pz)/n)}{n\Gamma((\mu-pz)/n)+1+z)} \end{aligned} \tag{3}$$

By means of the inversion formula 3.7.2, we can now write

$$y^\mu(x) = \frac{1}{2\pi i}\int_{c-i\infty}^{c+i\infty} Y(z)x^{-z}\,dz, \qquad 0 < c < \frac{\mu}{p} \tag{4}$$

The choice of c is suggested by the fact that in order to perform the integration in (2) it was necessary to keep z in this strip. In order to close the path of integration over the left half plane we use Stirling's formula to verify that the contribution from a large semicircle vanishes. A detailed verification shows that we require that

$$|x| < \frac{1}{(p/n)^{p/n}(1-p/n)^{1-p/n}}$$

which is, of course, equivalent to saying that x must stay sufficiently close to the origin to be sure that (1) has n distinct solutions. Using residue integration we find that

$$y^\mu(x) = \frac{\mu}{n}\sum_{r=0}^{\infty}\frac{(-1)^r\Gamma((\mu+pr)/n)x^r}{r!\,\Gamma((\mu+rp)/n)+1-r)} \tag{5}$$

In particular for $\mu = 1$ we have

$$y(x) = \frac{1}{n}\sum_{r=0}^{\infty}\frac{(-1)^r\Gamma((1+pr)/n)x^r}{r!\,\Gamma((1+rp)/n)+1-r)} \tag{6}$$

and all other solutions are found by evaluating $\varepsilon^k y(\varepsilon^{pk}x)$.

Lastly we shall show that the solution (5) satisfies certain linear differential equations. We shall define the following symbol, often referred to as the Barnes symbol.

$$(a)_n = a(a+1)\cdots(a+n-1) = \frac{\Gamma(a+n)}{\Gamma(a)}.$$

The last equality can be proved using the recurrence formula for the Gamma function. Using this notation, we have, from (3)

$$\frac{Y(z+n)}{Y(z)} = \frac{\Gamma(z+n)}{\Gamma(z)} \cdot \frac{\Gamma((\mu - pz)/n) - p)}{\Gamma((\mu - pz)/n)} \cdot \frac{\Gamma((\mu - pz)/n) + 1 + z)}{\Gamma((\mu - pz)/n) + 1 + z + n - p)}$$

$$= \frac{(z)_n}{((\mu - pz)/n) - p)_p((\mu - pz)/n) + 1 + z)_{n-q}} \equiv \frac{(z)_n}{(pz)} \tag{7}$$

The right side of (7) is a rational function of z, where both numerator and denominator are polynomials of degree n.

From (4) we obtain by application of successive differentiations and multiplications by x

$$-x \frac{d}{dx} y^\mu(x) = \frac{1}{2\pi i} \int_{c-i\infty}^{c+i\infty} zY(z)x^{-z}\, dz$$

$$x^2 \left(\frac{d}{dx}\right)^2 y^\mu(x) = \frac{1}{2\pi i} \int_{c-i\infty}^{c+i\infty} z(z+1)Y(z)x^{-z}\, dz$$

$$\vdots$$

$$(-x)^n \left(\frac{d}{dx}\right)^n y^\mu(x) = \frac{1}{2\pi i} \int_{c-i\infty}^{c+i\infty} (z)_n Y(z)x^{-z}\, dz$$

$$= \frac{1}{2\pi i} \int_{c-i\infty}^{c+i\infty} p(z)Y(z+n)x^{-z}\, dz$$

Letting $\xi = z + n$ we find

$$\left(-\frac{d}{dx}\right)^n y^\mu(x) = \frac{1}{2\pi i} \int_{c+n-i\infty}^{c+n+i\infty} p(z-n)Y(z)x^{-z}\, dx.$$

$$= \frac{1}{2\pi i} \int_{c-i\infty}^{c+i\infty} p(z-n)Y(z)x^{-z}\, dz \tag{8}$$

since in the strip $c < \operatorname{Re} z < n + c$ the integrand has no poles.

Similarly

$$-x\frac{d}{dx}\,y^{\mu}(x)=\frac{1}{2\pi i}\int_{c-i\infty}^{c+i\infty} zY(z)x^{-z}\,dz$$

$$\left(-x\frac{d}{dx}\right)^{2} y^{\mu}(x)=\frac{1}{2\pi i}\int_{c-i\infty}^{c+i\infty} z^{2}Y(z)x^{-z}\,dz$$

$$\vdots$$

$$\left(-x\frac{d}{dx}\right)^{n} y^{\mu}(x)=\frac{1}{2\pi i}\int_{c-i\infty}^{c+i\infty} z^{n}Y(z)x^{-z}\,dz$$

and more generally for an arbitrary polynomial

$$\phi\left(-x\frac{d}{dx}\right)y^{\mu}(x)=\frac{1}{2\pi i}\int_{c-i\infty}^{c+i\infty} \phi(z)Y(z)x^{-z}\,dz \tag{9}$$

Combining (8) and (9) we have

$$\left(-\frac{d}{dx}\right)^{n} y^{\mu}(x)=p\left(-x\frac{d}{dx}-n\right)y^{\mu}(x)$$

which is an nth order-linear differential equation for $y^{\mu}(x)$.

Exercises

1. Show that

$$\Gamma(z)=\int_0^{\infty}\left[e^{-t}-\sum_{k=0}^{n}\frac{(-t)^k}{k!}\right]t^{z-1}\,dt,\quad -n-1<\operatorname{Re} z<-n$$

2. Show that

$$\gamma=-\int_0^{\infty} e^{-t}\log t\,dt,$$

where γ is the Euler-Mascheroni constant.

3. Apply Watson's Lemma to obtain an asympotic series for

$$\int_0^{\infty} e^{-st}\sqrt{1+t^2}\,dt$$

for large s. Demonstrate explicitly that the resulting series is not convergent.

4. In Stirling's formula only the first two terms in the expansion $1+1/12z+\cdots$ were calculated. Find the next term.

5. Show that

$$\prod_{k=0}^{m-1}\sin\pi\left(z+\frac{k}{m}\right)=2^{1-m}\sin\pi mz.$$

6. Evaluate

$$\int_0^1 t^{x-1}(1-t)^{y-1}\frac{dt}{(t+p)^{x+y}},\ p>0.$$

7. Evaluate

$$\int_z^{z+1} \ln \Gamma(\zeta)\, d\zeta.$$

8. Show that

$$|\Gamma(iy)|^2 = \frac{\pi}{y \sinh \pi y}.$$

9. Evaluate

$$\int_0^{\pi/2} \sin^{3/2} t\, dt.$$

10. Show that

$$\int_0^\infty \frac{\cosh 2yt}{(\cosh t)^{2x}}\, dt = 2^{2x-2}\frac{\Gamma(x+y)\,\Gamma(x-y)}{\Gamma(2x)}$$

where $\operatorname{Re} x > |\operatorname{Re} y|$, $\operatorname{Re} x > 0$.

11. Show that

$$\int_{-\infty}^{\infty} \frac{e^{its}\, ds}{(\Gamma u+s)\Gamma(\gamma-s)} = \frac{2[\cos \frac{1}{2} t]^{\mu+\gamma-2} e^{1/2\, it(\gamma-\mu)}}{\Gamma(\mu+\nu-1)}, \qquad |t|<\pi$$

$$= 0 \qquad |t|>\pi.$$

12. Apply the method of section 8 to the equation

$$y^3 + xy - 1 = 0.$$

Find the series representation for the solution and the differential equation satisfied by it.

CHAPTER 4

Hypergeometric Functions

1. Review of Linear Differential Equations with Regular Singular Points

Many of the differential equations of mathematical physics have regular singular points that are sometimes known as Fuchsian singularities. If the equation can be written as a system it has the following form. Without loss of generality the singular point can be taken as the origin.

$$\frac{d}{dz} X = \frac{1}{z} A(z)X. \tag{1}$$

In the above X is a vector in an n-dimensional space and $A(z)$ is an $n \times n$ matrix whose entries are analytic functions. Then $A(z)$ must have a series expansion of the form

$$A(z) = A_0 + zA_1 + z^2A_2 + \cdots = \sum_{k=0}^{\infty} z^k A_k,$$

where the A_k are constant matrices. We can suppose that $A_0 \neq 0$, otherwise the singularity is removable. We can show that (1) must have solutions of the form

$$X = z^{\mu} \sum_{k=0}^{\infty} z^k X_k, \tag{2}$$

where the X_k are constant vectors, if the parameter μ satisfies certain conditions. We can insert (2) in (1) multiply the resultant series and obtain

$$z^{\mu-1} \sum_{k=0}^{\infty} (\mu + k) z^k X_k = z^{\mu-1} \sum_{k=0}^{\infty} z^k \sum_{l=0}^{k} A_{k-l} X_l. \tag{3}$$

By comparing coefficients of corresponding terms we have

$$\begin{aligned}(\mu I - A_0)X_0 &= 0\\ ((\mu+1)I - A_0)X_1 &= A_1X_0\\ &\vdots\\ ((\mu+k)I - A_0)X_k &= \sum_{l=0}^{k-1} A_{k-l}X_l\\ &\vdots\end{aligned} \tag{4}$$

If μ is selected as an eigenvalue of A_0 the first of the above equations will have a solution X_0. If $\mu + 1$ is not an eigenvalue the second equation will have a solution X_1, and more generally if for no positive integer k, $\mu + k$ is an eigenvalue then we will be able to solve for X_k recursively. The resultant series (3) can be shown to converge in the same z-domain in which the series for $A(z)$ converges.

If we let $z = 1/t$ in (1) then

$$\frac{d}{dt}X = -\frac{1}{t}A\left(\frac{1}{t}\right)X. \tag{5}$$

If the latter has a regular singular point at $t = 0$ then we say that (1) has a regular singular point at $z = \infty$.

Suppose now that (1) has precisely one regular singular point at $z = 0$. Then (5) must have a point of regularity at $t = 0$. But for this to be the case we require that $A(1/t) \equiv 0$. In this case we see that no equation such as (1) has precisely one regular singular point. Similarly if (1) is to have precisely two regular singular points it is necessary that A be a constant matrix.

Lastly let us consider the case where (1) has precisely three singular points. Suppose these are located, without loss of generality, at $z = 0, 1, a$. Then

$$X' = \frac{A(z)}{z(z-1)(z-a)}X. \tag{6}$$

If $z = \infty$ is to be regular we require that

$$A(z) = A_0 + A_1 z,$$

so that $A(z)$ is linear. To see this we let $z = 1/t$ so that

$$\frac{dX}{dt} = \frac{-tA(1/t)}{(1-t)(1-at)}X.$$

If $A(1/t)$ were not linear in the above, $tA(1/t)$ would be singular at $t = 0$.

By means of the bilinear transformation

$$z = \frac{at}{t + (a-1)}$$

applied to (6) we can place the singular points at $t = 0, 1, \infty$. Then

$$\frac{dX}{dt} = \frac{A_0(a-1) + (A_0(a-1) + A_1 a)t}{t(t-1)} X \tag{7}$$

and we will use (7) as the standard form of an equation with precisely three regular singular points.

2. The Hypergeometric Differential Equation

We will now specialize the analysis of the previous section to an equation of the second order. If we consider

$$X' = \frac{1}{z} A(z) X$$

where X has the components y and u and where $A(z)$ is a 2×2 matrix, we can eliminate one of the variables, say u. Then y satisfies a second order differential equation

$$y'' + \frac{1}{z} p(z) y' + \frac{1}{z^2} q(z) y = 0$$

where $p(z)$ and $q(z)$ are analytic functions, that depend on the entries in $A(z)$. More generally if the equation has Fuchsian singularities at $z = 0$ and $z = 1$ we have

$$y'' + \left(\frac{p_0(z)}{z} + \frac{p_1(z)}{z-1}\right) y' + \left(\frac{q_0(z)}{z^2} + \frac{q_1(z)}{(z-1)^2} + \frac{q_2(z)}{z(z-1)}\right) y = 0. \tag{1}$$

If the above is to have no other singularities in the finite part of the plane the functions p_i and q_i must be entire functions.

We now wish to determine conditions under which (1) has a Fuchsian singularity at ∞. As before we let $z = 1/t$ and (1) takes the form

$$\frac{d^2 y}{dt^2} + \frac{1}{t}\left[2 - p_0\left(\frac{1}{t}\right) - \frac{p_1(1/t)}{1-t}\right] \frac{dy}{dt} + \frac{1}{t^2}\left[q_0\left(\frac{1}{t}\right) + \frac{q_1(1/t)}{(1-t)^2} + \frac{q_2(1/t)}{1-t}\right] y = 0. \tag{2}$$

From (2) we observe that in order for $t = 0$ to be a regular singular point all p_i and q_i must be constants.

In analogy to IV.1.2 we can now search for solutions of the form

$$y = z^{\mu} \sum_{0}^{\infty} a_n z^n \qquad \text{near} \quad z = 0$$

$$y = (z-1)^{\mu} \sum_{0}^{\infty} b_n (z-1)^n \qquad \text{near} \quad z = 1 \tag{3}$$

$$y = z^{-\mu} \sum_{0}^{\infty} c_n z^{-n} \qquad \text{near} \quad z = \infty.$$

To determine the appropriate values of μ we insert the trial solutions (3) in (1) and by comparing coefficients of corresponding powers of z we obtain the following three equations

$$\mu(\mu-1) + p_0\mu + q_0 = 0 \qquad \text{with roots } \mu_1 = \alpha_1, \mu_2 = \alpha_2$$

$$\mu(\mu-1) + p_1\mu + q_1 = 0 \qquad \text{with roots } \mu_1 = \beta_1, \mu_2 = \beta_2$$

$$\mu(\mu-1) + (2 - p_0 - p_1)\mu + q_0 + q_1 + q_2 = 0 \qquad \text{with roots } \mu_1 = \gamma_1, \mu_2 = \gamma_2. \tag{4}$$

The six quantities $\alpha_1, \alpha_2, \beta_1, \beta_2, \gamma_1, \gamma_2$ are known as the indicial roots. They are not independent of one another. Note that

$$\alpha_1 + \alpha_2 = 1 - p_0$$

$$\beta_1 + \beta_2 = 1 - p_1$$

$$\gamma_1 + \gamma_2 = -1 + p_0 + p_1$$

so that

$$\alpha_1 + \alpha_2 + \beta_1 + \beta_2 + \gamma_1 + \gamma_2 = 1.$$

A knowledge of the indicial roots allows us to deduce the five quantities p_0, p_1, q_0, q_1, q_2 via (4). These values coupled with the fact that the three singular points are located at $z = 0, 1, \infty$ fully determine the differential equation

$$y'' + \left[\frac{p_0}{z} + \frac{p_1}{z-1}\right] y' + \left[\frac{q_0}{z^2} + \frac{q_1}{(z-1)^2} + \frac{q_2}{z(z-1)}\right] y = 0. \tag{5}$$

It is often convenient to associate with the differential equation (5) the *Riemann-Papperitz* symbol

$$P\begin{Bmatrix} 0 & 1 & \infty & \\ \alpha_1 & \beta_1 & \gamma_1 & z \\ \alpha_2 & \beta_2 & \gamma_2 & \end{Bmatrix}$$

which carries all the information necessary to construct (5). It tells us where the singular points are and what the indices associated with them are. We can easily show that with a proper understanding of this symbol many transformations of the equation can be carried out by inspection.

For example

$$z^{\lambda}(1-z)^{\nu} P \begin{Bmatrix} 0 & 1 & \infty & \\ \alpha_1 & \beta_1 & \gamma_1 & z \\ \alpha_2 & \beta_2 & \gamma_2 & \end{Bmatrix} = P \begin{Bmatrix} 0 & 1 & \infty & \\ \alpha_1+\lambda & \beta_1+\nu & \gamma_1-\lambda-\nu & z \\ \alpha_2+\lambda & \beta_2+\nu & \gamma_2-\lambda-\nu & \end{Bmatrix}. \tag{6}$$

The above essentially tells us that if we consider the equation associated with the symbol on the left and introduce a new independent variable u via the substitution

$$u = z^{\lambda}(1-z)^{\nu} y \tag{7}$$

then u satisfies the equation associated with the symbol on the right. If y has the form

$$y = z^{\alpha_1} \sum a_n z^n$$

then

$$u = z^{\alpha_1+\lambda} \sum a_n z^n$$

showing that the index α_1 is transformed into $\alpha_1 + \lambda$. The coefficients $\bar{p}_i, \bar{q}_i$ associated with the right side of (6) can be calculated from (4)

$$\begin{aligned} \bar{p}_0 &= p_0 - 2\lambda, & \bar{p}_1 &= p_1 - 2\nu \\ \bar{q}_0 &= \lambda(\lambda+1) - p_0\lambda + q_0, & \bar{q}_1 &= \nu(\nu+1) - p_1\nu + q_1 \\ \bar{q}_2 &= q_2 + 2\lambda\nu - \lambda p_1 - \nu p_0 \end{aligned}$$

A direct substitution of (7) in (5) would lead to

$$u'' + \left[\frac{p_0 - 2\lambda}{z} + \frac{p_1 - 2\nu}{z-1}\right]u' + \left[\frac{\lambda(\lambda+1) - p_0\lambda + q_0}{z^2} + \frac{\nu(\nu+1) - p_1\nu + q_1}{(z-1)^2}\right.$$
$$\left. + \frac{q_2 + 2\lambda\nu - \lambda p_1 - \nu p_0}{z(z-1)}\right]u = 0.$$

It is particularly convenient to let $\lambda = -\alpha_1$ and $\nu = -\beta_1$. Then we have near $z = 0$ one solution that is a pure power series, and similarly another one near $z = 1$. We now label the indices at ∞ a and b and the second index at $z = 0$, $1 - c$. These are conventional designations. The resultant Riemann–Papperitz symbol becomes

$$P \begin{Bmatrix} 0 & 1 & \infty & \\ 0 & 0 & a & z \\ 1-c & c-a-b & b & \end{Bmatrix}.$$

The associated equation becomes

$$z(1-z)y'' + [c - (1 + a + b)z]y' - aby = 0 \tag{8}$$

which is the canonical form in which we shall study the hypergeometric equation.

3. The Hypergeometric Function

We saw in the last section that the hypergeometric differential equation can be put into the form

$$z(1-z)y'' + [c - (1 + a + b)z]y' - aby = 0. \tag{1}$$

It must have a solution of the form

$$y = \sum_{n=0}^{\infty} a_n z^n.$$

Inserting the latter in (1) we find that a_n satisfies the recurrence formula

$$a_{n+1} = \frac{(a+n)(b+n)}{(n+1)(c+n)} a_n.$$

If we let $a_0 = 1$ we can show from the above that

$$a_n = \frac{(a)_n(b)_n}{n!(c)_n}.$$

It is customary to represent this function by the symbol

$$F(a, b; c; z) = \sum_{n=0}^{\infty} \frac{(a)_n(b)_n}{n!(c)_n} z_n. \tag{2}$$

From the general theory of equations with regular singular points we know that (2) must converge up to the next singularity, which is at $z = 1$. This also follows from the ratio test. For $|z| < 1$ we have

$$\lim_{n\to\infty} \left| \frac{(a)_{n+1}(b)_{n+1} z^{n+1} n!(c)_n}{(n+1)!(c)_{n+1}(a)_n(b)_n z^n} \right| = \lim_{n\to\infty} \left| \frac{(a+n)(b+n)z}{(n+1)(c+n)} \right| < 1.$$

On the circle of convergence the situation is more delicate. We have the following expansion for the above ratio.

$$\frac{(a+n)(b+n)}{(n+1)(c+n)} = 1 - \frac{1 + c - a - b}{n} + \cdots.$$

By Raabe's test, which says that the real part of the coefficient of $-1/n$ must exceed unity for convergence, we have that (2) converges on $|z| = 1$ if

$$\operatorname{Re}(c - a - b) > 0.$$

Many familiar functions are special cases of hypergeometric functions. For example, the elliptic integrals

$$K(k) = \frac{\pi}{2} F(\tfrac{1}{2}, \tfrac{1}{2}; 1; k^2) = \int_0^{\pi/2} \frac{d\phi}{\sqrt{1 - k^2 \sin^2 \phi}}$$

$$E(k) = \frac{\pi}{2} F(-\tfrac{1}{2}, \tfrac{1}{2}; 1, k^2) = \int_0^{\pi/2} \sqrt{1 - k^2 \sin^2 \phi}\, d\phi$$

the associated Legendre polynomials

$$P_n^{\mu}(x) = \frac{1}{\Gamma(1 - \mu)} \left(\frac{x + 1}{x - 1}\right)^{\mu/2} F(-n, n + 1; 1 - \mu; \tfrac{1}{2} - \tfrac{1}{2}x)$$

the Gegenbauer polynomials

$$C_n^{\nu}(x) = F(-n, n + 2\nu; \nu + \tfrac{1}{2}; \tfrac{1}{2} - \tfrac{1}{2}x)$$

$$(1 + x)^n = F(-n, 1; 1; -x)$$

$$(1 + x)^n - (1 - x)^n = 2F(-\tfrac{1}{2}n, -\tfrac{1}{2}n + \tfrac{1}{2}; \tfrac{1}{2}; x^2)$$

$$\ln(1 + x) = xF(1, 1; 2; -x).$$

The solution (2) of (1) is of course only one solution and we should like to find another linearly independent one. One such can be associated with the index $1 - c$. We see that

$$z^{c-1} P \begin{Bmatrix} 0 & 1 & \infty & \\ 0 & 0 & a & z \\ 1 - c & c - a - b & b & \end{Bmatrix} = P \begin{Bmatrix} 0 & 1 & \infty & \\ c - 1 & 0 & a + 1 - c & z \\ 0 & c - a - b & b + 1 - c & \end{Bmatrix}$$

and the solution corresponding to the index 0 at $z = 0$ associated with the right side is

$$F(a + 1 - c, b + 1 - c; 2 - c; z).$$

It follows that

$$y = z^{1-c} F(a + 1 - c, b + 1 - c; 2 - c; z) \tag{3}$$

satisfies (1). To test whether (2) and (3) are linearly independent we examine their Wronskian.

In general if y_1 and y_2 are two solution of

$$y'' + p(z)y' + q(z)y = 0$$

then

$$W(y_1, y_2) = \begin{vmatrix} y_1 & y_2 \\ y_1' & y_2' \end{vmatrix}$$

satisfies

$$W' = -p(z)W$$

so that

$$W(z) = ke^{-\int p(z)\,dz}.$$

For (1) we find

$$W(z) = kz^{-c}(1-z)^{c-1-a-b}.$$

To deduce the value of k if y_1 and y_2 are given by (2) and (3) we need work only with the leading terms

$$\begin{vmatrix} 1 + \dfrac{ab}{c} z + \cdots & z^{1-c}[1 + \cdots] \\ \dfrac{ab}{c} + \cdots & (1-c)z^{-c}[1 + \cdots] \end{vmatrix} = kz^{-c}[1 + \cdots]$$

so that $k = 1 - c$. Hence if $c \neq 1$ (2) and (3) are linearly independent; if $c = 1$ they are identical. In this case, by the method of variation of parameters, we can easily verify that

$$y = F(a, b; 1; z) \int^z \frac{d\zeta}{\zeta(1-\zeta)^{a+b} F^2(a, b; 1; \zeta)}$$

is a solution independent of (2). Observe, however, that it has a logarithmic singularity.

By means of the Riemann-Papperitz symbol we can generate numerous solutions. For example

$$(1-z)^{a+b-c} P \begin{Bmatrix} 0 & 1 & \infty & \\ 0 & 0 & a & z \\ 1-c & c-a-b & b & \end{Bmatrix} = P \begin{Bmatrix} 0 & 1 & \infty & \\ 0 & a+b-c & c-b & z \\ 1-c & 0 & c-a & \end{Bmatrix}$$

so that

$$y_3 = (1-z)^{c-a-b} F(c-a, c-b; c; z).$$

Clearly constants A and B must exist such that

$$(1-z)^{c-a-b} F(c-a, c-b; c; z)$$
$$= AF(a, b; c; z) + Bz^{1-c} F(a+1-c, b+1-c; 2-c; z).$$

To evaluate A and B we examine the neighborhood of $z = 0$. Two of the terms are regular there, but the third, for general values of c, has a branch point. Hence $B = 0$. Then letting $z = 0$ we find that $A = 1$ and we obtain the identity

$$(1 - z)^{c-a-b}F(c - a, c - b; c; z) = F(a, b; c; z).$$

We shall, in a subsequent section, study all solutions that can be obtained by such and related techniques.

Using Beta functions we can derive a useful integral representation. (2) can be rewritten in the form

$$\begin{aligned}
F(a, b; c; z) &= \sum_{n=0}^{\infty} \frac{\Gamma(a + n)\Gamma(b + n)\Gamma(c)}{\Gamma(a)\Gamma(b)\Gamma(c + n)} \times \frac{z}{n!} \\
&= \frac{\Gamma(c)}{\Gamma(a)\Gamma(c - a)} \sum_{n=0}^{\infty} \frac{\Gamma(a + n)\Gamma(c - a)}{\Gamma(c + n)} \times \frac{\Gamma(b + n)}{\Gamma(b)} \times \frac{z^n}{n!} \\
&= \frac{1}{B(a, c - a)} \sum_{n=0}^{\infty} B(a + n, c - a) \frac{\Gamma(b + n)}{\Gamma(b)} \times \frac{z^n}{n!} \\
&= \frac{1}{B(a, c - a)} \sum_{n=\infty}^{\infty} \int_0^1 t^{a+n-1}(1 - t)^{c-a-1}\, dt \times \frac{\Gamma(b + n)}{\Gamma(b)} \times \frac{z^n}{n!} \\
&= \frac{1}{B(a, c - a)} \int_0^1 t^{a-1} \frac{(1 - t)^{c-a-1}}{(1 - tz)^b}\, dt. \qquad (4)
\end{aligned}$$

The last step is accomplished by noting that upon interchange of summation and integration we obtain a binomial series.

It was shown earlier that if Re $(c - a - b) > 0$ then (2) converged for $|z| = 1$. In this case (4) yields

$$\begin{aligned}
F(a, b; c; 1) &= \frac{1}{B(a, c - a)} \int_0^1 t^{a-1}(1 - t)^{c-a-b-1}\, dt \\
&= \frac{B(a, c - a - b)}{B(a, c - a)} \\
&= \frac{\Gamma(c - a - b)\Gamma(c)}{\Gamma(c - a)\Gamma(c - b)}. \qquad (5)
\end{aligned}$$

The representation (4) is restricted to those cases where Re $a > 0$ and Re $(c - a - b) > 0$. Yet the hypergeometric function will exist in more general situations. Exactly as was done in the case of the Gamma functions and Beta functions we can find certain contour integrals that exist in more general cases. For example (4) can be generalized to

$$F(a, b; c; z) = \frac{-\Gamma(c)e^{-i\pi c}}{\Gamma(a)\Gamma(c-a)4 \sin \pi a \sin \pi(c-a)}$$

$$\times \int^{(1+,0+,1-,0-)} \frac{t^{a-1}(1-t)^{c-a-1}}{(1-tz)^b}\, dt. \quad (6)$$

Another generalized integral representation can be found via the Mellin-Barnes integral. Consider

$$F(a, b; c; z) = \frac{\Gamma(c)}{\Gamma(a)\Gamma(b)2\pi i} \int_{-i\infty}^{i\infty} \frac{\Gamma(a+s)\Gamma(b+s)\Gamma(-s)(-z)^s}{\Gamma(c+s)}\, ds. \quad (7)$$

The path is so chosen that all poles of $\Gamma(a+s)\ \Gamma(b+s)$ lie on the left and those of $\Gamma(-s)$ lie on the right. For Re $(c-a-b)$ we can close the contour on the right and recapture the power series (2) by means of residue integration.

Hypergeometric functions are related to derivatives of other hypergeometric functions. We see that

$$\frac{d}{dz} F(a, b; c; z) = \sum_{n=1}^{\infty} \frac{(a)_n(b)_n}{(c)_n} \times \frac{z^{n-1}}{(n-1)!}$$

$$= \frac{ab}{c} \sum_{n=1}^{\infty} \frac{(a+1)_{n-1}(b+1)_{n-1}}{(c+1)_{n-1}} \times \frac{z^{n-1}}{(n-1)!}$$

$$= \frac{ab}{c} F(a+1, b+1; c+1; z).$$

By repeating the above we find

$$\left(\frac{d}{dz}\right)^k F(a, b; c; z) = \frac{(a)_k(b)_k}{(c)_k} F(a+k, b+k; c+k; z). \quad (8)$$

Similarly we can show that

$$\left(\frac{d}{dz}\right)^k z^{a+k-1}F(a, b; c; z) = (a)_k z^{a-1} F(a+k, b; c; z). \quad (9)$$

Another set of interesting identities, known as the *contiguity relations*, can be derived as follows. In order to facilitate the notation we introduce the following

$$\delta \equiv z \frac{d}{dz}$$

$$F_{a\pm} \equiv F(a \pm 1, b; c; z).$$

By a comparison of power series we can easily demonstrate the validity of each of the following.

$$(\delta + a) = F_{a+}$$
$$(\delta + b)F = F_{b+}$$
$$(\delta + c - 1)F = (c-1)F_{c-}$$
$$(1-z)\delta F = (c-a)F_{a-} + (a - c + bz)F$$
$$(1-z)\delta F = (c-b)F_{b-} + (b - c + az)F$$
$$c(1-z)\delta F = z(c-a)(c-b)F_{c+} + zc(a+b-c)F.$$

The set of functions $F_{a\pm}$, $F_{b\pm}$, $F_{c\pm}$ consists of six elements. We can select two of these in $\binom{6}{2} = 15$ different ways. Between F and any two of these contiguous functions there must exist a relationship that can be found by means of the above identities, and it can be obtained by eliminating δ from some of them. For example, using the second and third we find

$$F_{b+} - (b - c + 1)F - (c-1)F_{c-} = 0. \tag{10}$$

Gauss first showed that one can often use such relations to develop hypergeometric functions in terms of continued fractions. One can easily verify that

$$\frac{(a)_n(b+1)_n}{n!(c+1)_n} - \frac{(a)_n(b)_n}{n!(c)_n} = \frac{a(c-b)}{c(c+1)} \times \frac{(a+1)_{n-1}(b+1)_{n-1}}{(n-1)!(c+2)_{n-1}}$$

from which we conclude that

$$F(a, b+1; c+1; z) - F(a, b; c; z) = \frac{a(c-b)}{c(c+1)} zF(a+1, b+1; c+2; z). \tag{11}$$

(11) can be rewritten as

$$\frac{F(a, b+1; c+1; z)}{F(a, b; c; z)} = 1 \Bigg/ 1 - \frac{a(c-b)}{c(c+1)} z \left\{\frac{F(a+1, b+1; c+2; z)}{F(a, b+1; c+1; z)}\right\}.$$

Similarly, by an interchange of symbols

$$\frac{F(a+1, b+1; c+2; z)}{F(a, b+1; c+1; z)} = 1 \Bigg/ 1 - \frac{(b+1)(c+1-a)}{(c+1)(c+2)} z \left\{\frac{F(a+1, b+2; c+3, z)}{F(a+1, b+1; c+2, z)}\right\}$$

and replacing this in the former we find

$$\frac{F(a, b+1; c+1; z)}{F(a, b; c; z)} = \cfrac{1}{1 - \cfrac{\dfrac{a(c-b)}{c(c+1)} z}{1 - \dfrac{(b+1)(c+1-a)}{(c+1)(c+2)} z \left\{\dfrac{F(a+1, b+2; c+3; z)}{F(a+1, b+1; c+2; z)}\right\}}}.$$

By repeating the above the following is obtained

$$\frac{F(a, b+1; c+1; z)}{F(a, b; c; z)} = \frac{1|}{|1} - \frac{(a(c-b)/c(c+1))z|}{|\quad 1} - \frac{((b+1)(c+1-a)/(c+1)(c+2))z|}{|\quad 1}$$
$$- \frac{((a+n-1)(c-b+n-1)/(c+2n-2)(c+2n-1))z|}{|\quad 1}$$
$$- \frac{(b+n)(c-a+n)}{(c+2n-1)(c+2n)} z \frac{F(a+n, b+n+1; c+2n+1; z)}{F(a+n, b+n; c+2n; z)}. \tag{12}$$

The case $b = 0$ is of particular interest since in that case

$$F(a, 0; c; z) = 1.$$

(12) now reduces to

$$F(a, 1; c+1; z) = \frac{1|}{1} - \frac{(ac/c(c+1))z|}{|\quad 1} - \frac{((c-a+1)/(c+1)(c+2))z|}{|\quad 1}$$
$$- \frac{((a+n-1)(c+n-1)/(c+2n-2)(c+2n-1))z|}{|\quad 1}$$
$$- \frac{n(c-a+n)}{(c+2n-1)(c+2n)} z \frac{F(a+n, n+1; c+2n+1; z)}{F(a+n, n; c+2n; z)} \tag{13}$$

Note that for $a = c = 1$

$$F(1, 1; 2; z) = \frac{1}{z} \ln \frac{1}{1-z}$$

so that

$$\frac{1}{z}\ln\frac{1}{1-z} = \frac{1|}{1} - \frac{z/2|}{|1} - \frac{z/6|}{|1} - \cdots \frac{(n/(4n-2))z}{|\quad 1} - \frac{(n/(4n+2))z}{1}\cdots. \tag{14}$$

These continued fraction expansions can be used to derive similar relations for confluent hypergeometric functions. For example

$$\lim_{a\to\infty} F\left(a, 1; 1; \frac{z}{a}\right) = \lim_{a\to\infty} \sum_{n=0}^{\infty} \frac{(a)_n(z/a)^n}{n!} = e^z.$$

Letting $c = 0$ in (13), replacing z by z/a and taking the limit we find

$$e^z = \frac{1|}{|1} - \frac{z|}{|1} + \frac{z/2|}{|1} - \cdots \frac{z/4n-2|}{|\quad 1} + \frac{z/4n+2|}{|\quad 1} - \cdots. \tag{15}$$

The preceding discussion omitted a number of so-called degenerate cases. For example we saw that

$$c_1 F(a, b; c; z) + c_2 z^{1-c} F(a+1-c; b+1-c; 2-c; z)$$

represents a general solution of (1). But when c is a negative integer the solution $F(a, b; c; z)$ may fail to exist, since the term $(c)_n$ in the denominator of the expansion will vanish for some n. If, however, either a or b is a negative integer and larger than c then this function reduces to a polynomial. For example, let $a = -3$, $b = 2$, $c = -7$. Then

$$z(1-z)y'' - 7y' + 6y = 0$$

and

$$y = 1 + \tfrac{6}{7}z + \tfrac{3}{7}z^2 + \tfrac{4}{35}z^3$$

is a solution, in spite of the fact that $c = -7$. For a full tabulation of all the special cases that may arise the reader should consult *Higher Transcendental Functions* by Erdelyi et al.

4. A General Method for Finding Integral Representations

In studying the properties of solutions of differential equations integral representations are often far more useful than other representations, such as series expansions. We shall now consider a very general technique for obtaining such integral representations. We define an n'th order differential operator L_z as follows

$$L_z[y(z)] \equiv \sum_{k=0}^{n} l_k(z) y^{(k)}(z) \tag{1}$$

and the subscript in L_z indicates the fact that the independent variable is z. Suppose that $K(z, t)$ is some given function of the variables z and t, and sufficiently differentiable so that all subsequent operations are permissible.

We now suppose that we have another differential operator M_t defined by

$$M_t[u(t)] \equiv \sum_{k=0}^{n} m_k(t)u^{(k)}(t). \tag{2}$$

With M_t we associate a second operator M_t^*, said to be the adjoint of M_t. It is defined by

$$M_t^*[v(t)] \equiv \sum_{k=0}^{n} (-1)^k[m_k(t)u(t)]^{(k)}. \tag{3}$$

By use of the identity

$$\frac{d}{dt}\sum_{\tau=0}^{k-1} (-1)^\tau v^{(\tau)}(t)u^{(k-\tau-1)}(t) = v(t)u^{(k)}(t) - (-1)^k u(t)v^k\ (t)$$

we obtain (with $v(t)$ replaced by $m_k(t)\ v(t)$)

$$v(t)M_t[u(t)] - u(t)M_t^*[v(t)]$$
$$= \frac{d}{dt}\sum_{k=1}^{n}\sum_{\tau=0}^{k-1} (-1)^\tau[m_k(t)v(t)]^{(\tau)}u^{(k-\tau-1)}(t) \equiv \frac{d}{dt}P(u, v). \tag{4}$$

The term $P(u, v)$ defined in (4) is sometimes known as the bilinear concommitant.

We now consider the differential equation

$$L_x[y(x)] = 0 \tag{5}$$

and, for a given function $K(x, t)$ and suitable limits α, β, seek a solution in the form

$$y = \int_\alpha^\beta K(x, t)v(t)\, dt. \tag{6}$$

In order for (6) to satisfy (5) it is sufficient that

$$\int_\alpha^\beta L_x[K(x, t)]v(t)\, dt = 0 \tag{7}$$

and this can be considered as an integral equation for a suitable function $v(t)$. We now suppose that a differential operator M_t exists such that

$$L_x[K(x, t)] = M_t[K(x, t)]. \tag{8}$$

If so we can replace (7) by

$$\int_a^\beta M_t[K(x, t)]v(t)\, dt = 0 \tag{9}$$

and by the use of (4), (9) can be rewritten as

$$\int_{\alpha}^{\beta} [K(x, t)M_t^*[v(t)] + \frac{\partial}{\partial t} P(K(x, t), v(t))]\, dt = 0. \tag{10}$$

If $v(t)$ can be so selected that

$$M_t^*[v(t)] = 0$$

(10) can be integrated exactly.

$$P(K(x, t), v(t))\Big|_{\alpha}^{\beta} = 0 \tag{11}$$

We now seek α and β so as to satisfy (11).

EXAMPLE We will consider the particular equation

$$L_x[y(x)] \equiv y'' + xy = 0 \tag{12}$$

and select

$$K(x, t) = e^{xt}.$$

The latter choice is suggested by the fact that this type of kernel arises in the Laplace transform.

$$L_x e^{xt} = (t^2 + x)e^{xt} = \left(t^2 + \frac{d}{dt}\right)e^{xt} \equiv M_t e^{xt}.$$

Thus we see that

$$M_t \equiv \frac{d}{dt} + t^2$$

is of the first order whereas L_x is of the second. Note that

$$M_t^* = -\frac{d}{dt} + t^2$$

and

$$vM_t[u] - uM_t^*[v] = \frac{d}{dt} uv.$$

We now seek a solution of (12) in the form

$$y = \int_{\alpha}^{\beta} e^{xt} v(t)\, dt.$$

Following the previously outlined steps we have

$$L_x[y] = \int_\alpha^\beta (t^2 + x)e^{xt}v(t)\,dt$$

$$= \int_\alpha^\beta v(t)M_t[e^{xt}]\,dt$$

$$= \int_\alpha^\beta \left\{e^{xt}M_t^*[v(t)] + \frac{d}{dt}\,e^{xt}\,v(t)\right\} dt = 0$$

To satisfy the above we seek v so that

$$M_t^*[v] = -v' + t^2 v = 0.$$

Clearly $v = e^{t^3/3}$. To find α and β

$$e^{xt+t^3/3}\Big|_\alpha^\beta = 0$$

and we can select for our path of integration in the t plane any path on which Re $t^3 \to -\infty$ at the end of the path. Three such paths are indicated in the following diagram.

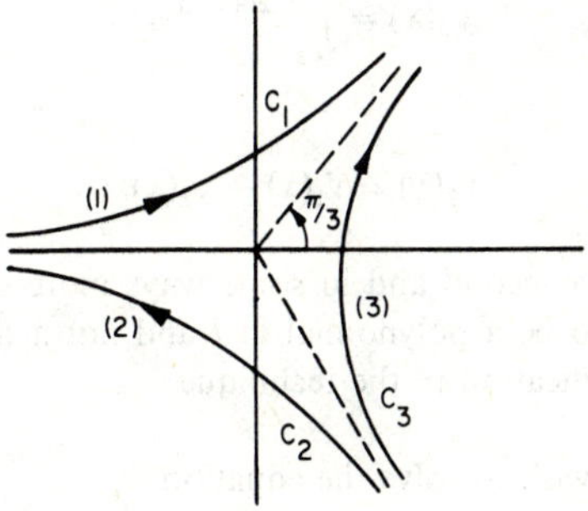

Two solutions will be

$$y_1(x) = \int_{c_1} e^{xt+t^3/3}\,dt$$

$$y_2(x) = \int_{c_2} e^{xt+t^3/3}\,dt$$

We can easily convince ourselves that these are independent. Note that

$$\int_{c_1} t^n e^{t^3/3}\,dt = \int_{-\infty}^{0} t^n e^{t^3/3}\,dt + e^{i(n+1)\pi/3} \int_0^\infty t^n e^{-t^3/3}\,dt$$

$$= [(-1)^n + e^{i(n+1)\pi/3}] \int_0^\infty t^n e^{-t^3/3}\,dt$$

$$= [(-1)^n + e^{i(n+1)\pi/3}]3^{(n-2)/3}\Gamma\left(\frac{n}{3}+\frac{1}{3}\right).$$

Using the latter we see that

$$y_1(x) = \sum_{n=0}^{\infty} \frac{[(-1)^n + e^{i(n+1)\pi/3}]3^{(n-2)/3}\Gamma(n/3+1/3)}{n!}\,x^n$$

$$y_2(x) = \sum_{n=0}^{\infty} \frac{[(-1)^n + e^{-i(n+1)\pi/3}]3^{(n-2)/3}\Gamma(n/3+1/3)}{n!}\,x^n.$$

It follows that

$$W(y_1, y_2) = y_1(0)y_2'(0) - y_1'(0)y_2(0) = -i\sqrt{3}\Gamma(\tfrac{1}{3})\Gamma(\tfrac{2}{3}) = -2\pi i \neq 0,$$

the last step, by means of the multiplication theorem for the Gamma function.
Note that a third possible solution is

$$y_3(x) = \int_{c_3} e^{xt+t^3/3}\,dt$$

but

$$y_3(x) = y_1(x) - y_2(x).$$

We now consider a second and in some ways more elementary example. Here M_t turns out to be a polynomial in t and not a differential operator. This requires a modification in the technique.

EXAMPLE We wish to solve the equation

$$L_x[y] = y'' + 3y' + 2y = 0.$$

Again we let $K(x, t) = e^{xt}$. Clearly

$$L_x[e^{xt}] = (t^2 + 3t + 2)e^{xt} = M_t e^{xt}$$

where $M_t = t^2 + 3t + 2 = M_t^*$. If

$$y = \int_\alpha^\beta e^{xt} v(t)\,dt$$

then

$$L_x[y] = \int_\alpha^\beta (t^2 + 3t + 2)e^{xt}v(t)\,dt = 0.$$

In order to select $v(t)$ we must proceed differently than before since v need not satisfy any particular differential equation. We let $\alpha = \beta$ and integrate over a closed contour. Then

$$L_x[y] = \int_c (t^2 + 3t + 2)e^{xt}v(t)\,dt = 0$$

for all $v(t) = (at + b)/(t^2 + 3t + 2)$, even if the poles of $v(t)$ are inside c. In that case

$$\begin{aligned} y &= \int_c \frac{at + b}{t^2 + 3t + 2}\, e^{xt}\,dt \\ &= 2\pi i(-a + b)e^{-t} + 2\pi i(2a - b)e^{-2t}. \end{aligned}$$

5. Integral Representations for the Hypergeometric Function

We shall now apply this technique to the hypergeometric equation

$$L_z[y] = z(1 - z)y'' + [c - (a + b + 1)z]y' - aby = 0 \tag{1}$$

For our kernel we let

$$K(z, t) = K(zt)$$

where $K(\rho)$ satisfies

$$\rho(1 - \rho)K'' + [d - (a + b + 1)\rho]K' - abK = 0. \tag{2}$$

A computation shows that

$$\begin{aligned} zL_z[K(zt)] &= z^2(1 - z)t^2K'' + z[c - (a + b + 1)z]tK' - zabK \\ &= \left[-t(1 - t)\frac{\partial^2 K}{\partial t^2} + (ct - d)\frac{\partial K}{\partial t}\right] = M_t[K]. \end{aligned}$$

(It is convenient to work with zL_z rather than L_z). We can observe that

$$M_t^*[v(t)] = (-t(1 - t)v)'' - ((ct - d)v)'.$$

In order for

$$y = \int_\alpha^\beta K(zt)v(t)\,dt$$

to satisfy (1), $v(t)$ must satisfy

$$M_t^*[v(t)] = 0$$

and it follows that a particular solution is

$$v(t) = t^{d-1}(1-t)^{c-d-1}.$$

From (2) we see that

$$K(zt) = k\,F(a, b; d; zt) \tag{3}$$

where k is an arbitrary constant. To select α and β we examine IV.4.11. We require that

$$P(K, v)\Big|_\alpha^\beta = -t^d(1-t)^{c-d}\frac{\partial K}{\partial t}\Big|_\alpha^\beta = 0$$

For Re $d > 0$ and Re $(c-d) > 0$ we can let $\alpha = 0$ and $\beta = 1$. We now find

$$y = k\int_0^1 F(a, b; d; zt)t^{d-1}(1-t)^{c-d-1}\,dt.$$

For $y(0) = 1$, k in (3) must be so chosen that

$$1 = k\int_0^1 t^{d-1}(1-t)^{c-d-1}\,dt = kB(d, c-d)$$

Then

$$F(a, b; c; z) = \frac{\Gamma(c)}{\Gamma(d)\Gamma(c-d)}\int_0^1 F(a, b; d; zt)t^{d-1}(1-t)^{c-d-1}\,dt. \tag{4}$$

(4) is a generalization of IV.3.4. If we let $a = d$

$$F(a, b; a, zt)^{-b} = (1-zt)^{-b}$$

Again

$$F(a, b; c; z) = \frac{\Gamma(c)}{\Gamma(a)\Gamma(c-a)}\int_0^1 \frac{t^{d-1}(1-t)^{c-d-1}}{(1-zt)^b}\,dt.$$

6. The Twenty-four Solutions of the Hypergeometric Equation

In section 3 it was seen that by means of the Riemann-Papperitz symbol various solutions could be obtained. That there ought to be a system of 24 solutions can be seen as follows. The three singular points 0, 1 ∞ can be permuted in $3! = 6$ ways. For each of these cases, by means of substitutions of the type

$$y = z^\lambda(1-z)^\nu u$$

we can modify one index at the origin and another at $z = 1$ so as to make them vanish. This can be done in 4 ways and altogether we can produce $4 \times 6 = 24$ solutions. We shall now discuss a technique that will be useful in other contexts as well, and one that will lead us to the afore-mentioned 24 solutions systematically.

The general second order, linear, homogeneous differential equation

$$y'' + p(z)y' + q(z)y = 0 \tag{1}$$

can, by means of the substitution

$$y = e^{-\frac{1}{2}\int^z p(\zeta)\, d\zeta}\, V, \tag{2}$$

be reduced to the form

$$V'' + J(z)\, V = 0 \tag{3}$$

where

$$J(z) = q(z) - \tfrac{1}{2}p'(z) - \tfrac{1}{4}p^2(z),$$

provided of course that $p(z)$ is sufficiently nice to permit the necessary operations. Suppose that we have a second equation

$$u'' + I(x)u = 0 \tag{4}$$

and that the independent variables x and z are somehow related, via

$$\phi(x, z) = 0. \tag{5}$$

Letting $z' = dz/dx$, $z'' = d^2z/dx^2$ and using z as a suitable independent variable in (4) we find

$$z'^2 \frac{d^2u}{dz^2} + z'' \frac{du}{dz} + I(x(z))u = 0.$$

Letting

$$u = \exp\left(-\frac{1}{2}\int \frac{z''}{z'^2}\, dz\right) w = \frac{1}{\sqrt{z'}}\, w$$

we find that $w(z)$ satisfies

$$w'' + G(z)w = 0 \tag{6}$$

where

$$G(z) = \frac{I(x) - \frac{1}{2}\{z, x\}}{z'^2} \tag{7}$$

and

$$\{z, x\} = \frac{z'''}{z'} - \frac{3}{2}\left(\frac{z''}{z'}\right)^2 = \left(\frac{z''}{z'}\right)' - \frac{1}{2}\left(\frac{z''}{z'}\right)^2. \tag{8}$$

The expression $\{z, x\}$ defined in (8) is known as the Schwarzian derivative.

We can now make the following observation. If the independent variables x and z of (4) and (6) are related in terms of (5) and if $I(x)$ and $G(z)$ satisfy (7) then if we can solve either (4) or (6) we can certainly solve the other. In particular we have the following lemma.

LEMMA If u_1 and u_2 are linearly independent solutions of (4) then

$$\left\{\frac{u_2}{u_1}, x\right\} = 2I(x).$$

Conversely if, for some function $z(x)$,

$$\{z, x\} = 2I(x)$$

then $z(x)$ is the ratio of two solutions of

$$u'' + I(x)u = 0.$$

Proof. Suppose

$$\{z, x\} = 2I(x).$$

Then from (7) we observe that $G(z) = 0$, and

$$w'' = 0.$$

Let $w_1 = 1$, $w_2 = z$, so that

$$u_1 = \frac{w_1}{\sqrt{z'}} = \frac{1}{\sqrt{z'}}$$

$$u_2 = \frac{w_2}{\sqrt{z'}} = \frac{z}{\sqrt{z'}}$$

and

$$z = \frac{u_2}{u_1}.$$

Now suppose that u_1 and u_2 are solution (4). Let

$$z = \frac{u_2}{u_1}.$$

Then a simple computation shows that

$$\frac{z''}{z'} = -2\frac{u_1'}{u_1}$$

and

$$\{z, x\} = 2I(x). \quad ■$$

The third order differential equation

$$\{z, x\} = 2I(x) \tag{9}$$

does not possess unique solutions in the following sense. If $z(x)$ satisfies (9), so does

$$\zeta = \frac{az + b}{cz + d}, \ (ad - bc \neq 0).$$

Obviously

$$\zeta' = \frac{ad - bc}{(cz + d)^2} z'$$

$$\frac{\zeta''}{\zeta'} = \frac{z''}{z'} - \frac{2cz'}{cz + d}$$

so that

$$\{\zeta, x\} = \{z, x\} = 2I(x).$$

We now return to the hypergeometric equation

$$z(1 - z)y'' + [c - (1 + a + b)z]y' - aby = 0,$$

and let

$$y = z^{-c/2}(1 - z)^{(c-a-b-1)/2} V.$$

Then V satisfies

$$V'' + J(z)V = 0 \tag{10}$$

where

$$J(z) = \frac{q_0}{z^2} + \frac{q_1}{(z - 1)^2} + \frac{q_2}{z(z - 1)}.$$

A simple calculation verifies that

$$q_0 = \tfrac{1}{4}[1 - (1 - c)^2]$$

$$q_1 = \tfrac{1}{4}[1 - (a + b - c)^2]$$

$$q_2 = \tfrac{1}{4}[(1 - c)^2 + (a + b - c)^2 - (a - b)^2 - 1)].$$

Our task is to find all equations

$$u'' + I(x)u = 0 \tag{11}$$

where

$$I(x) = \frac{p_0}{x^2} + \frac{p_1}{(x - 1)^2} + \frac{p_2}{x(x - 1)}$$

and

$$I(x) - \tfrac{1}{2}\{z, x\} = z'^2 J(z). \tag{12}$$

This is a formidable task and we shall restrict ourselves to some special cases. In particular we are now interested in those transformations $z(x)$ that are independent of the parameters a, b, c. This will certainly be the case if we let

$$\{z, x\} = 0.$$

In order for this to hold we see that

$$z = \frac{\alpha x + \beta}{\gamma x + \delta}. \tag{13}$$

Then (12) can be rewritten as

$$\frac{Ax^2 + Bx + C}{x^2(x-1)^2} = z'^2 \frac{A'z^2 + B'z + C'}{z^2(z-1)^2}, \tag{14}$$

where A, B, C depend of course on a, b, c, and similarly A', B', C' depend on a', b', c'.

Noting that (14) must be an identity for arbitrary a, b, c, and expressing z in terms of (13) we find

$$\begin{aligned} m(Ax^2 + Bx + C) = (\alpha\delta - \beta\gamma)^2[A'(\alpha x + \beta)^2 + B'(\alpha x + \beta)(\gamma x + \delta) \\ + C'(\gamma x + \delta)^2] \end{aligned} \tag{15}$$

$$mx^2(x-1)^2 = (\alpha x + \beta)^2(\gamma x + \delta)^2[(\alpha x + \beta) - (\gamma x + \delta)]^2.$$

The latter is obtained by making the numerators in (14) proportional, and also the denominators, m being the constant of proportionality. The second of the above leads to an identity for a polynomial of degree 6. The coefficients depend on α, β, γ, δ and m. An explicit calculation shows us that there are only six essentially different cases for which this equation reduces to an identity. These are tabulated as follows

1. $\gamma = \beta = a - \delta = 0$, $\quad m = \alpha^6$, $z = x$
2. $\gamma = \delta - \beta = \alpha + \beta = 0$, $\quad m = \alpha^6$, $z = 1 - x$
3. $\alpha = \delta = \beta - \gamma - 0$, $\quad m = \beta^6$, $z = \dfrac{1}{x}$
4. $\alpha = \delta - \beta = \gamma + \delta = 0$, $\quad m = \beta^6$, $z = \dfrac{1}{1 - x}$
5. $\gamma - a = \beta = \delta + \gamma = 0$, $\quad m = \alpha^6$, $z = \dfrac{x}{x -- 1}$
6. $\gamma - \alpha = \delta = \beta + \alpha = 0$, $\quad m = \alpha^6$, $z = \dfrac{x - 1}{x}$

The above six possibilities take into account the six possible permutations of the three singular points 0, 1, ∞. To compute the possible values of a, b, c and a', b', c' we can either consider the first of the equations (15) or else return to (14) or (12). We consider the first of the above cases. Then from (12)

$$\frac{1-(1-c)^2}{x^2}+\frac{1-(a+b-c)^2}{(x-1)^2}+\frac{(1-c)^2+(a+b-c)^2-(a-b)^2-1}{x(x-1)}$$
$$=\frac{1-(1-c')^2}{x^2}+\frac{1-(a'+b'-c')^2}{(x-1)^2}$$
$$+\frac{(1-c')^2+(a'+b'-c')^2-(a'-b')^2-1}{x(x-1)} \tag{16}$$

By comparing the first terms on the left and right in (16) we find that either

$$c=c' \quad \text{or} \quad c'=2-c.$$

The first of these then shows, using the second terms that either

$$a+b=a'+b' \quad \text{or} \quad a+b+a'+b'=c'+c=2c$$

and the last terms show that

$$a-b=a'-b' \quad \text{or} \quad a-b=-a'+b'.$$

By examining all possible solutions we are led to the following four cases.

1. $a'=a,\ b'=b,\ c'=c$
2. $a'=c-a,\ b'=c-b,\ c'=c$
3. $a'=a-c+1,\ b'=b-c+1,\ c'=2-c$
4. $a'=1-a,\ b'=1-b,\ c'=2-c.$

These correspond to the four cases

$$P\begin{Bmatrix} 0 & 1 & \infty & \\ 0 & 0 & a & z \\ 1-c & c-a-b & b & \end{Bmatrix}$$

$$(1-z)^{a+b-c}\,P\begin{Bmatrix} 0 & 1 & \infty & \\ 0 & 0 & a & z \\ 1-c & c-a-b & b & \end{Bmatrix}=P\begin{Bmatrix} 0 & 1 & \infty & \\ 0 & 0 & c-a & z \\ 1-c & a+b-c & c-b & \end{Bmatrix}$$

$$z^{c-1}P\begin{Bmatrix} 0 & 1 & \infty & \\ 0 & 0 & a & z \\ 1-c & c-a-b & b & \end{Bmatrix}=P\begin{Bmatrix} 0 & 1 & \infty & \\ 0 & 0 & a+1-c & z \\ c-1 & c-a-b & b+1-c & \end{Bmatrix}$$

$$z^{c-1}(1-z)^{a+b-c}P\begin{Bmatrix} 0 & 1 & \infty & \\ 0 & 0 & a & z \\ 1-c & c-a-b & b & \end{Bmatrix}$$

$$= P\begin{Bmatrix} 0 & 1 & \infty & \\ 0 & 0 & 1-a & z \\ c-1 & a+b-c & 1-b & \end{Bmatrix}.$$

An analysis of all possible cases arising from (15) leads to 24 different solutions of the hypergeometric equation.

7. The Schwarz-Christoffel Transformation

According to the Riemann mapping theorem one can always map simply connected domains with at least two boundary points into a unit circle or equivalently into a half-plane. The general proof is purely existential however. But in special cases explicit mappings can, of course, often be produced. In particular for polygonal regions this can be done. We shall first examine the case where the polygon has n given vertices and its sides are straight.

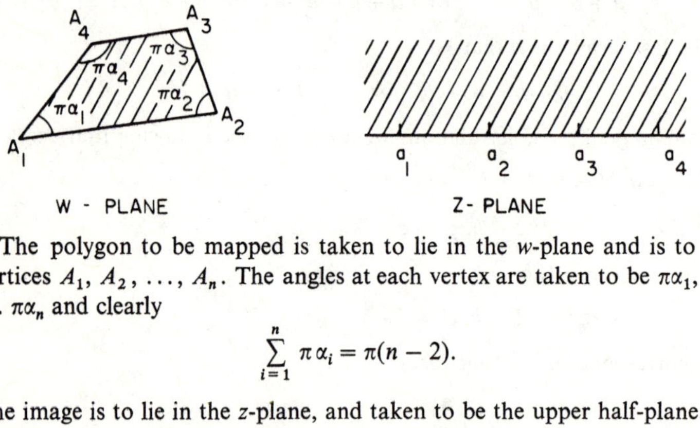

The polygon to be mapped is taken to lie in the w-plane and is to have vertices $A_1, A_2, \ldots, A_n$. The angles at each vertex are taken to be $\pi\alpha_1, \pi\alpha_2, \ldots \pi\alpha_n$ and clearly

$$\sum_{i=1}^{n} \pi\alpha_i = \pi(n-2). \tag{1}$$

The image is to lie in the z-plane, and taken to be the upper half-plane. The images of the vertices are $a_1, a_2, \ldots, a_n$ respectively. It will prove to be convenient to study the function $f(z)$, which maps the region $\operatorname{Im}(z) > 0$ into the polygon rather than its inverse.

In general we cannot expect $f(z)$ to be a single-valued function. To see this we note that near a vertex, say a_1, the mapping has to behave like a mapping of the domain $0 \le \arg w \le \pi\alpha_1$ into the half-plane $\operatorname{Im} z > 0$. Such a mapping is of the form $w = (z - a_1)^{\alpha_1}$. It follows that

$$f(z) = A_1 + (z - a_1)^{\alpha_1} h(z) \tag{2}$$

where $h(z)$ is analytic near $z = a_1$. The effect of this transformation is to deform the angle π at $z = a_1$ into the angle $\pi\alpha_1$ at $w = A_1$. The above clearly has a branch point at $z = a_1$, and similarly $f(z)$ has branch points at all a_i.

Suppose the edge A_1A_2 of the polygon is defined by

$$w = A_1 + e^{i\theta}t$$

where t is a real parameter. Then the function

$$F(z) = (f(z) - A_1)\, e^{-i\theta}$$

is real for z real and $a_1 < z < a_2$. According to the Schwarz reflection principle $F(z)$ can be continued across (a_1, a_2) into the lower half plane by means of the formula

$$F(z) = \overline{F(\bar{z})}, \qquad \text{Im } z < 0.$$

The appropriate continuation for $f(z)$ is then given by

$$f(z) = A_1 + (\overline{f(\bar{z})} - \bar{A}_1)e^{2i\theta}, \qquad \text{Im } z < 0.$$

Similar continuations are possible across every one of the intervals $(-\infty, a_1)$, $(a_1, a_2) \cdots (a_{n-1}, a_n)$, (a_n, ∞), but because of the multi-valued character of $f(z)$ each of these leads to a different branch of $f(z)$. Any one of them can now be continued back into the upper half plane across any one of these intervals.

What is the effect of these in the w plane? If, for example, we continue $f(z)$ across (a_1, a_2) into the lower plane the resultant function maps Im $z < 0$ into the polygon obtained by reflecting the original polygon across side (A_1, A_2).

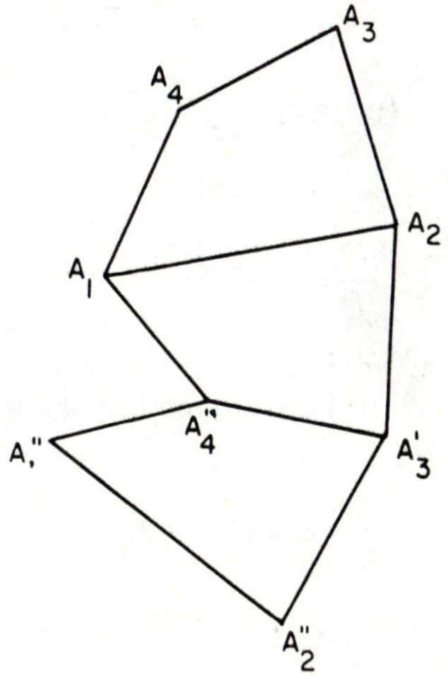

If we now continue this function back into Im $z > 0$ across (a_3, a_4) the new polygon is reflected across (A'_3, A'_4) and we obtain as the image of the upper half plane a polygon congruent to the original one. In the above figure this process is illustrated for $n = 4$. If we call the resultant mapping $f_1(z)$ we have

$$f_1(z) = Af(z) + B$$

since $f_1(z)$ and $f(z)$ differ by at most a translation and a rotation. By elimination of A and B we have

$$\frac{f_1''}{f_1'} = \frac{f''}{f'}$$

from the above. If $f_k(z)$ is any branch of $f(z)$ we see that necessarily

$$\frac{f_k''}{f_k'} = \frac{f''}{f'}$$

and necessarily the form $f''(z)/f'(z)$ is single valued.

From (2) we have that near $z = a_1$

$$\frac{f''(z)}{f'(z)} = \frac{\alpha_1 - 1}{z - a_1} + m(z)$$

where $m(z)$ is analytic near $z = a_1$. More generally

$$\frac{f''(z)}{f'(z)} = \sum_{i=1}^{n} \frac{\alpha_i - 1}{z - a_i} + g(z). \tag{3}$$

In view of the fact that $f''(z)/f'(z)$ is single valued and has singularities only at $a_1, a_2, \ldots a_n$ $g(z)$ must be an entire function. Since we are dealing with a finite polygon $f(z)$ must be regular near $z = \infty$. Then

$$f(z) = c_1 + \frac{c_2}{z} + \cdots$$

for large z and

$$\frac{f''}{f'} = \frac{-2}{z} + \cdots.$$

Using the above we see that $g(z)$ is bounded and hence a constant. On closer inspection we see that $g(z) \equiv 0$. Furthermore comparing (3) and the above we find

$$\sum_{i=1}^{n} (\alpha_i - 1) = -2$$

which agrees with (1).

Finally, by integration

$$f(z) = c_1 \int_0^z \prod_{i=1}^{n} (\zeta - a_i)^{\alpha_i - 1} \, d\zeta + c_2 \tag{4}$$

where c_1 and c_2 are constants of integration.

EXAMPLE Consider the isosceles triangle where $A_1 = -1$, $A_2 = 1$ and we require that $f(1) = 1, f(-1) = -1, f(\tau) = A_3$.

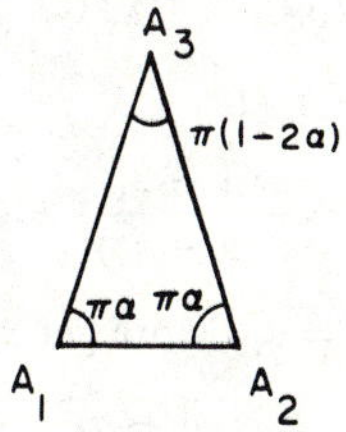

Using (4) we find

$$f(z) = c_1 \int_{-1}^{z} \frac{dz}{(1 - z^2)^{1-\alpha}(\tau - z)^{2\alpha}} + c_2.$$

Since $f(1) = 1, f(-1) = -1$ we see that

$$f(z) = 2 \frac{\int_{-1}^{z} (dz/(1 - z^2)^{1-\alpha}(\tau - z)^{2\alpha})}{\int_{-1}^{1} (dz/(1 - z^2)^{1-\alpha}(\tau - z)^{2\alpha})} - 1.$$

As an interesting special case we let $\tau \to \infty$ and $\alpha = 1/2$. In this case the triangle degenerates into a semi-infinite rectangle and $f(\infty) = \infty$. Then

$$\frac{2 \int_{-1}^{z} (dz/\sqrt{1 - z^2})}{\int_{-1}^{1} (dz/\sqrt{1 - z^2})} - 1 = \frac{2}{\pi} \sin^{-1} z.$$

We now consider a generalization of the aforegoing procedure for the case where the sides of the polygon are not straight, but are circular arcs. The preceding development can be followed with a few modifications. As before let $A_1, A_2, \ldots, A_n$ denote the vertices of the polygon and $\pi\alpha_1, \pi\alpha_2, \ldots, \pi a_n$ the included angles. $a_1, a_2, \ldots a_n$, as before, denote the image points of the vertices in the z plane. (2) still holds near each vertex

$$f(z) = A_1 + (z - a_1)^{\pi\alpha_1} h(z).$$

We can again continue $f(z)$ into the lower half plane, across some interval (a_i, a_{i+1}). The corresponding effect in the w plane is to reflect the polygon in one of its sides. But since we are now dealing with a reflection in a circular arc we use an inversion. After an even number of reflections in the z plane we will have performed an even number of inversions in the w plane. Such an even number of inversions is necessarily a bilinear transformation so that if $f_1(z)$ is the function obtained from $f(z)$ by such a continuation process we have

$$f_1(z) = \frac{Af(z) + B}{Cf(z) + D}.$$

We can eliminate A, B, C, D using the Schwarzian derivative. Then

$$\{f_1(z), z\} = \{f(z), z\}.$$

Use of (2) shows that near $z = a_1$

$$\{f(z), z\} = \frac{1}{2}\frac{1 - \alpha_1^2}{(z - a_1)^2} + \frac{\beta_1}{z - a_1} + h(z)$$

where $h(z)$ is regular near $z = a_1$. More generally

$$\{f(z), z\} = \sum_{i=1}^{n} \frac{1}{2}\frac{1 - \alpha_i^2}{(z - a_i)^2} + \frac{\beta_i}{z - a_i} + g(z). \tag{5}$$

The expression $\{f(z), z\}$ is necessarily single valued since it is invariant under all bilinear transformations of $f(z)$. It can have singularities only at the vertices $z = a_i$. Hence $g(z)$ must be entire.

Since $f(z)$ is taken to be regular at $z = \infty$ we have for large $|z|$

$$f(z) = c_1 + \frac{c_2}{z} + \cdots$$

and a calculation shows that

$$\{f(z), z\} = \frac{k}{z^4} + \cdots.$$

Using the last result in (5) we conclude that $g(z) \equiv 0$ and by expanding the remaining terms in powers of $1/z$ we also find that

$$\begin{aligned} &\sum_{i=1}^{n} \beta_i = 0 \\ &\sum_{i=1}^{n} [2a_i\beta_i + 1 - \alpha_i^2] = 0 \\ &\sum_{i=1}^{n} [\beta_i a_i^2 + a_i(1 - \alpha_i^2)] = 0 \end{aligned} \tag{6}$$

We shall summarize this discussion in the form of two theorems.

THEOREM The function $f(z)$, mapping the half plane $\operatorname{Im} z > 0$ onto a polygon of n sides with interior angles $\pi\alpha_1, \pi\alpha_2, \ldots, \pi a_n$ at vertices corresponding to $a_1, a_2, \ldots, a_n$ satisfies

$$\frac{f''(z)}{f'(z)} = \sum_{i=1}^{n} \frac{\alpha_i - 1}{z - a_i}. \tag{7}$$

The explicit solution of the above differential equation is

$$f(z) = c_1 \int_0^z \prod_{i=1}^{n} (\zeta - a_i)^{\alpha_i - 1}\, d\zeta + c_2. \tag{8}$$

A second theorem relates to curvilinear polygons

THEOREM The function $f(z)$ mapping the half plane $\operatorname{Im} z > 0$ onto a curvilinear polygon with vertices $A_1, A_2, \ldots, A_n$ bounded by circular arcs which form the angles $\pi\alpha_1, \pi\alpha_2, \ldots, \pi\alpha_n$ at the vertices, and such that the vertices correspond to $a_1, a_2, \ldots, a_n$ satisfies

$$\{f(z), z\} = \sum_{i=1}^{n} \frac{1}{2}\frac{1 - \alpha_i^2}{(z - a_i)^2} + \frac{\beta_i}{z - a_i}. \tag{9}$$

The parameters in the above must also satisfy

$$\sum_{i=1}^{n} \beta_i = 0$$

$$\sum_{i=1}^{n} [2a_i\beta_i + 1 - \alpha_i^2] = 0$$

$$\sum_{i=1}^{n} [\beta_i a_i^2 + a_i(1 - \alpha_i^2)] = 0.$$

One big difference between these two cases to be noted is the following. Equation (7) can always be solved, as shown by (8). However (9) is a third order non-linear differential equation and a general solution can in general not be produced. By the use of the lemma of the preceding section we can make the following assertion. If we know the general solution of the equation

$$u'' + \left[\sum_{i=1}^{n} \frac{1}{4}\frac{1 - \alpha_i^2}{(z - a_i)^2} + \frac{1}{2}\frac{\beta_i}{z - a_i}\right] u = 0$$

and if u_1, u_2 are independent solutions of the above then

$$f(z) = \frac{u_1}{u_2}$$

satisfies (9).

Another question to be resolved is the following. In order to define the polygon we require n vertices and n angles, or a total of $3n$ real parameters. But since the mapping is unique only up to a bilinear map, depending on three complex numbers we have a total of $3n - 6$ parameters determining the mapping. In (9) we have $2n$ parameters α_i and β_i, that satisfy three conditions so that we have $2n - 3$ parameters. We therefore must still be able to determine

$$(3n - 6) - (2n - 3) = n - 3$$

parameters from the given data. There are still the n points $a_1, a_2, \ldots, a_n$. Three of these are to be selected by us, the remaining $n - 3$ must be determined by means of the given data. These $n - 3$ parameters are called the accessory parameters and their determination is in general very difficult.

For $n = 3$, the case of a triangle, there are no accessory parameters. In this case we are free to choose a_1, a_2, a_3 as we wish, and no other vertex need to be found. If we let $a_1 = 0$, $a_2 = 1$ (9) reduces to

$$\{w, z\} = \frac{1 - \alpha_1^2}{2z^2} + \frac{1 - \alpha_2^2}{2(z - 1)^2} + \frac{1 - \alpha_3^2}{(z - a_3)^2} + \frac{\beta_1}{z} + \frac{\beta_2}{z - 1} + \frac{\beta_3}{z - a_3}$$

$$\beta_1 + \beta_2 + \beta_3 = 0$$

$$2\beta_2 + 2a_3\beta_3 - \alpha_1^2 - \alpha_2^2 - \alpha_3^2 + 3 = 0$$

$$\beta_2 + 1 - \alpha_2^2 + a_3^2\beta_3 + a_3(1 - \alpha_3^2) = 0.$$

The case where $a_3 \to \infty$ is, as will be clear shortly, of particular significance to us. A direct calculation shows that

$$\lim_{a_3 \to \infty} \beta_1 = -\tfrac{1}{2}[\alpha_1^2 + \alpha_2^2 - \alpha_3^2 - 1]$$

$$\lim_{a_3 \to \infty} \beta_2 = \tfrac{1}{2}[\alpha_1^2 + \alpha_2^2 - \alpha_3^2 - 1]$$

$$\lim_{a_3 \to \infty} \beta_3 = 0$$

and finally for (9) we find

$$\{w, z\} = \frac{1 - \alpha_1^2}{2z^2} + \frac{1 - \alpha_2^2}{2(z - 1)^2} + \frac{\alpha_1^2 + \alpha_2^2 - \alpha_3^2 - 1}{2z(z - 1)}. \tag{10}$$

The associated second order equation is

$$u'' + \left[\frac{1 - \alpha_1^2}{4z^2} + \frac{1 - \alpha_2^2}{4(z - 1)^2} + \frac{\alpha_1^2 + \alpha_2^2 - \alpha_3^2 - 1}{4z(z - 1)}\right]u = 0. \tag{11}$$

A comparison with 6.10 shows that (11) is a hypergeometric differential equation with the parameters

$$\alpha_1 = 1 - c, \qquad \alpha_2 = a + b - c, \qquad \alpha_3 = a - b$$

or equivalently

$$a = \tfrac{1}{2}(1 - \alpha_1 + \alpha_2 + \alpha_3), \qquad b = \tfrac{1}{2}(1 - \alpha_1 + \alpha_2 - \alpha_3), \qquad c = 1 - \alpha_1. \quad (12)$$

We see therefore that the theory of hypergeometric functions and the theory of conformal mappings of curvilinear triangles onto half planes are intimately related. Every such mapping must be a ratio of hypergeometric functions.

8. Mappings of Curvilinear Triangles

In the previous section we studied the function $f(z)$, which maps the half plane $\operatorname{Im} z > 0$ onto a curvilinear polygon. The case of a curvilinear triangle will now be discussed in more detail. As we saw $f(z)$ is necessarily a multil-valued function. We shall not discuss its inverse function denoted by $S(w)$. The latter may or may not be single valued depending on the particular character of the triangle.

Let us for the moment focus attention on one point of the triangle. After performing all possible reflections of the triangle, or equivalently after performing all possible continuations from the upper to the lower z half plane and back we may find that the point in question is covered by an infinite number of distinct triangles. In that case $S(w)$ cannot possibly be single valued and must have an infinite number of branches. If all points of the triangle are such that each belongs to at most a finite number of triangles then $S(w)$ has only a finite number of branches. If the triangle can never partially overlap with any other triangle then $S(w)$ is single valued.

Near a vertex we have

$$w = A_i + (z - a_i)^{\alpha_i} h(z)$$

so that

$$S(w) = z = a_i + (w - A_i)^{1/\alpha_i} H(w)$$

where $h(z)$ and $H(w)$ are regular near $z = a_i$ and $w = A_i$. A necessary condition for single valuedness of $S(w)$ is clearly that $1/\alpha_i$ be an integer. If $1/\alpha_i$ were irrational $S(w)$ would have an infinite number of branches.

We shall restrict our discussion to those triangles where all $1/\alpha_i$ are integers, in order to investigate all situations in which $S(w)$ is single valued. Furthermore it will prove to be advantageous to separate our discussion into three subcases, namely those where the sum of the interior angles is less than, equal to, or greater than π.

Case 1. Let $\alpha_1 = 1/n$, $a_2 = 1/m$, $\alpha_3 = 1/l$ and consider the case where

$$\pi\alpha_1 + \pi\alpha_2 + \pi\alpha_3 = \pi$$

or equivalently

$$\frac{1}{n}+\frac{1}{m}+\frac{1}{l}=1.$$

It is impossible for $n, m, l > 3$ simultaneously. There are at most a finite number of values n, m, l satisfying the above. The following table summarizes the only four possibilities

$1/\alpha_1$	$1/\alpha_2$	$1/\alpha_3$	c	$a+b$	a	b	ab	Figure
2	2	∞	$\frac{1}{2}$	1	$\frac{1}{2}$	$\frac{1}{2}$	$\frac{1}{4}$	semi-infinite rectangle
2	3	6	$\frac{1}{2}$	$\frac{5}{6}$	$\frac{1}{2}$	$\frac{1}{3}$	$\frac{1}{6}$	30, 60, 90 triangle
2	4	4	$\frac{1}{2}$	$\frac{3}{4}$	$\frac{1}{2}$	$\frac{1}{4}$	$\frac{1}{8}$	isosceles right triangle
3	3	3	$\frac{2}{3}$	1	$\frac{2}{3}$	$\frac{1}{3}$	$\frac{2}{9}$	equilateral triangle

To determine a, b, c we used Eq. 4.7.12.

In these cases we can find explicit solutions of the hypergeometric equation since we know how to construct the corresponding conformal mapping.

EXAMPLE 1 Consider the case where $a = b = c = \frac{1}{2}$, corresponding to an infinite rectangle. We have

$$z(1-z)y'' + [\tfrac{1}{2} - 2z]y' - \tfrac{1}{4}y = 0.$$

$$y_1 = F(\tfrac{1}{2}, \tfrac{1}{2}; \tfrac{1}{2}; z) = (1-z)^{-1/2}$$

$$y_2 = z^{1/2}F(1, 1; \tfrac{3}{2}; z) = \tfrac{1}{2}(1-z)^{-1/2}\int_0^z \frac{d\zeta}{\sqrt{\zeta-\zeta^2}} = \frac{\sin^{-1}(1-2z)-\pi/2}{2(1-z)^{1/2}}.$$

The mapping is given by

$$w = \frac{y_2}{y_1} = \tfrac{1}{2}\sin^{-1}(1-2z) - \pi/4$$

and its inverse is

$$S(w) = \frac{1-\cos 2w}{2} = \sin^2 w.$$

Note that whereas w is multivalued $S(w)$ is single valued and also periodic. The following figure indicates the images of Im $z > 0$ in the w plane

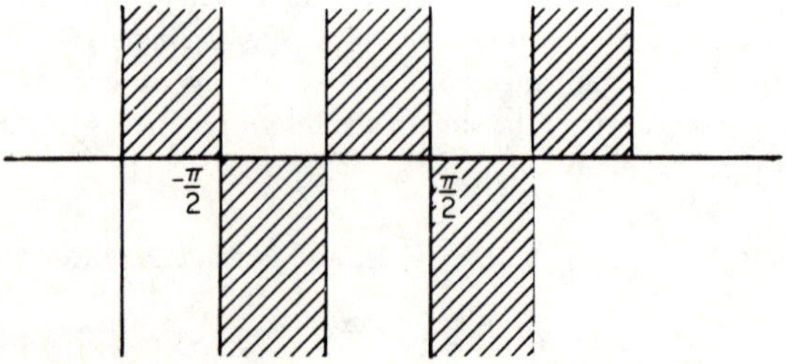

EXAMPLE 2 Consider the case where $a = c = \frac{1}{2}, b = \frac{1}{4}$, corresponding to an isosceles right triangle

$$z(1-z)y'' + [\tfrac{1}{2} - \tfrac{7}{4}z]y' - \tfrac{1}{8}y = 0$$

$$y_1 = F(\tfrac{1}{2}, \tfrac{1}{4}; \tfrac{1}{2}; z) = (1-z)^{-1/4}$$

$$y_2 = \sqrt{z}\,F(1, \tfrac{3}{4}; \tfrac{3}{2}; z) = \tfrac{1}{2}(1-z)^{-1/4}\int_0^z \zeta^{-1/2}(1-\zeta)^{-3/4}\,d\zeta.$$

The mapping is given by

$$w = \frac{y_2}{y_1} = \frac{1}{2}\int_0^z \zeta^{-1/2}(1-\zeta)^{-3/4}\,d\zeta.$$

The above integral no longer can be evaluated in terms of elementary functions, but can be evaluated in terms of elliptic integrals. To find the image points corresponding to $z = 0$, $z = 1$, $z = \infty$ we note that

$$w(0) = 0$$

$$w(1) = \int_0^1 \zeta^{-1/2}(1-\zeta)^{-3/4}\,d\zeta = \frac{\Gamma(1/2)\Gamma(1/4)}{\Gamma(3/4)}$$

$$w(\infty) = \int_0^{-\infty} \zeta^{-1/2}(1-\zeta)^{-3/4}\,d\zeta = i\int_0^{\infty} \frac{t^{-1/2}}{(1+t)^{3/4}}\,dt = \frac{i\Gamma(1/2)\Gamma(1/4)}{\Gamma(3/4)}.$$

The images of Im $z > 0$ are shown in the following figure.

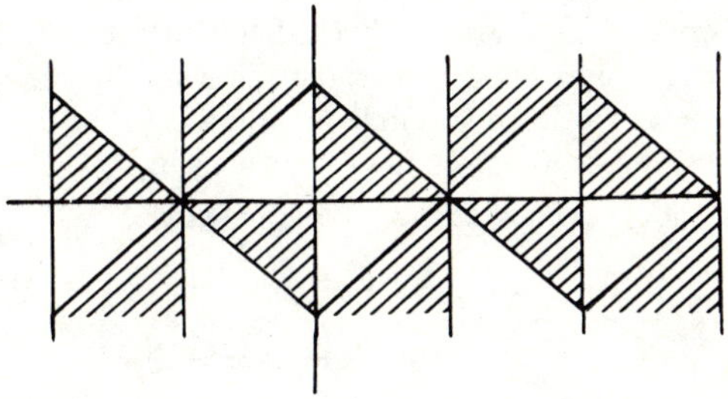

If we denote the inverse function of $f(z)$ by $S(w)$, then $S(w)$ maps all the shaded triangles onto Im $z > 0$. We see that

$$w = \frac{1}{2}\int_0^{S(w)} \zeta^{-1/2}(1-\zeta)^{-3/4}\,d\zeta.$$

The above is an elliptic integral and $S(w)$ is known as an elliptic function. Let

$$p = 2\,\frac{\Gamma(1/2)\Gamma(1/4)}{\Gamma(3/4)}.$$

Then, we see from the diagram that

$$S(w + nip + mp) = S(w)$$

for arbitrary integers n and m. It is clear therefore that $S(w)$ is a doubly periodic function. In example 1 we note that $S(w)$ is simply periodic.

Case 2. We now turn to the case where

$$\pi\alpha_1 + \pi\alpha_2 + \pi\alpha_3 < \pi$$

or equivalently

$$\frac{1}{n} + \frac{1}{m} + \frac{1}{l} < 1.$$

Unlike Case 1, we see that there will be an infinite number of possibilities for n, m and l satisfying the above conditions. Another difference between these two cases in the following. In Case 1 we saw that the images of $\text{Im } z > 0$ and their reflections cover the w plane. In Case 2 the images of $\text{Im } z > 0$ and their reflections (inversions) cover a finite circle known as the orthogonal circle.

Suppose the two circles whose arcs form the two sides of the triangle at the vertex of angle $\pi\alpha_1$ meet at P_1 and P_2. Let P_1 denote the vertex. By means of a bilinear mapping we can map P_1 into the origin and P_2 into ∞. Then the two sides in question become straight lines. We can, accordingly, without loss of generality suppose that one vertex of the triangle is at the origin and the corresponding two sides are straight lines. Let Γ denote the circle corresponding to the third side, as shown in the following figure. We can now construct the so-called orthogonal circle as follows. From the vertex 0 we draw a tangent to Γ and denote the point of tangency by P. The circle centered at 0 and of radius $\overline{OP}$ is defined as the orthogonal circle.

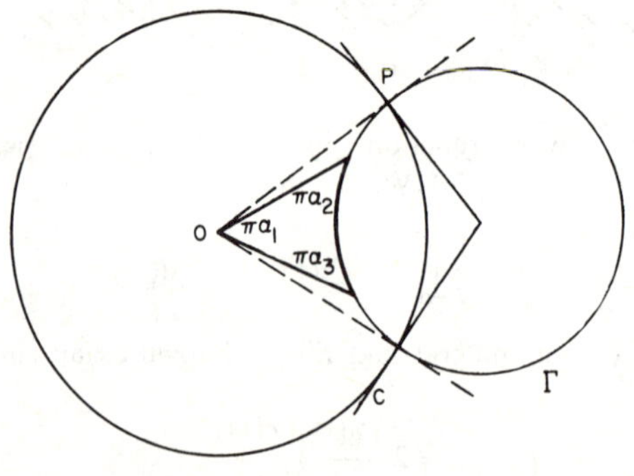

All reflections of the triangle in any one of its sides lead to new triangles whose sides are orthogonal to C. This follows from the fact that inversions are angle preserving. Furthermore all inversions are with respect to circles orthogonal to C so that C has to remain invariant. It also follows that all triangles constructed from the original one by successive inversions are in the interior of C. For example under inversion in Γ the region interior to C as well as Γ and the region interior to C and exterior to Γ are interchanged. But clearly every point interior to C remains interior to C. The following figure illustrates this situation for the case where $n = 7$, $m = 2$, $l = 3$. The shaded regions are the images of the upper half plane $\text{Im } z > 0$.

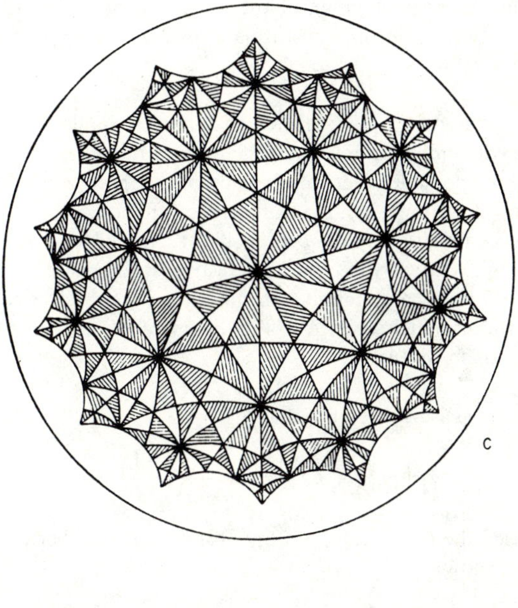

We can show that every point interior to C must belong to one of the triangles obtained from all the inversions of the original triangle. We suppose that some point P is a boundary point of the domain in question and that P is interior to C. P must be a vertex of one of these triangles or lie on an arc bounding a triangle. In either case it lies on a circular arc. If that arc had a finite curvature we could enlarge the domain by another inversion. Hence the arc has zero curvature. But every such arc must lie on a circle orthogonal to C. These conditions are contradictory unless P belongs to the boundary of C.

We see then that $S(w)$ maps the original triangle and all others obtained from it by an even number of reflections onto $\text{Im } z > 0$. If

$$\zeta_1(w) = \frac{a_1 w + b_1}{c_1 w + d_1}$$

$$\zeta_2(w) = \frac{a_2 w + b_2}{c_2 w + d_2}$$

are two bilinear mappings of the triangle onto one of the other triangles corresponding to $\text{Im } z > 0$, then

$$\zeta_1(\zeta_2(w))$$

will be another bilinear mapping of the initial triangle onto another image of $\text{Im } z > 0$. All such bilinear transformations form a group. Furthermore

$$S(\zeta(w)) = S(w)$$

for all $\zeta(w)$ in the group. We see then that $S(w)$ is single valued inside the orthogonal circle and invariant under the group of all even inversions generated by the three possible inversions in each of the three sides of the triangle. Such a function is known as an automorphic function.

In our preceding discussion we took it for granted that n, m and l are positive, but finite integers. Certain interesting special cases arise if one of them is allowed to become infinite. In this case one vertex becomes a cusp. These cases are intimately related to the theory of elliptic functions, and we shall not take them up further.

Case 3. In this case we require that

$$\pi\alpha_1 + \pi\alpha_2 + \pi\alpha_3 = \pi\left(\frac{1}{n} + \frac{1}{m} + \frac{1}{l}\right) > \pi,$$

and as in Case I the above can hold only for a finite number of cases. These are tabulated below.

$1/\alpha_1$	$1/\alpha_2$	$1/\alpha_3$	c	a	b	Group
2	2	n	$\frac{1}{2}$	$\frac{n+1}{2n}$	$\frac{n-1}{2n}$	Dihedral
2	3	3	$\frac{1}{2}$	$\frac{7}{8}$	$\frac{1}{4}$	tetrahedral
2	3	4	$\frac{1}{2}$	$\frac{13}{24}$	$\frac{7}{24}$	octohedral
2	3	5	$\frac{1}{2}$	$\frac{31}{60}$	$\frac{19}{60}$	Isosahedral

Geometrically these are best visualized not in the w plane, but on a Riemann sphere onto which the w plane is mapped stereographically. With each of these we can then associate a rotation group of the sphere that leaves it invariant. In other words the various images of $\operatorname{Im} z > 0$ on the sphere are simply permuted in some way.

Strictly speaking the first of the tabulated cases corresponds to an infinity of situations, depending on the natural number n. But as will be seen they are quite similar. We take the triangle to have a vertex at the origin and the corresponding angle to be π/n, as is shown in the following figure.

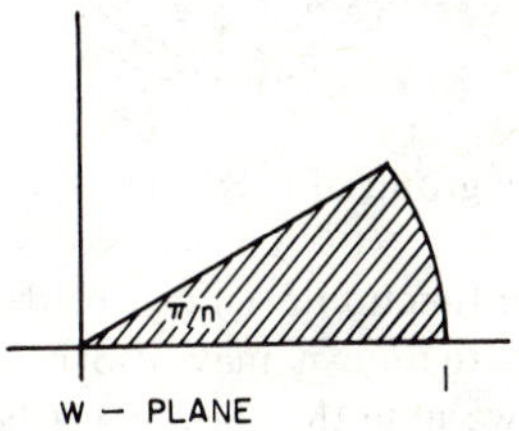

The third side has to be a circular arc centered at the origin. A simple calculation shows that

$$z = S(w) = \frac{1}{4}\frac{(w^n - 1)^2}{w^n}$$

is the mapping that maps the triangle onto $\operatorname{Im} z > 0$. The totality of all images of the half plane for the case $n = 8$ is shown below.

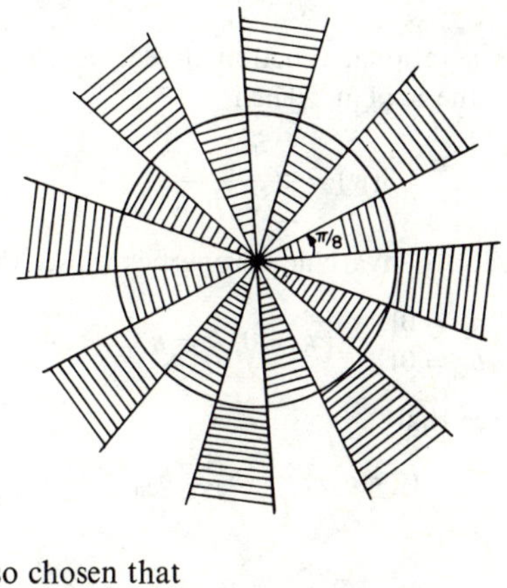

The mapping is so chosen that

$$S(0) = \infty, \qquad S(1) = 0, \qquad S(e^{i\pi/n}) = 1.$$

If we map the above stereographically onto the Riemann sphere we obtain the following,

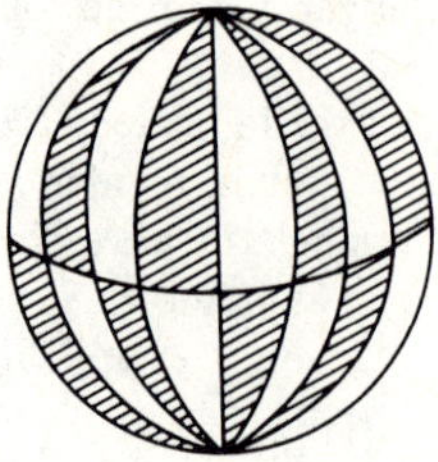

The dihedral group is the group of rotations about the axis, and is generated by rotations of $2\pi/n$.

Although the mapping function $S(w)$ can be determined without recourse to the geometry of the situation it may also be discovered on purely geometrical grounds. First we note that $S(w)$ must be invariant under all rotations about the origin of multiples of the angle $2\pi/n$. Analytically this means that

$$S(e^{i2\pi k/n}w) = S(w).$$

We also note that under all reflections in $|w| = 1$ the figure remains invariant. That is

$$S\left(\frac{1}{w}\right) = S(w).$$

Lastly, $S(w)$ must be a rational function of degree $2n$, since there are $2n$ images of $\operatorname{Im} z > 0$ in the w plane. Then

$$S(w) = \frac{\sum_{k=0}^{2n} a_k w^k}{\sum_{k=0}^{2n} b_k w^k}.$$

From the first of the above invariance properties we see that

$$\left.\begin{aligned} a_k &= 0 \\ b_k &= 0 \end{aligned}\right\}, \qquad k \neq 0,\ k \neq n,\ k \neq 2n.$$

From the second we see that

$$a_0 = a_{2n}, \qquad b_0 = b_{2n}.$$

Then

$$S(w) = \frac{a_0(1 + w^{2n}) + a_n w^n}{b_0(1 + w^{2n}) + b_n w^n}.$$

If we now stipulate that $S(0) = \infty$ we have $b_0 = 0$. From $S(1) = 0$ we have $a_n = -2a_0$. Lastly $S(e^{i\pi/n}) = 1$ shows that $a_0/b_n = -\frac{1}{4}$, which yields the mapping given earlier.

The hypergeometric equation associated with this case is given by

$$z(1-z)y'' + \left(\frac{1}{2} - 2z\right)y' - \frac{1}{4}\left(1 - \frac{1}{n^2}\right)y = 0.$$

Two solutions of the above are

$$F\left(\frac{1}{2} + \frac{1}{2n}, \frac{1}{2} - \frac{1}{2n}; \frac{1}{2}; z\right)$$

$$z^{1/2}F\left(1 + \frac{1}{2n}, 1 - \frac{1}{2n}; \frac{3}{2}; z\right).$$

Also, using 7.10 and the Lemma of section 6, we see that

$$y_1 = \frac{z^{-1/4}(1-z)^{-3/4}}{(w')^{1/2}}$$

$$y_2 = y_1 w$$

are also solutions. w is the inverse of $S(w)$, and is given by

$$w^n = 1 - 2z \pm 2\sqrt{z^2 - z}.$$

To determine the branch we note that we defined $S(w)$ so that

$$S(1) = 0, \qquad S(e^{i\pi/n}) = 1, \qquad S(0) = \infty.$$

Then

$$w = [1 - 2z + 2(z^2 - z)^{1/2}]^{1/n}$$

and the branches can be selected accordingly. We can easily verify that

$$F\left(\frac{1}{2} + \frac{1}{2n}, \frac{1}{2} - \frac{1}{2n}; \frac{1}{2}; z\right) = \frac{1}{2}(y_1 + y_2) = \frac{(1-z)^{-1/2}}{2}[w^{1/2} + w^{-1/2}].$$

To prove the above we note that at $z = 0$ both sides agree. Also if z traverses a small circle about the origin

$$w = [1 - 2z + 2(z^2 - z)^{1/2}] \to [1 - 2z - 2(z^2 - z)^{1/2}]^{1/n} = \frac{1}{w}$$

so that

$$w^{1/2} + w^{-1/2} \to w^{1/2} + w^{-1/2}$$

and the latter is single valued, so that the identity is established. Similarly it can be verified that

$$z^{1/2}F\left(1+\frac{1}{2n}, 1-\frac{1}{2n}; \frac{3}{2}; z\right) = \frac{n}{2i}(1-z)^{-1/2}[w^{1/2} - w^{-1/2}].$$

For the next case $\alpha_1 = \frac{1}{2}$, $\alpha_2 = \alpha_3 = \frac{1}{3}$. We place the origin so that the angle there is $\pi/2$. The fundamental triangle is shown in the following figure.

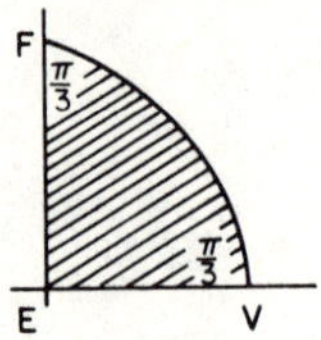

If we perform all possible reflections and then map onto the Riemann sphere we obtain the following.

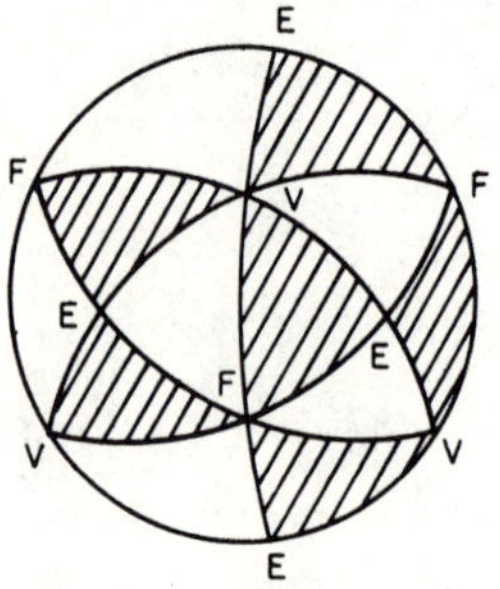

With the above figure we can associate a tetrahedron. The points labeled V are vertices, those labeled E are midpoints of the six edges, and the F correspond to the centers of the four faces. A single face is shown in the following.

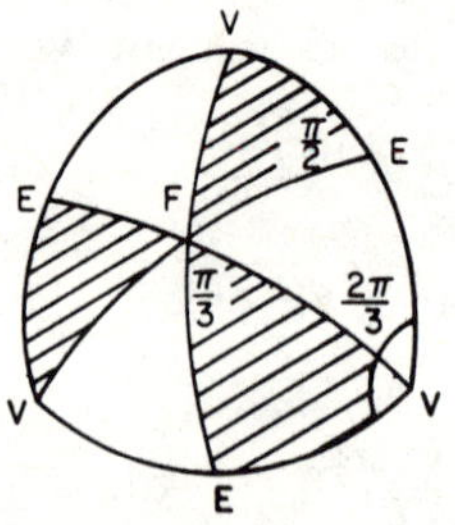

The mapping can be constructed, as before, from the geometry and can be shown to be

$$S(w) = \left[\frac{w^4 + 2w^2\sqrt{3} - 1}{w^4 - 2w^2\sqrt{3} - 1}\right]^3.$$

The above shows that for every z there will be 12 values of w, and indeed the tetrahedron has four faces and on each there are three images of $\text{Im } z > 0$. The tetrahedral group is of order 12 and consists of those 12 rotations which rotate the sphere in such a way as to map each of the 12 triangles onto some one of the remaining ones.

We can now consider the case where $\alpha_1 = \frac{1}{2}$, $\alpha_2 = \frac{1}{3}$, $\alpha_3 = \frac{1}{4}$. In this case it develops that the triangles on the Riemann sphere can be associated with an octahedron. Such an octahedron has 8 faces, 12 edges, and 6 vertices. Every face will have the structure indicated in the following figure.

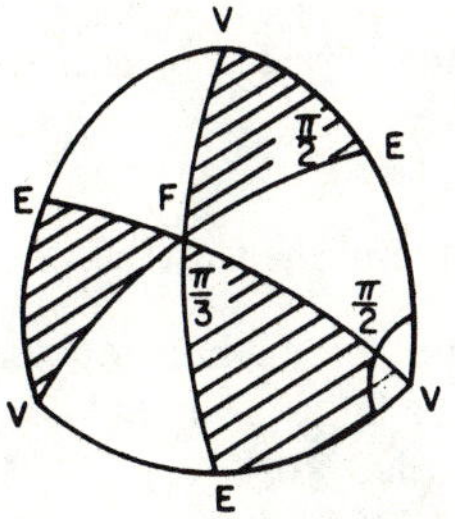

Every face contains three images of $\text{Im } z > 0$. Therefore $S(w)$ must be of degree 24. One can show that

$$S(w) = \frac{(w^8 + 14w^4 + 1)^3}{108w^4(w^4 - 1)}.$$

Lastly we have $\alpha_1 = \frac{1}{2}$, $\alpha_2 = \frac{1}{3}$, $\alpha_3 = \frac{1}{5}$. In this case we can associate a dodecahedron with the mapping. The dodecahedron has 12 faces, 30 edges and 20 vertices. On each face there will be five images of $\text{Im } z > 0$ so that $S(w)$ will have to be of degree 60. Each face is as follows.

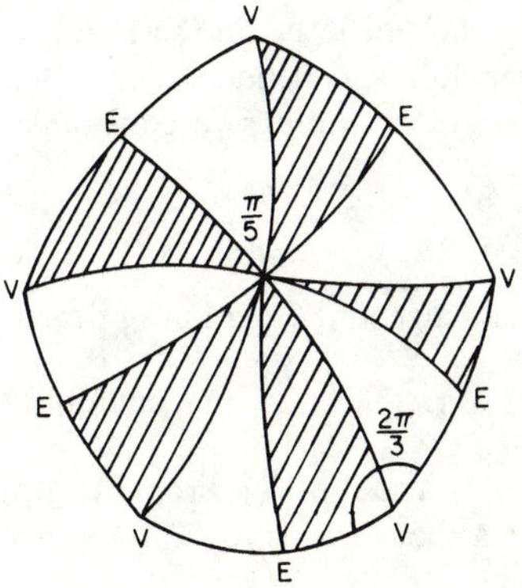

The appropriate mapping function is given by

$$S(w) = \frac{(w^{20} - 228w^{15} + 494w^{10} + 228w^5 + 1)^3}{-1728w^5(w^{10} + 11w^5 - 1)^5}.$$

To verify that the previously presented $S(w)$ are indeed correct we would have to show that $\{w, z\}$ has the correct form. But an explicit determination of $w(z)$ is clearly out of the question. But we have the identity

$$\{w, z\} = -\{z, w\}/\left(\frac{dz}{dw}\right)^2$$

and since $z(w)$ is known the verification is a tedious, but straightforward business.

9. Group Theoretic Discussion of the Case $\pi(\alpha_1 + \alpha_2 + \alpha_3) > \pi$

We shall now discuss another method for analyzing Case 3 of the preceding section. Suppose G is a finite group of bilinear mappings. Let N denote the order of the group, and let

$$I = w = \psi_0(w), \psi_1(w), \ldots, \psi_{N-1}(w)$$

denote the group elements.

We now select two complex numbers, a, b and suppose that they are such that

$$\psi_i(b) \neq a \qquad \text{for all } i.$$

In view of the finiteness of the group such a choice can be made in an infinity of ways. The reason for this choice is that in the expression

$$X = \prod_{i=0}^{N-1} \frac{\psi_i(w) - a}{\psi_i(w) - b} \tag{1}$$

there can be no cancellation of terms in the rational function on the right. After multiplying numerator and denominator by the denominators of all $\psi_i(w)$ we can rewrite (1) as the ratio of two polynomials of degree N,

$$X = \frac{G(w, a)}{G(w, b)} \tag{2}$$

For a given X (2) is an algebraic equation of degree N for w.

LEMMA All roots of (2) have the same multiplicity.

Proof. Suppose $w = \sigma$ is a root of (2). From the group structure it follows that $\psi_i(\sigma)$ must also be a root for every i. Those elements ψ_i of G for which

$\psi_i(\sigma)=\sigma$ form a subgroup, which we denote by H. Let the order of H be V. According to the Lagrange decomposition theorem we can decompose G into N/V distinct cosets so that

$$G = H + \psi_k H + \psi_l H + \cdots.$$

The set H consists of all ψ_i for which $\psi_i(\sigma)=\sigma$. $\psi_k H$ consists of all ψ_i for which $\psi_i(\sigma)=\psi_k(\sigma)\neq\sigma$, etc. It follows that $\psi_k(\sigma)$ is distinct from σ and has the same multiplicity and similarly for all other roots. ■

The expression

$$\Delta(w) = G(w, a)G'(w, b) - G'(w, a)G(w, b)$$

is a polynomial of degree $2N-2$, and σ is a root of (2) of multiplicity V. Then σ is a zero of $\Delta(w)$ of multiplicity $V-1$ providing $V>1$. As X in (2) is allowed to range over the complex plane the multiple roots of (2) must also be zeros of $\Delta(w)$. But the latter can have only a finite number of zeros and they are independent of X. Conversely every zero of $\Delta(w)$ can be associated with some value of X in (2). For every possible situation we find a set of N/V_i distinct roots of (2). Each of these must be a zero of multiplicity V_i-1 of $\Delta(w)$, which is of degree $2N-2$. Summing over all possibilities we find

$$\sum_{i=1}^{k}\frac{N}{V_i}(V_i-1) = 2N-2$$

or equivalently

$$\sum_{i=1}^{k}\left(1-\frac{1}{V_i}\right) = 2-\frac{2}{N}, \qquad V_i \geq 2 \quad \text{for} \quad \text{all } i. \tag{3}$$

Since $V_i \geq 2$ we find that

$$2 > \sum_{i=1}^{k}\left(1-\frac{1}{V}\right) \geq \frac{k}{2}$$

so that $k<4$. Hence $k=1, 2,$ or 3. But an inspection of (3) shows that $k=1$ is impossible. For $k=2$ we have

$$\frac{1}{V_1}+\frac{1}{V_2}=\frac{2}{N}$$

but $V_i \leq N$ for all i so that necessarily $V_1=V_2=N$. This corresponds to rotations of the plane through multiples of π/N.

For $k=3$ we find

$$\frac{1}{V_1}+\frac{1}{V_2}+\frac{1}{V_3}=1+\frac{2}{N}. \tag{4}$$

There is only a limited number of possibilities satisfying (4). These are tabulated below

V_1	V_2	V_3	N
2	2	$N/3$	even, unrestricted otherwise
2	3	3	12
2	3	4	24
2	3	5	60

The above cases correspond to the dihedral, tetrahedral, octahedral and dodecahedral groups discussed earlier. We thus see that the number of finite groups of bilinear mappings is an essentially limited one.

Another point of view in regard to this problem may be found in Zassenhaus' The Theory of Groups. There he works directly with the finite rotation groups of spheres.

10. Nonlinear Transformations of Hypergeometric Functions

In section 6 of the present chapter the 24 different solutions of the hypergeometric equation were discussed. These represent the totality of all bilinear transformations of the independent variable under which the hypergeometric equation is transformed into another equation of hypergeometric type. One can accordingly inquire into the existence of more general rational transformations of the independent variable under which the differential equation is transformed into another one of hypergeometric type. In 1881 Goursat made an exhaustive study of all such transformations. The details of his analysis are rather formidable, and we shall content ourselves with the analysis of a few special cases.

Suppose we consider the transformation

$$z = 4x(1 - x). \tag{1}$$

We return to 6.12, and set up a condition analogous to 6.14. Since

$$\{z, x\} = -\frac{6}{(1 - 2x)^2}$$

we obtain

$$\frac{Ax^2 + Bx + C}{x^2(x - 1)^2} + \frac{3}{(1 - 2x)^2} = \frac{A'z^2 + B'z + C'}{z^2(z - 1)^2}\, 16(1 - 2x)^2 \tag{2}$$

Replacing z by (1) in (2) and regrouping terms we obtain an identity of two polynomials. A comparison of coefficients then leads to

$$\begin{aligned} C &= C' \\ B &= 4B' + 4C' \\ A &= -4B' - 4C' \\ A' + B' + C' &= \tfrac{3}{16} \end{aligned} \tag{3}$$

In view of the fact that

$$A' = \frac{1 - (a' - b')^2}{4}$$

$$B' = \frac{(1 - c')^2 - (a' + b' - c')^2 + (a' - b')^2 - 1}{4}$$

$$C' = \frac{1 - (1 - c')^2}{4}$$

we find from the last equation in (3) that

$$(a' + b' - c')^2 = \tfrac{1}{4} \quad \text{or} \quad c' = a' + b' + \tfrac{1}{2}.$$

Note that this implies that a', b', c' cannot be three independent quantities. Using (3) we obtain

$$c = c' \qquad a = 2a', \qquad b = 2b' \tag{4}$$

Use of the fundamental theory developed in section 6 now shows that if w satisfies

$$\frac{d^2w}{dz^2} + \frac{A'z^2 + B'z + C'}{z^2(z-1)^2}\, w = 0$$

then

$$u = \frac{w}{\sqrt{z'}},$$

where z satisfies (1), satisfies

$$\frac{d^2u}{dx^2} + \frac{Ax^2 + Bx + C}{x^2(x-1)^2}\, u = 0 \tag{5}$$

if (2) holds. But we know that

$$w = z^{c'/2}(1-z)^{(1+a'+b'-c')/2}F(a', b'; c'; z). \tag{6}$$

Then, using the above and (4) we find

$$u_1 = \frac{w}{\sqrt{z'}} = \frac{1}{2}[4x(1-x)]^{(a+b+1)/4}F\left(\frac{a}{2}, \frac{b}{2}; \frac{a+b+1}{2}; 4x(1-x)\right).$$

satisfies (5).

In analogy to (6) we find that another solution of (5) is given by

$$u_2 = x^{(a+b+1)/4}(1-x)^{(a+b+1)/4}F\left(a, b; \frac{a+b+1}{2}; x\right).$$

We now note that u_1 and u_2 both satisfy (5) and both have the same index at $x = 0$. Hence they differ by at most a constant. It follows that

$$F\left(a, b; \frac{a+b+1}{2}; x\right) = F\left(\frac{a}{2}, \frac{b}{2}; \frac{a+b+1}{2}; 4x(1-x)\right). \tag{7}$$

(7) can be rewritten in a different form, by letting $a \to 2a$, $b \to 2b$, $x \to \frac{1}{2}(1-(1-x)^{1/2})$. Then

$$F\left(2a, 2b; a+b+\frac{1}{2}; \frac{1-(1-x)^{1/2}}{2}\right) = F(a, b; a+b+\tfrac{1}{2}; x). \tag{8}$$

(8) could of course be verified directly using the differential equation, but the necessary calculations are very tedious.

When combined with certain linear transformations (8) can be used to obtain further nonlinear transformations. For example from

$$(1-x)^a P\begin{Bmatrix} 0 & 1 & \infty & \\ 0 & 0 & a & x \\ 1-c & c-a-b & b & \end{Bmatrix} = P\begin{Bmatrix} 0 & 1 & \infty & \\ 0 & a & 0 & x \\ 1-c & c-b & b-a & \end{Bmatrix}$$

$$= P\begin{Bmatrix} 0 & 1 & \infty & \\ 0 & 0 & a & \dfrac{x}{x-1} \\ 1-c & b-a & c-b & \end{Bmatrix}$$

we can read off the identity

$$F(a, b; c; x) = (1-x)^{-a}F\left(a, c-b; c; \frac{x}{x-1}\right). \tag{9}$$

Equation (9) is generally known as Kummer's relation.

By applying the above identity to the left side of (8) we obtain immediately

$$F(a, b; a + b + \tfrac{1}{2}; x) = 2^{2a}[1 + (1 - x)^{1/2}]^{-2a} F\left(2a, a - b + \frac{1}{2}; a + b + \frac{1}{2}; \frac{(1 - x)^{1/2} - 1}{(1 - x)^{1/2} + 1}\right).$$

In this case again a direct verification using the differential equation would prove to be formidable.

Exercises

1. Evaluate $F(a,b; a - b + 1; -1)$.
2. Show that

$$(1 - t)^{b-c}(1 - t + zt)^{-b} = \sum_{n=0}^{\infty} \frac{(c)_n}{n!} F(-n, b; c; z)t^n$$

if c is not a non-positive integer and $|t|$ is sufficiently small.

3. Verify 3.7 by explicitly intergrating the right side.
4. Show that

$$c[F_{b-} - F_{a-}] = (b - a)z\, F_{c+}$$

5. Show that

$$c(c + 1)[F - F_{c+}] = abz\, F(a + 1, b + 1; c + 2; z).$$

6. Show that

$$F(a, b; c; z) = \frac{\Gamma(c)\Gamma(b - a)}{\Gamma(b)\Gamma(c - a)} (-z)^{-a} F\left(a, 1 - c + a; 1 - b + a; \frac{1}{z}\right)$$
$$+ \frac{\Gamma(c)\Gamma(a - b)}{\Gamma(a)\Gamma(c - b)} (-z)^{-b} F\left(b, 1 - c + b; 1 - a + b; \frac{1}{z}\right).$$

7. Show that

$$\left(\frac{d}{dz}\right)^n [z^{c-a+n-1}(1 - z)^{a+b-c}F(a, b; c; z)]$$
$$= (c - a)_n z^{c-a-1}(1 - z)^{a+b-c-n}F(a - n, b; c; z).$$

8. Show that the function

$$W = \tfrac{1}{3}z\, F(\tfrac{1}{3}, \tfrac{2}{3}, \tfrac{3}{2}, -\tfrac{1}{4}z^2)$$

satisfies

$$W^3 + 3W - z = 0.$$

9. Use the method of section 4 to find an integral representation for the solution of

$$y'' + \frac{2m}{x} y' + y = 0.$$

Select $K(x, t) = \cos xt$.

10. Consider the linear differential equation with constant coefficients

$$P(D)y = \sum_{k=0}^{m} a_k D^k y = \sum_{k=0}^{m} a_k y^{(k)} = 0$$

with initial conditions

$$y^{(n)}(0) = 0 \qquad n = 0, 1, 2, \dots, k-1, k+1, \dots m-1$$

$$y^{(k)}(0) = 1.$$

Let L be any closed simply connected Jordan curve in the t plane enclosing all zeros of $P(t)$. Show that

$$y = \frac{1}{2\pi i} \int_L \frac{e^{tx}(P(t) - \sum_{l=0}^{k} a_l t^l)\, dt}{t^{k+1}\, P(t)}$$

satisfies the problem.

11. Show that

$$\{z, x\} + \left(\frac{dz}{dx}\right)^2 \{x, z\} = 0.$$

12. Solve the equation

$$x^2 \{z, x\} + a = 0.$$

13. Show that formula 7.7 also applies to the case where $f(z)$ maps the polygon in question into the unit circle in the z-plane. The integration is now restricted to paths lying in the unit circle.

14. Consider a polygon of n sides, each vertex lying on the unit circle, at the nth roots of unity, say $1, e^{2\pi i/n}, e^{4\pi i/n} \dots$. The sides are circular arcs, of equal length, such that the interior angle at each vertex is $\alpha\pi$, where $0 \leq \alpha \leq 2$. Show that the mapping $f(z)$ which maps the unit circle in the z-plane into the polygon such that $f(0) = 0$, $f'(0) > 0$ is given by

$$f(z) = z \frac{\Gamma(1 - 1/n)\Gamma(\frac{1}{2}(1 + \alpha) + 1/n) F(\frac{1}{2}(1 - \alpha) + 1/n, \frac{1}{2}(1 - \alpha); 1 + 1/n; z^n)}{\Gamma(1 + 1/n)\Gamma(\frac{1}{2}(1 + \alpha) - 1/n) F(\frac{1}{2}(1 - \alpha) - 1/n, \frac{1}{2}(1 - \alpha); 1 - 1/n; z^n)}.$$

[*Hint:* From the symmetry of the situation we see that every ray of the form $re^{2k\pi i/n}$ must be mapped into itself. Then $f(z) = zg(z^n)$. Now show that $\{f, z\} = \frac{1}{2}n^2(1 - \alpha^2)$ $z^{n-2}/(z^n - 1)^2$, and solve the associated second order differential equation.]

15. Show explicitly how the relations between α_i, β_i and a_i in 7.8 are derived.

16. Carry out the details of the limiting procedures used in deriving 7.9.

17. Show that

$$z^{1/2}F\left(1+\frac{1}{2n}, 1-\frac{1}{2n}; \frac{3}{2}; z\right) = \frac{n}{2i}(1-z)^{-1/2}[w^{1/2}-w^{-1/2}] \text{ where}$$

$w = [1 - 2z + 2(z^2 - z)^{1/2}]^{1/n}]$ and $w(0) = 1$.

18. Show that

$$\cos a\, z = F(\tfrac{1}{2}a, -\tfrac{1}{2}\, a; \tfrac{1}{2}; \sin^2 z).$$

CHAPTER 5

The Legendre Functions

1. Laplace's Differential Equation

Although it is certainly possible to study Legendre functions without ever referring to Laplace's differential equation, we prefer to do otherwise. Historically this is where Legendre functions first arose. If one wishes to study solutions of Laplace's equation in spherical co-ordinates these functions arise naturally. But the chief motivation for following this approach is that many properties are suggested in a natural manner from this point of view.

We now consider the equation

$$\Delta V = V_{xx} + V_{yy} + V_{zz} = 0 \tag{1}$$

(The subscripts denote partial derivatives; e.g., $V_{xx} = \partial^2 V/\partial x^2$.)
If the polar coordinates

$$x = r \sin \theta \cos \phi,\ y = r \sin \theta \sin \phi,\ z = r \cos \theta$$

are introduced into (1) one obtains

$$V_{rr} + \frac{2}{r} V_r + \frac{1}{r^2 \sin \theta} \frac{\partial}{\partial \theta} (\sin \theta V_\theta) + \frac{1}{r^2 \sin^2 \theta} V_{\phi\phi} = 0. \tag{2}$$

A standard method of tackling this equation is to use the method of separation of variables. That is we assume the solution to be a product of three functions, each of one variable. Then

$$V = R(r)\, \Theta\, (\theta)\, \Phi\, (\phi).$$

Insertion of this in (2) and subsequent division by V leads to

$$\left[\frac{R'' + (2/r)R'}{R}\right] + \frac{1}{r^2}\left[\frac{\Theta'' + \cot \theta \Theta'}{\Theta}\right] + \frac{1}{r^2 \sin^2 \theta}\left[\frac{\Phi''}{\Phi}\right] = 0. \tag{3}$$

In view of the fact that r, θ and ϕ are independent variables we can conclude from (3) that

$$r^2R'' + 2rR' - n(n+1)R = 0$$

$$\Theta'' + \cot\theta\Theta' + \left[n(n+1) - \frac{m^2}{\sin^2\theta}\right]\Theta = 0$$

$$\Phi'' + m^2\Phi = 0$$

m^2 and $n(n+1)$ are the separation constants, and are selected in this form for future convenience. The first and third of the above have the obvious solutions

$$R = c_1 r^n + \frac{c_2}{r^{n+1}}$$

$$\Phi = c_3 e^{im\phi} + c_4 e^{-im\phi}$$

where the c_i are constants of integration.

To discuss the equation for Θ we let $t = \cos\theta$ and $\Theta = T(t)$. Then

$$(1-t^2)T'' - 2tT' + \left[n(n+1) - \frac{m^2}{1-t^2}\right]T = 0. \tag{4}$$

For the case where $m = 0$ and n an integer we see that $T = P_n(t)$, the Legendre polynomial of chapter 2. More generally we denote the so-called associated Legendre functions of the first kind by the symbol $P_n{}^m(t)$. To define these properly we proceed as follows. m fold differentiation of the equation

$$(1-t^2)P_n'' - 2tP_n' + n(n+1)P_n = 0$$

leads to

$$(1-t^2)D^{m+2}P_n - 2t(m+1)D^{m+1}P_n + [n(n+1) - m(m+1)]D^mP_n = 0 \tag{5}$$

(In the above D stands for the operation d/dt.) Next we let

$$D^mP_n = (1-t^2)^{-m/2}\,T$$

and find that under this substitution in (5) we again obtain (4). The solution of (4) denoted by $P_n{}^m(t)$ is now defined as

$$P_n{}^m(t) = (-1)^m(1-t^2)^{m/2}D^mP_n(t). \tag{6}$$

A solution of (1) is given by

$$V = \left(c_1r^n + \frac{c_2}{r^{n+1}}\right)P_n{}^m(\cos\theta)(c_3 e^{im\phi} + c_4 e^{-im\phi}). \tag{7}$$

2. Maxwell's Theory of Poles

From 1.7 it follows that for $n = m = 0$

$$V = \frac{1}{r}$$

satisfies 1.1. By repeated differentiation of 1.1 with respect to x, y, z it follows that

$$\tilde{V} = \frac{\partial^{i+j+k}}{\partial x^i\, \partial y^j\, \partial z^k}\frac{1}{r}, \qquad r = \sqrt{x^2 + y^2 + z^2}$$

also satisfies 1.1.

As a special case consider

$$\frac{\partial}{\partial z}\frac{1}{r} = -\frac{z}{r^3} = -\frac{1}{r^2}P_1\left(\frac{z}{r}\right) = -\frac{1}{r^2}P_1(\cos\theta).$$

A comparison with 1.7 shows that the above corresponding to the case $m = 0$, $n = 1$. More generally we note that

$$r^{n+1}\frac{\partial^n}{\partial z^n}\frac{1}{r} = F(z, r)$$

is homogeneous in z and r. That is,

$$F(tz, tr) = F(z, r)$$

for arbitrary values of the parameter t. Hence

$$F(z, r) = f\left(\frac{z}{r}\right) = f(\cos\theta).$$

Then, corresponding to $m = 0$ in 1.7, we have

$$\frac{\partial^n}{\partial z^n}\frac{1}{r} = \frac{f(\cos\theta)}{r^{n+1}} = \frac{cP_n(\cos\theta)}{r^{n+1}} \tag{1}$$

for a suitable constant c. We could, at the same time, use the above as a definition for the function $P_n(\cos\theta)$. It follows immediately that $P_n(z/r)$ is a polynomial of degree n in the variable z/r.

To discuss the cases for which $m \neq 0$, we consider

$$\left(\frac{\partial}{\partial x} + i\frac{\partial}{\partial y}\right)\frac{1}{r} = -\frac{x + iy}{r^3} = -\frac{\sin\theta e^{i\phi}}{r^2} = -\frac{P_1{}^1(\cos\theta)e^{i\phi}}{r^2}$$

and more generally, one can show, by the method of mathematical induction that

$$\left(\frac{\partial}{\partial x} + i\frac{\partial}{\partial y}\right)^m f(r) = \left(\frac{1}{r}\frac{\partial}{\partial r}\right)^m f(r)(x+iy)^m.$$

For the case, where $f(r) = 1/r$ the latter yields

$$\left(\frac{\partial}{\partial x} + i\frac{\partial}{\partial y}\right)^m \frac{1}{r} = \frac{(-1)^m(2m)!(x+iy)^m}{2^m m! r^{2m+1}} = \frac{(-1)^m(2m)! \sin^m \theta e^{im\phi}}{2^m m! r^{m+1}}.$$

By combining these two operators it follows that

$$\left(\frac{\partial}{\partial z}\right)^{n-m}\left(\frac{\partial}{\partial x} + i\frac{\partial}{\partial y}\right)^m \frac{1}{r} = \frac{ce^{im\phi}P_n{}^m(\cos\theta)}{r^{n+1}}$$

where c is a suitable constant. The above depends of course on the definition of $P_n{}^m(\cos\theta)$, and can be most easily obtained using some of the recurrence formulas to be developed in subsequent sections. One can then show that

$$\left(\frac{\partial}{\partial z}\right)^{n-m}\left(\frac{\partial}{\partial x} + i\frac{\partial}{\partial y}\right)^m \frac{1}{r} = \frac{(-1)^{n-m}(n-m)! e^{im\phi}P_n{}^m(\cos\theta)}{r^{n+1}}.$$

The above could be used also as a definition of $P_n{}^m(\cos\theta)$. These formulas have certain physical interpretations that first led Maxwell to investigate them. $1/r$ represents the potential of a point charge, placed at $r = 0$. The function

$$V = \frac{\partial}{\partial z}\frac{1}{r} = \lim_{\zeta \to 0} \frac{[x^2+y^2+(z+\zeta)^2]^{-1/2} - [x^2+y^2+(z-\zeta)^2]^{-1/2}}{2\zeta}$$

can be interpreted as the potential resulting from the confluence of two point charges, approaching each other. The higher derivatives similarly are related to the potential due to the confluence of numerous point charges approaching each other from different directions.

3. Relationship to the Hypergeometric Functions

We return to the equation

$$(1-t^2)T'' - 2tT' + n(n+1)T = 0 \tag{1}$$

and note that it has three Fuchsian singularities at $t = 1$, $t = -1$ and $t = \infty$. A simple calculation shows that the indices at these singularities are given by

$t = 1$	$0, 0$
$t = -1$	$0, 0$
$t = \infty$	$-n, n+1$

We can therefore associate with the equation the following symbol

$$P\begin{Bmatrix} 1 & -1 & \infty & \\ 0 & 0 & -n & t \\ 0 & 0 & n+1 & \end{Bmatrix}.$$

Using a suitable bilinear substitution we can shift the singular points to 0, 1, ∞ so that the above is equivalent to

$$P\begin{Bmatrix} 0 & 1 & \infty & \\ 0 & 0 & -n & 1-2\tau \\ 0 & 0 & n+1 & \end{Bmatrix}.$$

so that $a = -n$, $b = n + 1$, $c = 1$. Then we can write

$$P_n(t) = F\left(-n, n+1; 1; \frac{1-t}{2}\right). \tag{2}$$

(2) satisfies (1) for arbitrary n, but will not converge at $t = 1$ unless n is an integer. In that case the function reduces to a polynomial of degree n. To show $P_n(t)$ in that case is indeed given by (2) we need to show that they agree at a point. But at $t = 1$

$$1 = F(-n, n+1; 1; 0) = P_n(1)$$

which can easily be shown to be so, using the generating function of the Legendre polynomials.

In view of the fact that the hypergeometric function is symmetric in the first two parameters, we see that it is invariant under the substitution

$$n \to -n - 1$$

Equation (1) is also invariant under the above. It follows that

$$P_n(t) = P_{-n-1}(t).$$

Using 1.6 we have more generally

$$P_n{}^m(t) = P^m_{-n-1}(t). \tag{3}$$

In view of the fact that the substitution $m \to -m$ leaves 1.4 invariant we expect that if $P_n{}^m(t)$ satisfies 1.4 so does $P_n{}^{-m}(t)$. But it need not be the same solution of course. To determine which it is we proceed as follows. For arbitrary u we can check easily that

$$V = (z + ix \cos u + iy \sin u)^n$$

satisfies

$$\Delta V = 0.$$

More generally, for sufficiently smooth f another solution is

$$V = \int_{-\pi}^{\pi} f(z + ix\cos u + iy\sin u, u)\,dy. \tag{4}$$

A special case of (4) is obviously

$$V = \int_{-\pi}^{\pi} [z + ix\cos u + iy\sin u]^n e^{imu}\,du. \tag{5}$$

If, in (5) we replace x, y, z by their equivalents in spherical coordinates we see that

$$V = r^n e^{im\phi} \int_{-\pi}^{\pi} [\cos\theta + i\sin\theta\cos u]^n e^{imu}\,du$$

A comparison with 1.7 shows that for a suitable constant $c_{m,n}$

$$P_n^m(\cos\theta) = c_{m,n} \int_{-\pi}^{\pi} [\cos\theta + i\sin\theta\cos u]^n \cos mu\,du \tag{6}$$

and using the substitution $n \to -n-1$

$$P_n^m(\cos\theta) = c'_{m,n} \int_{-\pi}^{\pi} \frac{\cos mu}{[\cos\theta + i\sin\theta\cos u]^{n+1}}\,du. \tag{7}$$

(6) and (7) may also be derived by another technique, which has the advantage that the constants are determined directly.

Using Rodrigues' formula and 1.6 we have

$$P_n^m(t) = (-1)^m(1-t^2)^{m/2}D^m P_n(t) = \frac{(-1)^m(1-t^2)^{m/2}}{2^n n!} D^{m+n}(t^2-1)^n. \tag{8}$$

If we use

$$f^{(n)}(z) = \frac{n!}{2\pi i}\int_c \frac{f(\zeta)}{(\zeta - z)^{n+1}}\,d\zeta$$

in (7) we obtain

$$P_n^m(t) = \frac{(-1)^m(1-t^2)^{m/2}}{2^n n!}\frac{(n+m)!}{2\pi i}\int_c \frac{(\zeta^2-1)^n}{(\zeta - t)^{n+m-1}}\,d\zeta$$

where for c we select the circle $|\zeta - t| = (1 - t^2)^{1/2}$. The change of variable

$$\zeta = t + i(1-t^2)^{1/2}e^{i\psi}$$

$$\zeta^2 - 1 = 2i(1-t^2)^{1/2}e^{i\psi}[t + i(1-t^2)^{1/2}\cos\psi]$$

leads to

$$P_n^m(t) = \frac{i^m(n+m)!}{2\pi n!} \int_{-\pi}^{\pi} [t + i(1-t^2)^{1/2} \cos\psi]^n \cos m\psi \, d\psi \tag{9}$$

and using the invariance under the substitution $n \to -n-1$, (9) yields

$$P_n^m(t) = \frac{(-i)^m n!}{2\pi(n-m)!} \int_{-\pi}^{\pi} \frac{\cos m\psi}{[t + i(1-t^2)^{1/2}\cos\psi]^{n+1}} \, d\psi \tag{10}$$

In the above we used the fact that

$$\frac{(n+m)!}{n!} = (n+m)(n+m-1)\cdots(n+1) \to (-1)\cdots(n-m+1)$$

$$= \frac{(-1)^m n!}{(n-m)!}.$$

Using either (9) or (10) we see that

$$P_n^{-m}(t) = (-1)^m \frac{(n-m)!}{(n+m)!} P_n^m(t). \tag{11}$$

For fixed m we can show that the set $\{P_n^m(t)\}$ is orthogonal over the interval $[-1, 1]$. To prove this we consider the integral

$$\int_{-1}^{1} P_n^m(t) P_k^m(t) \, dt = \frac{(-1)^m(n+m)!}{(n-m)!} \int_{-1}^{1} P_n^{-m}(t) P_k^m(t) \, dt$$

$$= \frac{(-1)^m(n+m)!}{2^{n+k} n! k! (n-m)!} \int_{-1}^{1} D^{n-m}(t^2-1)^n D^{k+m}(t^2-1)^k \, dt$$

where (9) was used. Repeated integration by parts in the last integral now shows that

$$\int_{-1}^{1} P_n^m(t) P_k^m(t) \, dt = \frac{(n+m)!}{2^{n+k} n! k! (n-m)!} \int_{-1}^{1} D^n(t^2-1)^n D^k(t^2-1)^k \, dt$$

$$= \frac{(n+m)!}{(n-m)!} \int_{-1}^{1} P_n(t) P_k(t) \, dt$$

$$= 0 \qquad n \neq k$$

$$= \frac{2(n+m)!}{(2n+1)(n-m)!}, \qquad n = k. \tag{12}$$

From (12) we can deduce the following fact immediately. Letting

$$P_{m+k}^m(t) = (1-t^2)^{m/2} \phi_k^m(t)$$

we see that $\phi_k^m(t)$ is a polynomial of degree k. For fixed m the set $\{\phi_k^m(t)\}$ is the set of orthogonal polynomials over $[-1, 1,]$ with weight function $(1 - t^2)^m$. Accordingly all theorems derived for orthogonal polynomials relating to completeness properties, recursion formulas etc. can be applied to the associated Legendre functions.

We can also relate the associated Legendre functions to the hypergeometric function. Clearly

$$\begin{aligned} P_n^m(t) &= (-1)^m(1 - t^2)^{m/2}D^m P_n(t) \\ &= (-1)^m(1 - t^2)^{m/2}D^m F\left(-n, n+1; 1; \frac{1-t}{2}\right) \\ &= \left(-\frac{1}{2}\right)^m \frac{(n+m)!}{(n-m)!m!}(1 - t^2)^{m/2} F\left(m-n, m+n+1; m+1; \frac{1-t}{2}\right) \end{aligned} \tag{13}$$

where 4.3.8 was used.

All the representations for $P_n^m(t)$ discussed so far can be used to find convergent series for $|(1 - t)/2| < 1$. To obtain their analytic continuations into the t plane we make use of some of the linear as well as nonlinear transformations that were derived for the hypergeometric function. Using

$$(1-z)^m\, P\begin{Bmatrix} 0 & 1 & \infty & \\ 0 & 0 & m-n & z \\ -m & -m & m+n+1 & \end{Bmatrix} = P\begin{Bmatrix} 0 & 1 & \infty & \\ 0 & 0 & -n & z \\ -m & m & n+1 & \end{Bmatrix}$$

in (13) with $z = (1 - t)/2$

$$P_n^m(t) = \frac{(-1)^m(n+m)!}{(n-m)!m!}\left(\frac{1-t}{1+t}\right)^{m/2} F\left(-n, n+1; m+1; \frac{1-t}{2}\right). \tag{14}$$

If we let

$$a = \frac{m-n}{2}$$

$$b = \frac{m+n+1}{2}$$

$$c = a + b + \tfrac{1}{2} = m + 1$$

and recall

$$F\left(2a, 2b; a+b+\tfrac{1}{2}; \frac{1-(1-z)^{1/2}}{2}\right) = F(a, b; a+b+\tfrac{1}{2}; z)$$

(13) can also be transformed into

$$P_n^m(t) = \left(-\frac{1}{2}\right)^m \frac{(n+m)!}{(n-m)!m!}(1-t^2)^{m/2} \times F\left(\frac{m-n}{2}, \frac{m+n+1}{2}; m+1; 1-t^2\right). \tag{15}$$

To the latter we apply the identity

$$F(a, b; c; z) = \frac{\Gamma(c)\Gamma(b-a)}{\Gamma(b)\Gamma(c-a)(1-z)^a} F\left(a, c-b; a-b+1; \frac{1}{1-z}\right) + \frac{\Gamma(c)\Gamma(a-b)}{\Gamma(a)\Gamma(c-b)(1-z)^b} F\left(b, c-a; b-a+1; \frac{1}{1-z}\right) \tag{16}$$

and obtain

$$P_n^m(t) = \frac{(2n)!(-1)^m}{2^n n!(n-m)!}(1-t^2)^{m/2} t^{n-m} F\left(\frac{m-n}{2}, \frac{m-n+1}{2}; \frac{1}{2}-n; \frac{1}{t^2}\right). \tag{17}$$

The second term from (17) dropped out by virtue of the fact that the denominator contains the term

$$\Gamma(a)\Gamma(c-b) = \Gamma\left(\frac{m-n}{2}\right)\Gamma\left(\frac{m-n+1}{2}\right).$$

Either $(m-n)/2$ or $(m-n+1)/2$ must be a negative integer so that the corresponding term in (16) vanishes.

If now we apply

$$F(a, b; c; z) = (1-z)^{-a} F\left(a, c-b; c; \frac{z}{z-1}\right) \tag{18}$$

to (17) we obtain

$$P_n^m(t) = \frac{(-1)^{(n+m)/2}(2n)!}{2^n n!(n-m)!}(1-t^2)^{n/2} F\left(\frac{m-n}{2}, -\frac{m+n}{2}; \frac{1}{2}-n; \frac{1}{1-t^2}\right) \tag{19}$$

If we again apply our quadratic transformation to the above we have

$$P_n^m(t) = \frac{(-1)^{(n+m)/2}(2n)!}{2^n n!(n-m)!}(1-t^2)^{n/2} \times F\left(m-n, -m-n, \frac{1}{2}-n; \frac{(t^2-1)^{1/2}-t}{2(t^2-1)^{1/2}}\right). \tag{20}$$

Another application of (18) to (20) finally yields

$$P_n^m(t) = \frac{(-1)^m (2n)!}{2^{2n-m} n!(n-m)!} (1-t^2)^{m/2}[(t^2-1)^{1/2}+t]^{n-m}$$

$$\times F\left(m-n, \frac{1}{2}+m; \frac{1}{2}-n; \frac{t-(t^2-1)^{1/2}}{t+(t^2-1)^{1/2}}\right). \tag{21}$$

The above formulas provide analytic continuations of the basic formula (13) into other regions of the t plane. This list is not exhaustive, but is a sampling of such. Repeated application of the many transformation formulas of hypergeometric functions will lead to many more.

4. Expansion Formulas

A typical boundary value problem encountered in mathematical physics is the following. We seek a solution V of

$$\Delta V = 0 \tag{1}$$

subject to the condition that

$$V = f(\theta, \phi) \qquad \text{on} \quad r = 1 \tag{2}$$

where $f(\theta, \phi)$ is a prescribed function of θ and ϕ. This function must be defined on the surface of the sphere, i.e. $0 \le \theta \le \pi$, $0 \le \phi \le 2\pi$. We can take the radius of the sphere to be unity, without loss of generality. We suppose $f(\theta, \phi)$ is such that

$$\int_0^{2\pi} \int_0^{\pi} f^2(\theta, \phi) \sin\theta \, d\theta \, d\phi < \infty.$$

To expand $f(\theta, \phi)$ we first write

$$f(\theta, \phi) = \sum_{-\infty}^{\infty} f_m(\theta) e^{im\phi}$$

Each of the Fourier coefficients $f_m(\theta)$ we expand in the terms of the orthogonal set $P_n^m(\cos\theta)$. Using 3.11 (with substitution $t = \cos\theta$) we have

$$f_m(\theta) = \sum_{n=|m|}^{\infty} A_{n,m} P_n^m(\cos\theta)$$

where

$$A_{n,m} = \frac{(2n+1)(n-m)!}{2(n+m)!} \int_0^{\pi} f_m(\theta) P_n^m(\cos\theta) \, d\theta$$

(Note that since $P_n^m(t)$ and $P_n^{-m}(t)$ are linearly dependent we need sum only over $(|m|, \infty)$.) Combining these we can write

$$f(\theta, \phi) = \sum_{n=0}^{\infty} \sum_{m=-n}^{n} A_{n,m} P_n^m(\cos\theta) e^{im\phi}$$

$$= \frac{1}{4\pi} \int_{-\pi}^{\pi} \int_0^{\pi} f(\theta', \phi') \sin\theta' \, d\theta' \, d\phi'$$

$$\times \sum_{n=0}^{\infty} (2n+1) \sum_{m=-n}^{n} \frac{(n-m)!}{(n+m)!} P_n^m(\cos\theta) P_n^m(\cos\theta') e^{im(\phi-\phi')}$$

$$= \frac{1}{4\pi} \int_{-\pi}^{\pi} \int_0^{\pi} f(\theta', \phi') \sin\theta' \, d\theta' \, d\phi'$$

$$\times \sum_{n=0}^{\infty} (2n+1) \sum_{m=0}^{n} \varepsilon_m \frac{(n-m)!}{(n+m)!} P_n^m(\cos\theta) P_n^m(\cos\theta') \cos m(\phi - \phi') \tag{3}$$

where

$$\varepsilon_m = \begin{matrix} 1 & m = 0 \\ 2 & m > 0 \end{matrix}$$

The validity of (3) follows from the completeness properties of Fourier series and the set $\{P_n^m(t)\}$.

To solve the boundary value problem (1), (2) we use the fact that the function

$$r^n P_n^m(\cos\theta) \cos m\phi$$

satisfies (1). It follows that

$$V = \frac{1}{4\pi} \int_{-\pi}^{\pi} f(\theta', \phi') \sin\theta' \, d\theta' \, d\phi$$

$$\times \sum_{n=0}^{\infty} (2n+1) r^n \sum_{m=0}^{n} \varepsilon_m \frac{(n-m)!}{(n+m)!} P_n^m(\cos\theta) P_n^m(\cos\theta') \cos m(\phi - \theta'). \tag{4}$$

satisfies (1) and reduces to $f(\theta, \phi)$ on $r = 1$. By virtue of the fact that at the north pole, defined by $\theta = 0$, we have

$$P_n(1) = 1$$

$$P_n^m(1) = 0 \qquad m > 0$$

(4) reduces at that point to

$$V(r, 0, \phi) = \frac{1}{4\pi} \int_{-\pi}^{\pi} f(\theta', \phi') \sin\theta' \, d\theta' \, d\phi' \sum_{n=0}^{\infty} (2n+1) r^n P_n(\cos\theta'). \tag{5}$$

The location of the axis $\theta = 0$ is purely a matter of convention. Any other point on the surface of the unit sphere could be designated as the north pole. It will prove to be convenient to select the direction (θ, ϕ) as that axis. The angle designated by θ' in (5) is the angle between the fixed direction (θ', ϕ) and (θ', ϕ'). The angle between the directions (θ, ϕ) and (θ', ϕ') will be designated by α. Clearly

$$\cos \alpha = \cos \theta \cos \theta' + \sin \theta \sin \theta' \cos (\phi - \phi').$$

This is evident from the following figure.

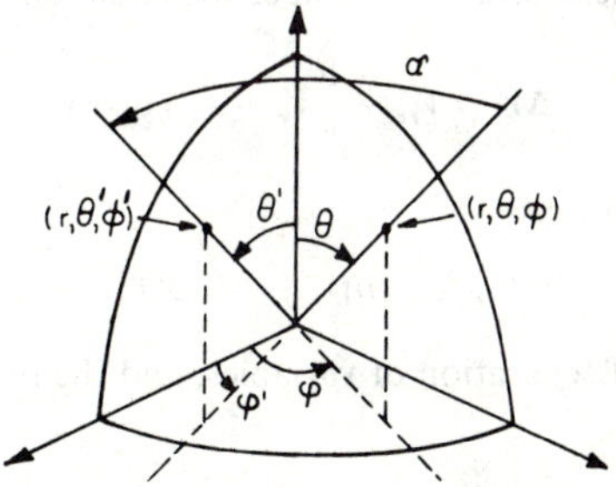

Relative to these directions and using (5) we can now write

$$V(r, \theta, \phi) = \frac{1}{4\pi} \int_{-\pi}^{\pi} \int_{0}^{\pi} f(\theta', \phi') \sin \theta' \, d\theta' \, d\phi' \sum_{n=0}^{\infty} (2n + 1) r^n P_n(\cos \alpha) \quad (6)$$

$$\cos \alpha = \cos \theta \cos \theta' + \sin \theta \sin \theta' \cos (\phi - \phi')$$

(4) and (6) are equivalent, but the advantage of (6) over (4) is that a simple sum is used rather than a double sum.

5. The Addition Theorem

A comparison of 4.4 and 4.6 shows that

$$\int_{-\pi}^{\pi} \int_{0}^{\pi} f(\theta', \phi') \sin \theta' \, d\theta' \, d\phi' \sum_{n=0}^{\infty} (2n + 1) r^n$$

$$\times \left\{ P_n(\cos \alpha) - \sum_{m=0}^{n} \varepsilon_m \frac{(n - m)!}{(n + m)!} P_n^m(\cos \theta) P_n^m(\cos \theta') \cos m(\phi - \phi') \right\} = 0$$

for all $f(\theta', \phi')$. By virtue of the completeness properties of the terms $P_n^m(\cos \theta)$ $e^{im\phi}$ the above implies that the bracketed term must vanish almost everywhere but since this term is continuous it vanishes everywhere. We conclude that

$$P_n(\cos\theta\cos\theta' + \sin\theta\sin\theta'\cos(\phi - \phi'))$$

$$= \sum_{m=0}^{n} \varepsilon_m \frac{(n-m)!}{(n+m)!} P_n^m(\cos\theta) P_n^m(\cos\theta')\cos m(\phi - \phi') \quad (1)$$

(1) is often referred to as the addition theorem for the Legendre polynomials. It shows us how the function $P_n(\cos\theta)$ can, when referred to a rotated coordinate system, be expressed in terms of the fundamental solutions $P_n^m(\cos\theta')e^{im\phi'}$.

It may be interesting to compare the above derivation to the comparable situation in two dimensions. We consider the equation

$$\Delta V = V_{rr} + \frac{1}{r}V_r + \frac{1}{r^2}V_{\theta\theta} = 0$$

with boundary conditions

$$V = f(\theta) \quad \text{on} \quad r = 1.$$

Using the method of separation of variables, and the resulting Fourier series we find that

$$V(r, \theta) = \frac{1}{2\pi}\int_{-\pi}^{\pi} f(\theta') \sum_{n=-\infty}^{\infty} r^{|n|}e^{in\theta}e^{-in\theta'}\, d\theta'. \quad (2)$$

As before we see that

$$V(r, 0) = \frac{1}{2\pi}\int_{-\pi}^{\pi} f(\theta') \sum_{n=-\infty}^{\infty} r^{|n|}e^{-in\theta'}\, d\theta'.$$

But the direction $\theta = 0$ is simply a selected direction. If we let θ be that preferred direction and refer θ' to that direction we have

$$\theta' - \theta = \alpha.$$

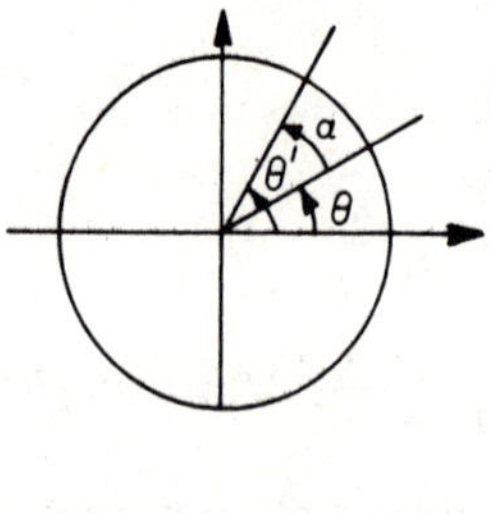

Then

$$V(r, \theta) = \frac{1}{2\pi}\int_{-\pi}^{\pi} f(\theta') \sum_{n=-\infty}^{\infty} r^{|n|}e^{-in\alpha}\, d\theta'. \quad (3)$$

A comparison of (2) and (3) now shows that

$$e^{-in\alpha} = e^{-in(\theta'-\theta)} = e^{in\theta}e^{-in\theta'}. \tag{4}$$

(4) is of course the standard addition theorem for exponential functions. But the proof offered here is based on the fact that certain basic solutions of Laplace's equation in one coordinate system may be expressed in terms of a similar set in a rotated coordinate system. In both cases treated what makes this feasible is the fact that Laplace's differential equation is invariant under rotations of the coordinate system. This is of course in consonance with the physical situation from which we can derive the equation. The corresponding physical phenomena do not depend on our choice of coordinate system. Hence any rigid transformation of coordinates, that is rotations or translations, must leave the equation invariant. Later we will see how the addition theorem for Bessel functions arises from a translation of coordinates.

We shall now discuss an alternative proof of (1). Using the aforegoing reasoning, it follows that a relationship of the type

$$r^n P_n(\cos\theta\cos\theta' + \sin\theta\sin\theta'\cos(\phi-\phi')) = r^n \sum_{m=-n}^{n} a_m P_n{}^m(\cos\theta)e^{im\phi}$$

must exist. Using the fact that the left side is symmetric in θ and θ' and even in $\phi - \phi'$ we conclude that

$$a_m = b_m P_n{}^m(\cos\theta')e^{-im\phi}$$

and b_m is so related to b_{-m} that we can write

$$P_n(\cos\theta\cos\theta' + \sin\theta\sin\theta'\cos(\phi-\phi'))$$
$$= \sum_{m=0}^{n} b_m P_n{}^m(\cos\theta)P_n{}^m(\cos\theta')\cos m(\phi-\phi'). \tag{5}$$

To determine b_m we shall use a limiting form of the above. Let $\cos\theta = x$. Then

$$\lim_{x\to\infty}\frac{\sin\theta}{x} = \lim_{x\to\infty}\frac{(1-x^2)^{1/2}}{x} = i$$

where we select a suitable branch of the radical. Similarly

$$\lim_{\substack{x\to\infty\\ x'\to\infty}} \frac{\cos\theta\cos\theta' + \sin\theta\sin\theta'\cos(\phi-\phi')}{xx'}$$
$$= \lim_{\substack{x\to\infty\\ x'\to\infty}} \frac{\cos\theta}{x}\frac{\cos\theta'}{x'} + \frac{\sin\theta}{x}\frac{\sin\theta'}{x'}\cos(\phi-\phi') = 1 - \cos(\phi-\phi').$$

But from Rodrigues' formula we find

$$P_n(x) = \frac{1}{2^n n!} D^n(x^2 - 1)^n = \frac{(2n)!}{2^n n!^2} x^n + \cdots$$

so that

$$\lim_{\substack{x \to \infty \\ x' \to \infty}} \frac{P_n(\cos\theta\cos\theta' + \sin\theta\sin\theta'\cos(\phi - \phi'))}{(xx')^n}$$

$$= \frac{(2n)!}{2^n n!^2}[1 - \cos(\phi - \phi')]^n = \frac{(2n)!}{n!^2} \sin^{2n}\frac{\phi - \phi'}{2}$$

$$= \frac{(2n)!(-1)^n}{2^{2n} n!^2}[e^{i(\phi-\phi')/2} - e^{-i(\phi-\phi')/2}]^{2n}$$

$$= \frac{(2n)!}{2^{2n} n!^2} \sum_{m=0}^{n} \varepsilon_m (-1)^m \binom{2n}{n-m} \cos m(\phi - \phi').$$

Similarly we find that

$$\lim_{\substack{x \to \infty \\ x' \to \infty}} \frac{P_n^m(\cos\theta)P_n^m(\cos\theta')}{(xx')^n} = \frac{(-1)^m(2n)!^2}{2^{2n}(n!)^2(n-m)!^2}$$

If we perform this limiting process in (5) we find

$$\frac{(2n!)}{2^{2n} n!^2} \sum_{m=0}^{n} \varepsilon_m (-1)^m \binom{2n}{n-m} \cos m(\phi - \phi')$$

$$= \sum_{m=0}^{n} b_m (-1)^m \frac{(2n)!^2}{2^{2n} n!^2 (n-m)!^2} \cos m(\phi - \phi')$$

a comparison of coefficients shows that

$$b_m = \varepsilon_m \frac{(n-m)!}{(n+m)!},$$

the latter of course agreeing with (1).

In comparing the two proofs of the addition theorem (1) we note that the latter requires more manipulation. The first proof, however, is in some sense deeper and uses the basic relationship with the Laplace equation. The second proof has the advantage over the first in that it lends itself more readily to extensions for nonintegral indices n. By its very nature we require n to be an integer in the first proof.

6. Green's Functions

In Eq. 5.4.5 we observed that it was possible to solve the boundary value problem 4.1, 4.2 in terms of a certain integral over the surface of the sphere. In the integrand the series

$$\sum_{n=0}^{\infty}(2n+1)r^n P_n(\cos\alpha)$$

appeared. This series can be summed explicitly as follows. We refer back to the generating function for the Legendre polynomials

$$\frac{1}{\sqrt{1-2r\cos\alpha+r^2}}=\sum_{n=0}^{\infty}r^n P_n(\cos\alpha).$$

By applying the operator $2\,r(d/dr)+1$ to both sides we immediately find

$$\sum_{n=0}^{\infty}(2n+1)r^n P_n(\cos\alpha)=\frac{1-r^2}{[1-2r\cos\alpha+r^2]^{3/2}}$$

By means of this we can rewrite 4.5 as

$$V(r,\theta,\phi)=\frac{1}{4\pi}\int_{-\pi}^{\pi}\int_{0}^{\pi}f(\theta',\phi')\frac{1-r^2}{[1-2r\cos\alpha+r^2]^{3/2}}\sin\theta'\,d\theta'\,d\phi'. \quad (1)$$

(1) can be interpreted in a physical sense if one introduces the concept of a Green's function. We will define such a function as follows. It is a solution of the equation

$$\Delta G=\delta(P,P_0) \quad (2)$$

$$G=0 \qquad \text{on} \qquad r=1$$

where $\delta(P, P_0)$ is a delta function. Here P and P_0 denote two points in space $P=(r,\theta,\phi)$, $P_0=(r_0,\theta_0,\phi_0)$. P_0 will be taken as a fixed point. The function $\delta(P, P_0)$ is so constituted that

$$\delta(P,P_0)=0 \qquad P\neq P_0$$

but

$$\int_D \delta(P,P_0)\,dv=1$$

where the above is a volume integral over any simple connected domain D having P_0 as an interior point. Physically G can be interpreted as the potential inside the sphere due to a point charge at P_0 with a perfectly conducting surface constituting the boundary of the sphere.

Using Green's theorem we have immediately

$$\int_D [V\,\Delta G - G\,\Delta V]\,dv = \int_{\partial D}\left[V\frac{\partial G}{\partial n} - G\frac{\partial V}{\partial n}\right] ds, \tag{3}$$

where ∂D denotes the boundary of D, $(\partial G/\partial n)$ the derivative of G in the exterior normal direction to ∂D and the integral on the right is a surface integral. In particular if D is a sphere then

$$\frac{\partial G}{\partial n} = \frac{\partial G}{\partial r}$$

and if $\Delta V = 0$ and G satisfies (2) we have, using (3),

$$\int_D V\,\delta(P, P_0)\,dv = \int_{-\pi}^{\pi}\int_0^{\pi} f(\theta', \phi')\frac{\partial G}{\partial r}\bigg|_{r=1} \sin\theta'\,d\theta'\,d\phi'.$$

Since V is a continuous function of P and $\delta(P, P_0) = 0$ for $P \neq P_0$ the integral on the left can be reduced to an integral over $|P - P_0| \leq \varepsilon$ for all $\varepsilon > 0$. It follows, from the definition of the function $\delta(P, P_0)$, that

$$\begin{aligned}\int_D V\,\delta(P, P_0)\,dv &= \int_{|P-P_0|\leq\varepsilon} V\,\delta(P, P_0)\,dv = V(P_0)\int_{|P-P_0|\leq\varepsilon}\delta(P, P_0)\,dv \\ &= V(P_0),\end{aligned}$$

so that

$$V(P_0) = \int_{-\pi}^{\pi}\int_0^{\pi} f(\theta', \phi')\frac{\partial G}{\partial r}\bigg|_{r=1} \sin\theta'\,d\theta'\,d\phi'. \tag{4}$$

We should like to show that (4) reduces to (1), but in order to do so we must find G explicitly. First we consider the potential of a point source

$$h(r, \theta, \phi) = \frac{-1}{4\pi R} = \frac{-1}{4\pi\,|P - P_0|}$$

$$R = \sqrt{r_0^2 - 2rr_0[\cos\theta\cos\theta_0 + \sin\theta\sin\theta_0\cos(\phi - \phi_0)] + r^2}.$$

Clearly

$$\Delta h = 0 \qquad P \neq P_0.$$

But

$$\begin{aligned}\int_D \Delta h\,dv &= \int_{|P-P_0|\leq\varepsilon}\Delta h\,dv = \int_{|P-P_0|\leq\varepsilon}\nabla\cdot\nabla h\,dv \\ &= \int_{|P-P_0|=\varepsilon}\frac{\partial h}{\partial n}\,ds = -\frac{1}{4\pi}\int_{|P-P_0|=\varepsilon}\frac{\partial}{\partial R}\frac{1}{R}\bigg|_{R=\varepsilon} ds \\ &= \frac{1}{4\pi}\int_{-\pi}^{\pi}\int_0^{\pi}\frac{1}{\varepsilon^2}\cdot\varepsilon^2\sin\theta\,d\theta\,d\phi \\ &= 1\end{aligned}$$

and it follows that $\Delta(-1/4\pi R) = \delta(P, P_0)$. From this we conclude that

$$G = -\frac{1}{4\pi R} + g \tag{5}$$

where g is a regular function which satisfies

$$\begin{aligned} \Delta g &= 0 \\ g &= \frac{1}{4\pi R} \qquad \text{on } r = 1 \end{aligned} \tag{6}$$

One can, in this relatively simple problem, find g by a technique known as the method of images. The point $P_0'(1/ro, \theta_0, \phi_0)$ is said to be the inverse point to $P_0'(r_0, \theta_0, \phi_0)$ with respect to the circle $r = 1$.

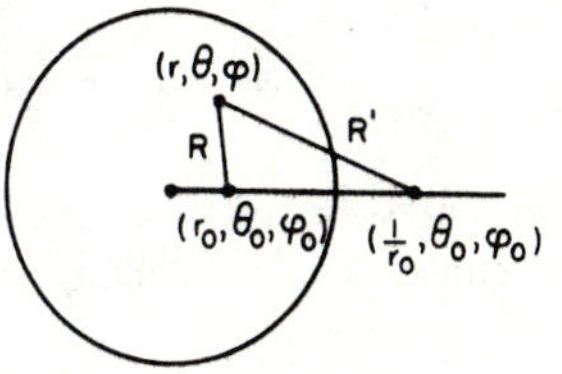

Clearly the function

$$g = \frac{1}{4\pi r_0 R'}$$

where

$$R' = \sqrt{\frac{1}{{r_0}^2} - 2\frac{r}{r_0}[\cos\theta\cos\theta_0 + \sin\theta\sin\theta_0\cos(\phi - \phi_0)] + r^2}$$

satisfies $\Delta g = 0$ inside $r \leq 1$, and a simple computation shows that all conditions in (6) hold. It follows that

$$G = -\frac{1}{4\pi\sqrt{{r_0}^2 - 2rr_0\cos\alpha + r^2}} + \frac{1}{4\pi\sqrt{1 - 2rr_0\cos\alpha + {r_0}^2r^2}} \tag{7}$$

is the appropriate Green's function. It can be interpreted as the potential resulting from placing one point charge at P_0 and a second one of suitable strength at P_0'. Their effects cancel on the surface $r = 1$. Finally a computation shows that

$$\left.\frac{\partial G}{\partial r}\right|_{r=1} = \frac{1 - {r_0}^2}{4\pi[1 - 2r_0\cos\alpha + {r_0}^2]^{3/2}}$$

which, when inserted in (4), yields (1) again.

7. The Complete Solution of Legendre's Differential Equation

In our discussion of the differential equation

$$L_t y = (1 - t^2)y'' - 2ty' + n(n + 1)y = 0 \tag{1}$$

up to now, we have concerned ourselves only with the solution

$$P_n(t) = F\left(-n, n + 1; 1; \frac{1 - t}{2}\right)$$

which is a polynomial for integral values of n. To find a second, linearly independent solution of (1) we shall seek an integral representation of the form

$$\mathcal{Q}_n(t) = \int_a^b \frac{V(\tau)}{t - \tau}\, d\tau \tag{2}$$

Here we again employ the method discussed in 4.4.

The operator L_t defined in (1) is formally self-adjoint so that

$$V(\tau)L_t \frac{1}{t - \tau} - \frac{1}{t - \tau} L_\tau V(\tau) = \frac{d}{d\tau}\left[\frac{1 - \tau^2}{(t - \tau)^2} V(\tau) - \frac{1 - \tau^2}{t - \tau} V'(\tau)\right].$$

We now require that

$$L_t \mathcal{Q}_n(t) = \int_a^b V(\tau) L_t \frac{1}{t - \tau}\, d\tau$$

$$= \int_a^b \frac{1}{t - \tau} L_\tau V(\tau)\, d\tau + \frac{(1 - \tau^2)V}{t - \tau} - \frac{1 - \tau^2}{t - \tau} V' \Bigg|_a^b = 0.$$

The integrated terms in the above will vanish if $a = -1$, $b = 1$, provided $V(\tau)$ is not too singular at $\tau = \pm 1$. To determine $V(\tau)$ we let

$$L_\tau V(\tau) = 0$$

In particular we let $V(\tau) = \frac{1}{2}P_n(\tau)$, in order to obtain a standard second solution. Then

$$\mathcal{Q}_n(t) = \frac{1}{2}\int_{-1}^{1} \frac{P_n(\tau)}{t - \tau}\, d\tau, \tag{3}$$

* From now on $Q_n(t)$ or $Q_n^m(t)$ will denote the value of the function for $-1 < t < 1$. For general complex values we reserve the symbols $\mathcal{Q}_n(t)$, $\mathcal{Q}_n^m(t)$.

provided t is not in the interval $[-1, 1]$. The contribution from the endpoints will vanish since $V(\tau)$ is a polynomial.

To study the analytic nature of $\mathcal{Q}_n(t)$ further we proceed as follows.

$$\mathcal{Q}_n(t) = \frac{1}{2}\int_{-1}^{1} \frac{P_n(t)}{t-\tau}\,d\tau - \frac{1}{2}\int_{-1}^{1} \frac{P_n(t) - P_n(\tau)}{t-\tau}\,d\tau.$$

The first term on the right can be integrated explicitly. The integrand in the second is clearly a polynomial in both t and τ. Then

$$\mathcal{Q}_n(t) = \frac{1}{2}\log\frac{t+1}{t-1}P_n(t) - W_{n-1}(t), \tag{4}$$

where $W_{n-1}(t)$ is a polynomial of degree $n-1$. The branch of the logarithm is so chosen that

$$\lim_{|t|\to\infty}\int_{-1}^{1}\frac{d\tau}{t-\tau} = \lim_{|t|\to\infty}\log\frac{t+1}{t-1} = 0.$$

(4) shows us quite simply that $\mathcal{Q}_n(t)$ has branch points at $t = \pm 1$. (3) as well as (4) define $\mathcal{Q}_n(t)$ in the complex plane cut along $[-1, 1]$.

To determine $W_{n-1}(t)$ explicitly we expand it in a (necessarily finite) series of Legendre polynomials. Then

$$W_{n-1}(t) = \sum_{k=0}^{n-1} A_k P_k(t)$$

and from (4)

$$L_t\mathcal{Q}_n(t) = 2P_n'(t) - L_t W_{n-1}(t) = 0.$$

But

$$L_t W_{n-1}(t) = \sum_{k=0}^{n-1} A_k L_t P_k(t) = \sum_{k=0}^{n-1} A_k(n-k)(n+k+1)P_k(t) = 2P_n'(t)$$

and by the orthogonality property

$$\begin{aligned}\frac{(n-k)(n+k+1)A_k}{2k+1} &= \int_{-1}^{1} P_n'(t)P_k(t)\,dt \\ &= P_n(t)P_k(t)\Big|_{-1}^{1} - \int_{-1}^{1} P_n(t)P_k'(t)\,dt \\ &= 1 - (-1)^{n+k}, \qquad \text{for} \quad n > k.\end{aligned}$$

Finally we have

$$\mathcal{Q}_n(t) = \frac{1}{2}\log\frac{t+1}{t-1}P_n(t) - \sum_{k=0}^{n-1}\frac{(2k+1)[1-(-1)^{n+k}]}{(n-k)(n+k+1)}P_k(t). \tag{5}$$

For $|t| > 1$

$$\frac{1}{t-\tau} = \frac{1}{t}\sum_{k=0}^{\infty}\left(\frac{\tau}{t}\right)^k$$

and using (3)

$$\mathcal{Q}_n(t) = \frac{1}{2t}\sum_{k=0}^{\infty}\frac{1}{t^k}\int_{-1}^{1}P_n(\tau)\tau^k\,d\tau$$

$$= \frac{1}{2t^{n+1}}\int_{-1}^{1}P_n(\tau)\tau^n\,d\tau + 0\left(\frac{1}{t^{n+2}}\right) \tag{6}$$

since the terms for $k < n$ vanish. To evaluate the above integral we see that

$$P_n(t) = \frac{(2n)!}{2^n(n!)^2}t^n + \sum_{k=0}^{n-1}A_k P_k(t).$$

$$\int_{-1}^{1}P_n^2(t)\,dt = \frac{(2n)!}{2^n(n!)^2}\int_{-1}^{1}P_n(t)t^n\,dt = \frac{2}{2n+1}$$

and

$$\mathcal{Q}_n(t) = \frac{2^n(n!)^2}{(2n+1)!t^{n+1}} + \cdots \tag{7}$$

The coefficients in (7) could be deduced by performing all necessary integrations in (6). There is a simpler technique. In 3.16 we had seen that

$$P_n(t) = \frac{(2n)!}{2^n(n!)^2}t^n F\left(-\frac{n}{2}, \frac{-n+1}{2}; \frac{1}{2}-n; \frac{1}{t^2}\right).$$

A second solution can be obtained by letting $n \to -n-1$. Then for suitable constants c_1 and c_2

$$c_1\mathcal{Q}_n(t) + c_2 P_n(t) = \frac{1}{t^{n+1}}F\left(\frac{n+1}{2}, \frac{n+2}{2}; n+\frac{3}{2}; \frac{1}{t^2}\right).$$

By comparing both sides for $|t|$ large and using (7) we find that $c_2 = 0$. After determining c_1

$$\mathcal{Q}_n(t) = \frac{2^n(n!)^2}{(2n+1)!t^{n+1}}F\left(\frac{n+1}{2}, \frac{n+2}{2}; n+\frac{3}{2}; \frac{1}{t^2}\right) \tag{8}$$

A bonus that is obtained follows by comparing (6) and (8).

$$\int_{-1}^{1} P_n(t)t^k\,dt = \begin{cases} 0, & n+k \text{ odd} \\ \dfrac{2^{n+1}(n!)^2\Gamma((1+k)/2)\Gamma((2+k)/2)\Gamma(n+3/2)}{(2n+1)!\Gamma((k-n+2)/2)\Gamma^2((1+n)/2)\Gamma((n+k+3)/2)}, & n+k \text{ even.} \end{cases}$$

Using (1) it follows that the Wronskian of any two solutions y_1 and y_2 is given by

$$W(y_1, y_2) = y_1 y_2' - y_1' y_2 = c\exp\left(\int \frac{2t}{1-t^2}\,dt\right) = \frac{c}{1-t^2}.$$

For $|t|$ large

$$P_n(t)\mathcal{Q}_n'(t) \approx \frac{-n-1}{2n+1}\cdot\frac{1}{t^2} + \cdots$$

$$P_n'(t)\mathcal{Q}_n(t) \approx \frac{n}{2n+1}\cdot\frac{1}{t^2} + \cdots,$$

so that $c = 1$ and

$$W(P_n(t), \mathcal{Q}_n(t)) = \frac{1}{1-t^2}. \tag{9}$$

We see that $\mathcal{Q}_n(t)$ has been defined in the t-plane except on the cut $[-1, 1]$. We now define

$$\mathcal{Q}_n(t \pm i0) = \lim_{\substack{\sigma\to 0\\ \sigma>0}} \mathcal{Q}_n(t \pm i\sigma) \qquad t \in [-1, 1]$$

and we are particularly interested in determining

$$\mathcal{Q}_n(t+i0) - \mathcal{Q}_n(t-i0).$$

To do so we consider the integral

$$I = \lim_{\substack{\sigma\to 0\\ \sigma>0}} \int_{-1}^{1} \frac{d\tau}{t + i\sigma - \tau}$$

and deform the path of integration as follows.

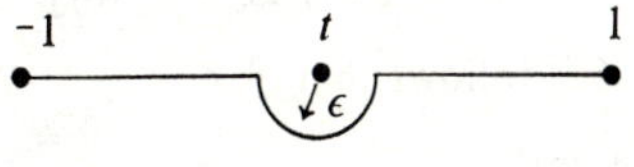

Then

$$I = \lim_{\varepsilon\to 0}\left\{\int_{-1}^{t-\varepsilon} + \int_{C_\varepsilon} + \int_{t+\varepsilon}^{1}\right\} = \int_{-1}^{1} \frac{d\tau}{t + i0 - \tau}$$

where c_ε denotes the semicircle shown in the figure.

$$\int_{-1}^{t-\varepsilon} \frac{d\tau}{t-\tau} + \int_{t+\varepsilon}^{1} \frac{d\tau}{t-\tau} = -\log(t-\tau)\Big|_{-1}^{t-\varepsilon} - \log(t-\tau)\Big|_{t+\varepsilon}^{1} = \log\frac{1+t}{1-t}$$

$$\int_{c_\varepsilon} \frac{d\tau}{t-\tau} = \int_{-\pi}^{0} \frac{i\varepsilon e^{i\theta}\,d\theta}{-i\varepsilon e^{i\theta}} = -i\pi$$

so that

$$\int_{-1}^{1} \frac{dt}{t+i0-\tau} = \log\frac{1+t}{1-t} - i\pi.$$

Similarly we find that

$$\int_{-1}^{1} \frac{dt}{t-i0-\tau} = \log\frac{1+t}{1-t} + i\pi.$$

By using these results it follows that

$$\begin{aligned} \mathcal{Q}_n(t+i0) &= \tfrac{1}{2}P_n(t)\log\frac{1+t}{1-t} - W_{n-1}(t) - \frac{\pi i}{2}P_n(t) \\ \mathcal{Q}_n(t-i0) &= \tfrac{1}{2}P_n(t)\log\frac{1+t}{1-t} - W_{n-1}(t) + \frac{\pi i}{2}P_n(t). \end{aligned} \tag{10}$$

The jump in $\mathcal{Q}_n(t)$ in crossing the cut is given by

$$\mathcal{Q}_n(t+i0) - \mathcal{Q}_n(t-i0) = -\pi i\, P_n(t).$$

For real t in $[-1, 1]$ we *define*

$$\begin{aligned} Q_n(t) &= \tfrac{1}{2}[\mathcal{Q}_n(t+i0) + \mathcal{Q}_n(t-i0)] \\ &= \tfrac{1}{2}P_n(t)\log\frac{1+t}{1-t} - W_{n-1}(t). \end{aligned} \tag{11}$$

Up to now we have concerned ourselves only with the Legendre functions, but by a very similar process we can construct all solutions for the associated Legendre equation as well. We now define

$$\mathcal{Q}_n^m(t) = (t^2-1)^{m/2} D^m \mathcal{Q}_n(t) \qquad t \notin [-1, 1] \tag{12}$$

and

$$\mathcal{Q}_n^m(t) = \tfrac{1}{2}[i^m \mathcal{Q}_n^m(t+i0) + (-i)^m \mathcal{Q}_n^m(t-i0)], \qquad t \in [-1, 1] \tag{13}$$

Using (8), (12), and 3.16 it follows that

$$\begin{aligned} \mathcal{Q}_n^m(t) = (-1)^m \frac{2^n n!(n+m)!}{(2n+1)!}(t^2-1)^{m/2} \\ \times t^{-(n+m+1)} F\left(\frac{n+m+1}{2}, \frac{n+m+2}{2}; n+\frac{3}{2}; \frac{1}{t^2}\right) \end{aligned} \tag{14}$$

In all the preceding we have still assumed that n and m are integers. But we can easily obtain representations that are valid for arbitrary values of n and m; in these cases we shall use the symbols ν and μ. Then by definition we say

$$\tilde{\mathcal{Q}}_\nu^\mu(t) = \frac{\sqrt{\pi}}{2^{\nu+1}\Gamma(\nu+3/2)}(t^2-1)^{\mu/2} \times t^{-(\nu+\mu+1)}F\left(\frac{\nu+\mu+1}{2}, \frac{\nu+\mu+2}{2}; \nu+\frac{3}{2}; \frac{1}{t^2}\right) \tag{15}$$

and we can easily verify that

$$\mathcal{Q}_n^m(t) = e^{im\pi}\Gamma(n+m+1)\tilde{\mathcal{Q}}_n^m(t). \tag{16}$$

Similarly we let

$$\mathcal{P}_\nu^\mu(t) = \frac{1}{\Gamma(1-\mu)}\left(\frac{t-1}{t+1}\right)^{-\mu/2} F\left(-\nu, \nu+1; 1-\mu; \frac{1-t}{2}\right). \tag{17}$$

When μ reduces to a positive integer (17) remains valid if we use a suitable limiting procedure. The function $P_\nu^\mu(t)$ is defined in the plane with a cut at $[-1, 1]$. One can show that

$$P_\nu^\mu(t) = e^{i\mu\pi/2}\mathcal{P}_\nu^\mu(t+i0) = e^{-i\mu\pi/2}\mathcal{P}_\nu^\mu(t-i0). \tag{18}$$

8. Asymptotic Formulas

Before plunging into a rigorous derivation of the asymptotic formulas for Legendre functions we shall describe a simpler, heuristic procedure. In the differential equation

$$P_n'' + \cot\theta P_n' + n(n+1)P_n = 0$$

we let

$$u = P_n(\cos\theta)\sqrt{\sin\theta}, \qquad 0 < \theta < \pi.$$

Then u satisfies

$$u'' + \left[\left(n+\frac{1}{2}\right)^2 + \frac{1}{4\sin^2\theta}\right]u = 0.$$

For $0 < \varepsilon < \theta \le \pi - \varepsilon < \pi$ and n sufficiently large we expect the second term to be negligible relative to the first term in the coefficient of u. In that case we expect that

$$u(\theta) \approx A_n \cos[(n+\tfrac{1}{2})\theta + \phi_n]$$

where A_n and ϕ_n are constants of integration. To determine A_n we note that

$$\frac{2}{2n+1} = \int_{-1}^{1} P_n^2(t)\,dt = \int_0^\pi P_n^2(\cos\theta)\sin\theta\,d\theta$$

$$= \int_0^\pi u^2(\theta)\,d\theta \approx A_n^2 \int_0^\pi \cos^2[(n+\tfrac{1}{2})\theta + \phi_n]\,d\theta = \frac{A_n^2\pi}{2}$$

so that $A_n \approx \sqrt{2/\pi n}$. The determination of ϕ_n is accomplished by recalling that $P_n(t)$ is an even or odd polynomial in t depending on the parity of n. In other words.

$$P_n(-t) = (-1)^n P_n(t).$$

or

$$P_n(\cos(\pi - \theta)) = (-1)^n P_n(\cos\theta)$$

Then

$$\cos[(n+\tfrac{1}{2})(\pi - \theta) + \phi_n] = (-1)^n \cos[(n+\tfrac{1}{2})\theta + \phi_n]$$

so that

$$(-1)^n \cos\left[\left(n+\frac{1}{2}\right)\theta - \phi_n - \frac{\pi}{2}\right] = (-1)^n \cos[(n+\tfrac{1}{2})\theta + \phi_n].$$

In order for the above to hold

$$-\phi_n - \frac{\pi}{2} = \phi_n$$

and

$$\phi_n = -\frac{\pi}{4}.$$

We conclude that for large n

$$P_n(\cos\theta) \approx \sqrt{\frac{2}{\pi n \sin\theta}} \cos\left[\left(n+\frac{1}{2}\right)\theta - \frac{\pi}{4}\right].$$

A rigorous development that leads to an asymptotic series can be accomplished by finding a suitable integral representation to which Watson's Lemma can be applied. We return to 3.8, in the slightly modified form

$$P_n^m(\cos\theta) = \frac{i^m(n+m)!}{2\pi n!} \int_{-\pi}^{\pi} [\cos\theta + i\sin\theta\cos\psi]^n e^{m\psi}\,d\psi, \qquad (1)$$

and introduce the new variable

$$w = e^{i\psi}.$$

(1) can be rewritten in the form

$$P_n^m(\cos\theta) = \frac{e^{im\pi/2}(n+m)!}{2\pi i n!} \int_{|w|=1} \frac{e^{n\log[\cos\theta + \frac{1}{2}i\sin\theta(w+1/w)]}}{w^{m+1}}\,dw \quad (2)$$

We now study the function

$$f(w) = \log\left[\cos\theta + \frac{1}{2}i\sin\theta\left(w + \frac{1}{w}\right)\right]$$

and find that

$$f'(\pm 1) = 0$$

$$f''(\pm 1) = \pm ie^{\mp i\theta}\sin\theta.$$

We concentrate our attention on $w = 1$. The case $w = -1$ can be studied in an analogous fashion. Near $w = 1$ we expand $f(w)$ in a Taylor series

$$f(w) = i\theta + \frac{i}{2}e^{-i\theta}\sin\theta(w-1)^2 + \cdots$$

and let

$$-\tau(w) = \frac{-2ie^{i\theta}[f(w) - i\theta]}{\sin\theta} = (w-1)^2 + \cdots$$

Then w as a function of τ has a branchpoint near $w = 1$. Solving for w we find

$$w_{1,2} = 1 \pm i\tau^{1/2} + \cdots.$$

We now consider the integral

$$I = \int_1^a \frac{e^{nf(w)}}{w^{m+1}}\,dw \quad (3)$$

where a is an arbitrary point in the first quadrant of the w-plane. By changing the integration to an integral in the τ-plane we obtain

$$I = \int_0^{\tau(a)} \frac{e^{nf(w_1(\tau))}}{w_1^{m+1}(\tau)} \frac{dw_1(\tau)}{d\tau}\,d\tau$$

$$= e^{in\theta}\int_0^{\tau(a)} e^{-\nu\tau}\sum_{k=0}^{\infty} a_k\tau^{k/2-1}\,d\tau$$

where

$$v = \frac{ine^{-i\theta}\sin\theta}{2}, \qquad \sum_{k=0}^{\infty} a_k \tau^{k/2-1} = w_1{}^{-m-1}\frac{dw_1}{d\tau}, \qquad a_0 = \frac{i}{2}.$$

For $0 < \varepsilon \leq \theta \leq \pi - \varepsilon < \pi$ we have arg $v > 0$ and I is a convergent integral. By Watson's Lemma we find

$$I \sim e^{in\theta} \sum_0^{\infty} a_k \int e^{-v\tau}\tau^{k/2-1}\, d\tau \sim \frac{i\sqrt{\pi}}{2\sqrt{v}} e^{in\theta} + \cdots$$

$$= \frac{e^{i[(n+1/2)\theta+\pi/4]}}{2}\sqrt{\frac{2\pi}{n\sin\theta}} + \cdots. \tag{4}$$

Integral (2) can be decomposed into four integrals of type (3); two near $w = -1$ and two near $w = 1$. Near $w = 1$ there will be two corresponding to the branches $w_{1,2}$. Combining these in (2) and letting

$$\frac{(n+m)!}{n!} \sim n^m \qquad (m \text{ fixed, } n \text{ large})$$

we finally obtain

$$P_n{}^m(\cos\theta) \sim n^m \sqrt{\frac{2}{\pi n \sin\theta}} \cos\left[\left(n + \frac{1}{2}\right)\theta - \frac{\pi}{4} + m\frac{\pi}{2}\right]. \tag{5}$$

For $m = 0$ the latter clearly reduces to the result found earlier.

Exercises

1. Verify by direct substitution that

$$P_n{}^m(t) = (-1)^m(1-t^2)^{m/2}D^m P_n(t)$$

satisfies

$$(1-t^2)y'' - 2ty' + \left[n(n+1) - \frac{m^2}{1-t^2}\right]y = 0.$$

2. Prove that if

$$F(tz, tr) = t^k F(z, r)$$

where t is a parameter independent of z and r, and if $F(z, r)$ is sufficiently differentiable that

$$F(z, r) = z^k f(z/r).$$

3. Verify directly that

$$V=\int_{-\pi}^{\pi} f(z+ix\cos u+iy\sin u, u)\,du$$

satisfies $\Delta V=0$, if f is sufficiently smooth.

4. Prove the validity of 3.15 and 3.17.

5. Solve the boundary value problem

$$\Delta V=0$$

$$\frac{\partial V}{\partial r}=f(\theta,\phi) \text{ on } r=1$$

where

$$\int_{-\pi}^{\pi}\int_{0}^{\pi} |f|^2(\theta,\phi)\sin\theta\,d\theta\,d\phi<\infty$$

and

$$\int_{-\pi}^{\pi}\int_{0}^{\pi} f(\theta,\phi)\sin\theta\,d\theta\,d\phi=0.$$

Why is the last condition necessary?

6. Show that

$$(1-t)^n P_n\left(\frac{1+t}{1-t}\right)=\sum_{k=0}^{n}\binom{n}{k}^2 t^k.$$

7. Show that

$$P_n(\cos\theta)=\frac{(2n)}{2^{2n}\,n!^2}\left[\cos n\theta-\frac{1}{1}\cdot\frac{n}{2n-1}\cos(n-2)\,\theta\right.$$
$$+\frac{1\times 3}{1\times 2}\;\frac{n(n-1)}{(2n-1)(2n-3)}\cos(n-4)\,\theta$$
$$\left.+\frac{1\times 3\times 5}{1\times 2\times 3}\;\frac{n(n-1)(n-2)}{(2n-1)(2n-3)(2n-5)}\cos(n-6)\,\theta+\cdots\right].$$

8. Use 7 to show that

$$|P_n(t)|\leqslant 1 \text{ for } |t|\leqslant 1.$$

9. Show that if

$$f(t)=\sum_{1}^{\infty} A_nP_n(t)$$

then

$$\int_1^t f(t)dt=-A_0-\tfrac{1}{3}A_1+\sum_{0}^{\infty}\left(\frac{A_{n-1}}{2n-1}-\frac{A_{n+1}}{2n+3}\right)P_n(t).$$

10. Show the function defined by

$$Q_n(t) = \frac{1}{2n+1} \int_{-1}^{1} \frac{(1-z^2)^n}{(t-z)}\, dz$$

where t does not lie in $[-1, 1]$, satisfies the Legendre equations.

11. Prove that

$$(2n+1)t\, P_n^m(t) - (n-m+1)P_{n+1}^m(t) - (n+m)P_{n-1}^m(t) = 0$$

$$P_n^{m+2}(t) + 2(m+1)\frac{t}{\sqrt{1-t^2}} P_n^{m+1}(t) + (n-m)(n+m+1)P_n^m(t) = 0$$

$$P_{n-1}^m(t) - P_{n+1}^m(t) = (2n+1)\sqrt{1-t^2}\, P_n^{m-1}(t).$$

12. Prove that

$$P_\nu(tt' - \sqrt{t^2-1}\, \sqrt{t'^2-1} \cos \phi)$$

$$= P_\nu(t)P_\nu(t') + 2\sum_{m=1}^{\infty} (-1)^m \frac{\Gamma(\nu-m+1)}{\mathrm{L}(\nu+m+1)} p_\nu^m(t)\, P_\nu^m(t') \cos m\phi$$

if $|\arg(t-1)| < \pi$, $|\arg(t'-1)| < \pi$, $\mathrm{Re}\, t > 0$, $\mathrm{Re}\, t' > 0$.

13. Show that

$$\int_{-1}^{1} \frac{P_{2n}(t)}{\sqrt{\cosh^2\alpha - t^2}}\, dt = 2iP_{2n}(0)\, Q_{2n}(i \sinh \alpha).$$

14. Solve the following boundary value problem. Find V such that

$$\Delta V = 0$$

$$V = f(\theta, \phi) \text{ on } = 1$$

where

$$f(\theta, \phi) = 1, \qquad 0 \le \theta < \alpha$$

$$= 0, \qquad \alpha < \theta \le \pi.$$

15. Show that

$$u_n = \frac{(2n)!}{2^{2n} n!^2}$$

is an increasing sequence and that

$$\lim_{n\to\infty} \sqrt{n}\, u_n = \frac{1}{\sqrt{\pi}}.$$

16. We know that $u(\theta) = \sqrt{\sin\theta}\, P_n(\cos\theta)$ satisfies

$$u'' + \left[\left(n + \frac{1}{2}\right)^2 + \frac{1}{4\sin^2\theta}\right]u = 0.$$

Use this fact and the results of 15 to prove that

$$\sqrt{\sin\theta}\,|P_n(\cos\theta)| < \sqrt{\frac{2}{\pi n}},\ 0 < \theta \leq \pi.$$

17. Show that

$$Q_n(t) = P_n(t) \int_t^\infty \frac{d\tau}{(\tau^2 - 1)P_n{}^2(\tau)}.$$

CHAPTER 6

Spherical Harmonics in p Dimensions

1. Homogeneous Polynomials

We now consider the Laplace operator in p dimensions

$$\Delta_p V = \sum_{i=1}^{p} \frac{\partial^2 V}{\partial x_i^2} \tag{1}$$

where $x_1, x_2, \ldots, x_p$ denote Cartesian coordinates in our p dimensional space. We had seen earlier that, for example, if $p = 2$ then

$$\Delta_2(x + iy)^n = 0.$$

and if $p = 3$ then

$$\Delta_3(x + iy \cos \theta + iz \sin \theta)^n = 0.$$

In each of the above examples we have solutions that are polynomials in the variables in question.

In particular we shall be interested in studying solutions of (1) that are homogeneous polynomials.

Definition. A polynomial $H_n(x_1, x_2, \ldots, x_p)$ will be said to be homogeneous of degree n in the p variables $x_1, x_2, \ldots, x_p$ if

$$H_n(tx_1, tx_2, \ldots, tx_p) = t^n H_n(x_1, x_2, \ldots, x_p).$$

For fixed values of n and p it is clear that there can exist only a finite number of linearly independent homogeneous polynomials.

THEOREM. Let $K(p, n)$ denote the number of linearly independent, homogeneous polynomials of degree n in p variables. Then

$$K(p, n) = \frac{(p + n - 1)!}{n!(p - 1)!}. \tag{2}$$

Proof. We can expand $H_n(x_1, x_2, \ldots, x_p)$ as a polynomial in one of the variables, say x_p, so that

$$H_n(x_1, x_2, \ldots, x_p) = \sum_{j=0}^{n} x_p{}^j A_{n-j}(x_1, x_2, \ldots, x_{p-1}) \tag{3}$$

where $A_{n-j}(x_1, x_2, \ldots, x_{p-1})$ denotes a homogeneous polynomial of degree $n-j$ in $p-1$ variables. For each of these there are $K(p-1, n-j)$ linearly independent choices so that

$$K(p, n) = \sum_{j=0}^{n} K(p-1, n-j) = \sum_{j=0}^{n} K(p-1, j). \tag{4}$$

(4) represents difference equation that the $K(p, n)$ must satisfy. We shall solve it by the method of generating functions. Let

$$M(p) = \sum_{n=0}^{\infty} r^n K(p, n). \tag{5}$$

If we substitute from (4) in (5) and interchange the n and j summations we obtain

$$\begin{aligned} M(p) &= \sum_{j=0}^{\infty} K(p-1, j) \sum_{n=j}^{\infty} r^n \\ &= \frac{1}{1-r} \sum_{j=0}^{\infty} r^j K(p-1, j) = \frac{M(p-1)}{1-r} \end{aligned} \tag{6}$$

From (6) we deduce readily that

$$M(p) = \frac{M(1)}{(1-r)^{p-1}}$$

But for $p = 1$ clearly $K(1, n) = 1$, since $H_n(x_1) = cx_1{}^n$, so that $M(1) = 1/(1-r)$ and

$$M(p) = (1-r)^{-p} = \sum_{n=0}^{\infty} \frac{(p+n-1)!}{n!(p-1)!} r^n. \tag{7}$$

Comparing (5) and (7) we can read off the value of $K(p, n)$. ■

Some, but not all of these polynomials will satisfy (1).

Definition. Any polynomial $R_n(x_1, \ldots x_p)$ that satisfies

$$\Delta_p R_n = 0$$

will be said to be a harmonic polynomial.

THEOREM. Let $N(p, n)$ denote the number of linearly independent, homogeneous harmonic polynomials of degree n in p variables. Then

$$N(p, n) = \frac{2n + p - 2}{n} \binom{n + p - 3}{n - 1}. \tag{8}$$

Proof. We shall decompose the operator Δ_p into two operators.

$$\Delta_p = \frac{\partial^2}{\partial x_p{}^2} + \Delta_{p-1}.$$

Here Δ_{p-1} denotes the $p - 1$ dimensional Laplace operator acting on functions of $x_1, x_2, \ldots, x_{p-1}$. Then let, using (3),

$$\begin{aligned} \Delta_p H_n &= \sum_{j=2}^{n} j(j-1)x_p{}^{j-2}A_{n-j} + \sum_{j=0}^{n} x_p{}^j \Delta_{p-1} A_{n-j} \\ &= \sum_{j=0}^{n} x_p{}^j[(j+1)(j+2)A_{n-j-2} + \Delta_{p-1}A_{n-j}] = 0. \end{aligned} \tag{9}$$

In the above $A_{-1} = A_{-2} = 0$. In order for (9) to hold we require that

$$\begin{aligned} \Delta_{p-1}A_n + 2A_{n-2} &= 0 \\ \Delta_{p-1}A_{n-1} + 6A_{n-3} &= 0 \\ \Delta_{p-1}A_{n-2} + 12A_{n-4} &= 0 \\ &\;\;\vdots \\ \Delta_{p-1}A_2 + n(n-1)A_0 &= 0 \\ \Delta_{p-1}A_1 &= 0 \\ \Delta_{p-1}A_0 &= 0. \end{aligned}$$

From the above we see that once A_n and A_{n-1} are selected all remaining A_{n-j} are determined recursively. But we can select A_n and A_{n-1} in $K(p - 1, n) + K(p - 1, n - 1)$ ways so that

$$\begin{aligned} N(p, n) &= K(p-1, n) + K(p-1, n-1) \\ &= \frac{2n + p - 2}{n} \binom{n + p - 3}{n - 1}. \end{aligned}$$ ■

For example for $p = 2$ we find that

$$N(2, n) = 2.$$

Two such polynomials are given by

$$(x+iy)^n + (x-iy)^n = 2\,r^n \cos n\theta$$
$$(x+iy)^n - (x-iy)^n = 2\,ir^n \sin n\theta.$$

For $p = 3$ it follows that

$$N(3, n) = 2n + 1.$$

These $(2n + 1)$ functions are given by, using spherical coordinates,

$$r^n P_n^{\,m}(\cos\theta)e^{im\phi},\ m = -n,\ -n+1, \ldots, n-1,\ n.$$

In general we write

$$H_n(X) = H_n(r\xi) = r^n S_n(\xi).$$

X is the vector $(x_1, x_2, \ldots, x_p)$ and ξ the unit vector $(\xi_1, \xi_2, \ldots, \xi_p)$. We shall refer to $S_n(\xi)$ as a spherical harmonic.

2. Orthogonality of Spherical Harmonics

Before we go on to study specific properties of the spherical harmonics, we shall find the surface area of a p dimensional unit sphere. Let $f(r)$ be any function of $r = \sqrt{\sum_1^p x_i^2}$ for which the integral

$$I_p = \int_{-\infty}^{\infty}\int_{-\infty}^{\infty}\cdots\int_{-\infty}^{\infty} f(r)\, dx_1\, dx_2 \cdots dx_p$$

exists. To perform the above integration we shall make use of the spherical symmetry of the function $f(r)$. Let ω_p denote the surface area of a unit sphere. Then we can evaluate I_p by integrating over spherical shells, so that

$$I_p = \omega_p \int_0^{\infty} f(r) r^{p-1}\, dr$$

since the surface area of a sphere or radius r is $\omega_p r^{p-1}$. It follows that

$$\omega_p = \frac{\int_{-\infty}^{\infty}\int_{-\infty}^{\infty}\cdots\int_{-\infty}^{\infty} f(r)\, dx_1\, dx_2 \cdots dx_p}{\int_0^{\infty} f(r) r^{p-1}\, dr}$$

The right side we note must be independent of the choice of $f(r)$. For the particular choice

$$f(r) = e^{-r^2} = e^{-(x_1^2 + x_2^2 + \cdots + x_p^2)}$$

we obtain

$$\omega_p = \frac{[\int_{-\infty}^{\infty} e^{-x^2}\, dx]^p}{\int_0^{\infty} e^{-r^2} r^{p-1}\, dr} = \frac{[\sqrt{\pi}]^p}{\frac{1}{2}\Gamma(p/2)} = \frac{2\pi^{p/2}}{\Gamma(p/2)}. \tag{1}$$

In particular we have

$$\omega_1 = 2, \qquad \omega_2 = 2\pi, \qquad \omega_3 = 4\pi, \qquad \omega_4 = 2\pi^2.$$

We are now in a position to prove the following lemma.

LEMMA. Any two real spherical harmonics of different degree are orthogonal over the sphere. That is

$$\int_{|\xi|=1} S_n(\xi) S_k(\xi)\, d\omega_p = 0, \qquad n \neq k. \tag{2}$$

The integration is carried out over the surface of the sphere $|\xi| = 1$.

Proof. Let $H_n(X)$ denote the homogeneous polynomial of degree n corresponding to $S_n(\xi)$. Then

$$H_n(tX) = t^n H_n(X)$$

and differentiation of the above with respect to t yields

$$\frac{d}{dt} H_n(tX) = nt^{n-1} H_n(X).$$

But

$$\frac{d}{dt} H_n(tX) = \sum_{i=1}^{p} \frac{\partial H_n(tX)}{\partial tx_i} \frac{dtx_i}{dt}$$

In particular for $t = 1$ we find by comparing the two above calculations

$$\sum \frac{\partial H_n(X)}{\partial x_i} x_i = nH_n(X). \tag{3}$$

Now let X be such that $|X| = 1$ so that $X = \xi$, and (3) reduces to

$$\sum_{i=1}^{p} \frac{\partial S_n(\xi)}{\partial \xi_i} \xi_i = nS_n(\xi). \tag{4}$$

Let v denote the exterior normal to the unit sphere $\xi = 1$. It follows that

$$\frac{\partial S_n(\xi)}{\partial v} = \xi \nabla S_n(\xi) = \sum_{i=1}^{p} \frac{\partial S_n(\xi)}{\partial \xi_i} \xi_i = nS_n(\xi). \tag{5}$$

With these preliminaries we can complete the proof. Using Green's theorem in p dimensions we have

$$0 = \int_{|X| \leq 1} [H_n(X)\Delta_p H_k(X) - H_k(X)\Delta_p H_n(X)]\, dx_1\, dx_2 \cdots dx_p$$

$$= \int_{|\xi|=1} \left[S_n(\xi)\frac{\partial S_k(\xi)}{\partial \nu} - S_k(\xi)\frac{\partial S_n(\xi)}{\partial \nu}\right] d\omega_p$$

$$= (k-n)\int_{|\xi|=1} S_n(\xi)S_k(\xi)\, d\omega_p$$

and for $n \neq k$ the result follows. ■

Suppose $\{S_n(\xi)\}$ is a set of $N(p, n)$ linearly independent real spherical harmonics. By the Gram-Schmidt process we can construct an orthonormal set

$$S_{n,1}(\xi),\ S_{n,2}(\xi), \ldots\ S_{n,\,N(p,\,n)}(\xi).$$

Then

$$\int_{|\xi|} S_{n,j}(\xi)S_{n,k}(\xi)\, d\omega_p = \delta_{j,k}. \tag{6}$$

Let A now denote an orthogonal matrix that represents a rotation of the coordinate system. It follows that

$$\int_{|\xi|=1} S_{n,j}(A\xi)S_{n,k}(A\xi)\, d\omega_p = \delta_{j,k} \tag{7}$$

for all such matrices. From the linear independence of the set $\{S_{n,j}(\xi)\}$, and also the fact that there are only $N(p, n)$ linearly independent spherical harmonics of degree n it follows that

$$S_{n,j}(A\xi) = \sum_{l=1}^{N(p,\,n)} C_{l,j}\, S_{n,l}(\xi) \tag{8}$$

By inserting (8) in (7) and using (6) we find that

$$\sum_{l=1}^{N(p,n)} C_{l,j}\, C_{l,k} = \delta_{j,k} \tag{9}$$

(9) expresses the fact that the matrix $C = (C_{ij})$, defined in (8) is also orthogonal.

We now examine the function

$$F(\xi, \eta) = \sum_{j=1}^{N(p,n)} S_{n,j}(\xi)S_{n,j}(\eta). \tag{10}$$

As we shall see shortly this function has similar properties and plays a similar role as does $K_n(x, y)$, defined by the Christoffel-Darboux formula in the theory of orthogonal polynomials. Using (8) and (9) we have

$$F(A\xi, A\eta) = \sum_{j=1}^{N(p,n)} \left(\sum_{l=1}^{N(p,n)} C_{l,j} S_{n,l}(\xi) \right) \left(\sum_{m=1}^{N(p,n)} C_{m,j} S_{n,m}(\eta) \right)$$

$$= \sum_{j=1}^{N(p,n)} S_{n,j}(\xi) S_{n,j}\, \eta) = F(\xi, \eta). \tag{11}$$

(11) expresses the fact that $F(\xi, \eta)$ is invariant under rotations of the coordinate system. We shall denote the inner product between two vectors by

$$(\xi, \eta) = \sum_{i=1}^{p} \xi_i \eta_i$$

and clearly for a rotation A

$$(A\xi, A\eta) = (\xi, A^T A\eta) = (\xi, \eta)$$

where A^T is the transpose of A. For a rotation $A^T A = I$.

LEMMA. The function $F(\xi, \eta)$ defined in (10) is a function of (ξ, η) only. That is

$$F(\xi, \eta) = \phi((\xi, \eta)). \tag{12}$$

Proof. We shall select our coordinate system in a special way. This can be done without loss of generality by a suitable rotation.

$$\eta = (1, 0, 0, \ldots, 0), \qquad \xi = (t, \sqrt{1 - t^2}, 0, \ldots, 0)$$

so that $(\eta, \xi) = t$. Now $F(\xi, \eta)$ will be a polynomial in the two variables t and $\sqrt{1 - t^2}$, that is

$$F(\xi, \eta) = P(t, \sqrt{1 - t^2}).$$

But there is certainly a rotation A such that

$$A\eta = \eta, \qquad A\xi = (t, -\sqrt{1 - t^2}, 0, \ldots, 0)$$

so that

$$F(A\xi, A\eta) = F(\xi, \eta)$$

which means that

$$P(t, -\sqrt{1 - t^2}) = P(t, \sqrt{1 - t^2}).$$

Then P is a polynomial in t and $1 - t^2$, which is a polynomial in t so that

$$F(\xi, \eta) = \phi(t) = \phi((\eta, \xi)). \qquad \blacksquare$$

3. Legendre Polynomials

We shall now define a set of polynomials that plays a similar role as the ordinary Legendre polynomials do for the case $p = 3$.

THEOREM. Consider the homogeneous and harmonic polynomial $L_n(X)$ characterized by the following properties.

1. Without loss of generality we let $\eta = (1, 0, 0 \ldots, 0)$. Then $L_n(\eta) = 1$.
2. Let A' be any rotation leaving η fixed, that is $A'\eta = \eta$. Then $L_n(A'X) = L_n(X)$.

The corresponding spherical harmonic $L_n(\xi)$ is uniquely defined by these properties and is a polynomial in $t = (\xi, \eta)$.

Proof. Let $r = \sqrt{x_1{}^2 + x_2{}^2 + \ldots + x_p{}^2}$, and ξ' be a unit vector orthogonal to η that is $(\xi', \eta) = 0$. We can express ξ in the form

$$\xi = t\eta + \sqrt{1 - t^2}\, \xi'$$

for suitable t and ξ'. Note that $t = (\xi, \eta)$.

We now expand $L_n(X)$ in the form

$$L_n(X) = \sum_{j=0}^{n} x_p{}^j A_{n-j}(x_1, x_2, \ldots, x_{p-1}).$$

In view of the fact that the transformation A' leaves $L_n(X)$ invariant, it must leave all $A_{n-j}(x_1, \ldots, x_{p-1})$ invariant. In other words, the A_{n-j} are invariant under all such rotations and it follows that

$$A_{n-j}(x_1, \ldots, x_{p-1}) = C_{n-j}(x_1{}^2 + x_2{}^2 + \cdots + x_{p-1}^2)^{(n-j)/2}.$$

so that $C_{n-j} = 0$ for all odd $n - j$. Otherwise these would not be polynomials. Hence either $A_n = 0$ if n is odd or $A_{n-1} = 0$ if n is even.

We had seen earlier that the specifications of A_n and A_{n-1} uniquely determines $L_n(X)$. In this case the specification of a single constant, either C_n or C_{n-1} depending on the parity of n, determines $L_n(X)$. We have now

$$L_n(\xi) = \sum_{j=0}^{n}{}' t^j(1 - t^2)^{(n-j)/2} C_{n-j}$$

where $\sum'$ denotes that we sum only over such values of j for which $n - j$ is even. By Property 1 with $t = 1$

$$L_n(\eta) = C_0 = 1.$$

Thus the single remaining constant is specified, and $L_n(\xi)$ uniquely determined. ■

The polynomial defined in the preceding theorem will be defined as the Legendre polynomial of degree n in p dimensions. Note that by definition.

$$P_n(1) = 1 \tag{13}$$

and from the homogeneity it follows that

$$P_n(-1) = (-1)^n. \tag{14}$$

We shall again use the symbol $P_n(t)$ without specifying p explicity, unless some ambiguity arises.

We now return to the function $F(\xi, \eta)$ defined in (10). If A' is any rotation leaving η fixed we have

$$F(A'\xi, A'\eta) = F(A'\xi, \eta) = F(\xi, \eta).$$

Denote the value $F(\eta, \eta)$ by C_n.
It follows that

$$F(\xi, \eta) = C_n P_n((\xi, \eta)). \tag{15}$$

To evaluate C_n we let $\xi = \eta$ and integrate (15) over the surface of the unit sphere in p dimensions

$$\int_{|\eta|=1} F(\eta, \eta)\, d\omega_p = \int_{|\eta|=1} \sum_{j=1}^{N} S_{n,j}^2(\eta)\, d\omega_p = N(p, n) = C_n \omega_p$$

so that $C_n = \dfrac{N(p, n)}{\omega_p}$. We have now an explicit expression for the Legendre polynomials in terms of a set of orthogonal spherical harmonics.

$$P_n((\xi, \eta)) = \frac{\omega_p}{N(p, n)} \sum_{j=1}^{N(p,n)} S_{n,j}(\xi) S_{n,j}(\eta). \tag{16}$$

(16) will be designated as the addition theorem for Legendre polynomials in p dimensions. For $p = 2$ and $p = 3$ we obtain the familiar expressions studied earlier. Consider $p = 2$ for example. Then $\omega_2 = 2\pi$, $N(2, n) = 2$ and we select

$$\xi = (\cos\theta, \sin\theta)$$
$$\eta = (\cos\phi, \sin\phi)$$

so that

$$t = (\xi, \eta) = \cos(\theta - \phi).$$

As our set of two linearly independent spherical harmonics we choose

$$S_{n,1}(\xi) = \frac{1}{\sqrt{\pi}} \cos n\theta, \qquad S_{n,2}(\eta) = \frac{1}{\sqrt{\pi}} \sin n\phi.$$

Then (16) becomes

$$P_n(\cos(\theta - \phi)) = \cos n\theta \cos n\phi + \sin n\theta \sin n\phi = \cos n(\theta - \phi)$$

or equivalently

$$P_n(t) = \cos(n \cos^{-1} t)$$

which shows that the Tchebicheff polynomials are the appropriate polynomials.

For $p = 3$ we have $\omega_3 = 4\pi$ and $N(3, n) = 2n + 1$, and we let

$$\xi = (\sin\theta \cos\phi, \sin\theta \sin\phi, \cos\theta)$$
$$\eta = (\sin\theta' \cos\phi', \sin\theta' \sin\phi', \cos\theta')$$

and so we obtain

$$t = (\xi, \eta) = \cos\theta \cos\theta' + \sin\theta \sin\theta' \cos(\phi - \phi') = \cos\alpha$$

For our orthonormal set we pick

$$\begin{aligned} S_{n,j} &= \sqrt{\frac{(2n+1)(n-j)!}{2\pi(n+j)!}}\, P_n^j(\cos\theta)\cos j\phi, \qquad j = 1, 2, \ldots, n \\ &= \sqrt{\frac{(2n+1)(n+j)!}{2\pi(n-j)!}}\, P_{n-j}(\cos\theta)\sin j\phi, \qquad j = -1, -2, \ldots, -n. \\ &= \sqrt{\frac{2n+1}{4\pi}}\, P_n^j(\cos\theta), \qquad j = 0. \end{aligned}$$

It now follows that

$$P_n(\cos\alpha) = P_n(\cos\theta)P_n(\cos\theta') + 2\sum_{j=1}^{n} \frac{(n-j)!}{(n+j)!} P_n^j(\cos\theta)P_n^j(\cos\theta')\cos j(\phi - \phi'),$$

which is the classical addition theorem for Legendre polynomials.

We shall, in what follows, require the following lemma.

LEMMA. Let $\{S_{n,j}(\eta)\}$ be a set of $k \le N(p, n)$ linearly independent spherical harmonics. Then there exists a set of k unit vectors $\{\eta_i\}$ such that the following $k \times k$ determinant does not vanish.

$$|S_{n,j}(\eta_i)| \ne 0.$$

Proof. We can certainly find η_1 so that $S_{n,1}(\eta_1) \ne 0$. Then we consider the spherical harmonic defined by

$$\begin{vmatrix} S_{n,1}(\eta_1) & S_{n,1}(\xi) \\ S_{n,2}(\eta_1) & S_{n,2}(\xi) \end{vmatrix}.$$

Since $S_{n,1}$ and $S_{n,2}$ are linearly independent the above does not vanish identically and we select η_2 so that it does not vanish. In a similar fashion we can select $\eta_3, \eta_4, \ldots, \eta_k$, by going on to higher determinants. ■

By means of the above lemma we can prove the following theorem. The key point of the theorem is that every spherical harmonic can be represented in terms of the basic Legendre polynomials.

THEOREM. Every spherical harmonic $S_n(\xi)$ can be represented in the form

$$S_n(\xi) = \sum_{k=1}^{N(p,n)} A_k P_n((\xi, \eta_k)) \tag{17}$$

where $P_n(t)$ is the Legendre polynomial of degree n in p dimensions and the η_k are a set of suitable unit vectors.

Proof. According to the previous lemma we can select $\{\eta_i\}$ so that

$$|S_{n,j}(\eta_i)| \neq 0$$

Then the system of equations

$$P_n((\xi, \eta_k)) = \frac{\omega_p}{N(p, n)} \sum_{j=1}^{N(p, n)} S_{n,j}(\xi) S_{n,j}(\eta_k) \qquad k = 1, 2, \ldots, N(p, n)$$

is invertible and

$$S_{n,j}(\xi) = \sum_{k=1}^{N(p, n)} A_k P_n((\xi, \eta_k)).$$

In view of the fact that an arbitrary spherical harmonic can be expressed as

$$S_n(\xi) = \sum_{j=1}^{N(p, n)} b_j S_{n,j}(\xi)$$

where

$$b_j = \int_{|\xi|=1} S_n(\xi) S_{n,j}(\xi)\, dp$$

we are done. ■

It was mentioned earlier that the function $P_n((\xi, \eta))$ had properties similar to the function $K_n(x, y)$ studied via the Christoffel-Darboux formula. This fact is encompassed in the following lemma.

LEMMA. For every spherical harmonic $S_n(\xi)$ we have

$$S_n(\xi) = \frac{N(p, n)}{\omega_p} \int_{|\eta|=1} S_n(\eta) P_n((\xi, \eta))\, d\omega_p. \tag{18}$$

Proof. Let

$$S_n(\xi) = \sum_{j=1}^{N(p,n)} b_j S_{n,j}(\xi).$$

Then the right side of (18) can be written as

$$\sum_{j=1}^{N(p,n)} S_{n,j}(\xi) \sum_{i=1}^{N(p,n)} b_i \int_{|\eta|=1} S_{n,j}(\eta) S_{n,i}(\eta)\, d\omega_p = \sum_{j=1}^{N(p,n)} b_j S_{n,j}(\xi) = S_n(\xi). \quad \blacksquare$$

The preceding results and ideas can be used to derive a number of estimates relating to spherical harmonics. These are formulated below.

THEOREM. Spherical harmonics and Legendre polynomials satisfy the following

$$|S_n(\xi)| \leq \left[\frac{N(p, n)}{\omega_p} \int_{|\eta|=1} S_n^{\,2}(\eta)\, d\omega_p\right]^{1/2} \tag{19}$$

$$|P_n(t)| \leq 1 \qquad \text{for} \quad |t| \leq 1 \tag{20}$$

$$\int_{|\xi|=1} P_n^{\,2}(\xi, \eta)\, d\omega_p = \frac{\omega_p}{N(p, n)} \tag{21}$$

Proof. Both (20) and (21) follow from the basic addition theorem (16). By the Cauchy-Schwarz inequality we have immediately

$$P_n^{\,2}((\xi, \eta)) \leq \left[\frac{\omega_p}{N(p, n)} \sum_{j=1}^{N(p,n)} S_{n,j}^2(\xi)\right]\left[\frac{\omega_p}{N(p, n)} \sum_{j=1}^{N(p,n)} S_{n,j}^2(\eta)\right] = P_n^{\,2}(1) = 1,$$

thus proving (20). To prove (21) we square (16) and integrate over ξ. Then

$$\int_{|\xi|} P_n^{\,2}((\xi, \eta))\, d\omega_p = \frac{\omega_p^{\,2}}{N^2(p, n)} \int_{|\xi|=1} \sum_{j,k=1}^{N(p,n)} S_{n,j}(\xi) S_{n,j}(\eta) S_{n,k}(\xi) S_{n,k}(\eta)\, d\omega_p$$

$$= \frac{\omega_p^{\,2}}{N^2(p, n)} \sum_{j=1}^{N(p,n)} S_{n,j}^2(\eta) = \frac{\omega_p}{N(p, n)} P_n(1) = \frac{\omega_p}{N(p, n)}$$

verifying (21). Finally (19) follows by applying the Cauchy-Schwarz inequality to (18) and using (21).

$$|S_n(\xi)|^2 \le \frac{N^2(p,n)}{\omega_p^{\,2}} \int_{|\eta|=1} S_n^{\,2}\ \eta)\, d\omega_p \int_{|\eta|=1} P_n^{\,2}((\xi,\eta))\, d\omega_p$$

$$= \frac{N(p,n)}{\omega_p} \int_{|\eta|=1} S_n^{\,2}(\eta)\, d\omega_p .$$

By taking square roots of the above (19) is obtained. ■

The integral in (21) can be reduced to a one-dimensional integral by using the symmetry of the integrand. Consider the following figure. The

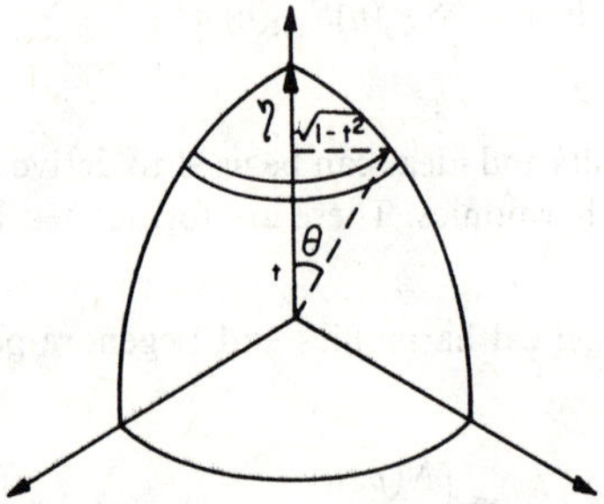

radius of the circle shown above is $\sqrt{1-t^2}$. It is the circle defined by $(\xi, \eta) = t$. Since it is a circle in a p dimensional space its circumference is given by $(1-t^2)^{(p-2)/2}\,\omega_{p-1}$. By the symmetry of $P_n((\xi,\eta))$ with respect to the η axis we can integrate about the circle keeping (ξ, η) constant. Thus the $p-1$ dimensional integral in (21) can be reduced to a one-dimensional integral.

$$\int_{|\xi|=1} P_n^{\,2}((\xi,\eta))\, d\omega_p = \omega_{p-1} \int_{-1}^{1} P_n^{\,2}(t)(1-t^2)^{(p-3)/2}\, dt$$

or equivalently

$$\int_{-1}^{1} P_n^{\,2}(t)(1-t^2)^{(p-3)/2}\, dt = \frac{\omega_p}{\omega_{p-1} N(p,n)} \tag{22}$$

Similarly we can show that

$$0 = \int_{-1}^{1} P_n((\xi,\eta))P_m((\xi,\eta))\, d\omega_p = \omega_{p-1} \int_{-1}^{1} P_n(t)P_m(t)(1-t^2)^{(p-3)/2}\, dt$$

and we see that the set of polynomials $P_n(t)$ is orthogonal over the interval $[-1, 1]$ with weight function $(1-t^2)^{(p-3)/2}$.

By referring to the results of Chapter 2 we see that

$$P_n(t) = \left(-\frac{1}{2}\right)^n \frac{\Gamma(p-1/2)}{\Gamma(n+(p-1)/2)} (1-t^2)^{(3-p)/2} D^n (1-t^2)^{n+(p-3)/2} \tag{23}$$

which is, of course, the Rodrigues formula. The coefficient is so adjusted that $P_n(1) = 1$. $P_n(t)$ satisfies the differential equation

$$(1 - t^2)P_n''(t) + (1 - p)tP_n'(t) + n(n + p - 2)P_n(t) = 0. \tag{24}$$

We also see that these polynomials are essentially the Gegenbauer polynomials and a simple comparison shows that

$$C_n^{(p-2)/2}(t) = \binom{n + p - 3}{n} P_n(t). \tag{25}$$

Knowing the recurrence formula for the Gegenbauer polynomials we obtain directly

$$(n + p - 2)P_{n+1}(t) - (2n + p - 2)tP_n(t) + nP_{n-1}(t) = 0. \tag{26}$$

The technique employed for reducing a surface integral to a one dimensional integral allows us to prove the following theorem.

HECKE-FUNK THEOREM.

$$\int_{|\eta|=1} f((\xi, \eta))S_n(\eta)\, d\omega_p = \omega_{p-1}S_n(\xi)\int_{-1}^{1} f(t)P_n(t)(1 - t^2)^{(p-3)/2}\, dt \tag{27}$$

Proof. We first let $S_n(\eta) = P((\eta, \zeta_k))$ where ζ_k is a fixed unit vector. Then the function

$$F(\xi, \zeta_k) = \int_{|\eta|=1} f((\xi, \eta))P_n((\eta, \zeta_k))\, d\omega_p \tag{28}$$

is invariant under all rotations. That is, for every orthogonal matrix A we find

$$\begin{aligned} F(A\xi, A\zeta_k) &= \int_{|\eta|=1} f((A\xi, \eta))P_n((\eta, A\zeta_k))\, d\omega_p \\ &= \int_{|\eta|=1} f((A\xi, A\eta))P_n((A\eta, A\zeta_k))\, d\omega_p \\ &= \int_{|\eta|=1} f((\xi, \eta))P_n((\eta, \zeta_k))\, d\omega_p = F(\xi, \zeta_k). \end{aligned}$$

We now select our coordinate system so that

$$\xi = (1, 0, 0, \ldots, 0), \qquad \zeta_k = (\tau, 1 - \tau^2, 0, \ldots, 0).$$

Then (28) defines a function of τ that can be shown to be a polynomial of degree n in τ, as has been done before. Then

$$F(\xi, \zeta_k) = cP_n(\tau)$$

for a suitable constant c. For $\tau = 1$ we obtain

$$\int_{|\eta|=1} f((\xi, \eta))P_n((\eta, \xi))\, d\omega_p = c = \omega_{p-1}\int_{-1}^{1} f(t)P_n(t)(1 - t^2)^{(p-3)/2}\, dt$$

where τ is the first component of η. Finally

$$\int_{|\eta|=1} f((\xi, \eta))P_n((\eta, \zeta_k))\, d\omega_p = \omega_{p-1} P_n((\xi, \zeta_k)) \int_{-1}^{1} f(t)P_n(t)(1-t^2)^{(p-3)/2}\, dt$$

The latter is (27) for the special form $S_n(\xi) = P_n((\xi, \zeta_n))$. But since every spherical harmonic can be expressed as a linear combination of such special cases (27) is established. ■

As an application of some of the ideas developed we can establish an integral representation

THEOREM

$$P_n(t) = \frac{\omega_{p-2}}{\omega_{p-1}} \int_{-1}^{1} [t + i\sqrt{1-t^2}\, s]^n (1-s^2)^{(p-4)/2}\, ds \tag{29}$$

Proof. Consider

$$L_n(X) = \frac{1}{\omega_{p-1}} \int_{|\eta'|=1} [(X, \zeta) + i(X, \eta')]^n\, d\omega_{p-1}$$

where

$$\zeta = (1, 0, 0, \ldots, 0), \qquad \eta' = \eta - \zeta(\eta, \zeta).$$

The integration is over the surface of a sphere in $p - 1$ dimensions. Evidently $L_n(X)$ is homogeneous of degree n, and by differentiation one can easily verify that it is harmonic as well. Also

$$L_n(\zeta) = 1$$

and if A is any rotation leaving ζ invariant we have

$$L_n(AX) = L_n(X)$$

Now let $X = t\zeta + \sqrt{1-t^2}\,\zeta'$, where ζ' is so that $(\zeta, \zeta') = 0$ and $(\zeta', \zeta') = 1$. Then using the results of section 3

$$L_n(X) = P_n(t)$$

and

$$P_n(t) = \frac{1}{\omega_{p-1}} \int_{|\eta'|=1} [t + i\sqrt{1-t^2}\,(\zeta', \eta')]^n\, d\omega_{p-1}.$$

As before we can reduce the above to one-dimensional integral. Then

$$P_n(t) = \frac{\omega_{p-2}}{\omega_{p-1}} \int_{-1}^{1} [t + i\sqrt{1-t^2}\, s]^n (1-s^2)^{(p-4)/2}\, ds.$$ ■

For $p = 3$ we obtain

$$P_n(t) = \frac{1}{\pi} \int_{-1}^{1} [t + i\sqrt{1-t^2}\, s]^n (1-s^2)^{-1/2}\, ds$$

a result obtained earlier.

As a final result we show that once the Legendre polynomials for the cases $p = 2$ and $p = 3$ are determined, all other can be expressed in terms of them.

THEOREM. We now denote the Legendre polynomial of degree n in p dimensions by the symbol $P_{n,p}(t)$. Then

$$P_{n-j,\,2j+p}(t) = \frac{((p-1)/2)_j}{(-n)_j(n+p-2)_j}\left(\frac{d}{dt}\right)^j P_{n,\,p}(t), \qquad j = 0, 1, \ldots, n. \tag{30}$$

Proof. Using the differential equation (24) we can easily show that

$$P_{n,\,p}(t) = F\left(-n, n+p-2; \frac{p-1}{2}; \frac{1-t}{2}\right).$$

The jth derivative of the above is given by

$$\left(\frac{d}{dt}\right)^j P_{n,\,p}(t) = \frac{(-n)_j(n+p-2)_j}{((p-1)/2)} F\left(-n+j, n+p+j-2; \frac{p-1}{2}+j; \frac{1-t}{2}\right)$$

$$= \frac{(-n)_j(n+p-2)_j}{((p-1)/2)_j} P_{n-j,\,p+2j}(t). \qquad \blacksquare$$

From (30) we see that all $P_{n,p}(t)$ for even p can be generated by successive differentiations of $P_{n,2}(t)$ and those for odd p from $P_{n,3}(t)$.

4. Applications to Boundary Value Problems

We can develop certain expansion theorems in connection with the orthonormal set $\{S_{n,j}(\xi)\}$ where $j = 1, 2, \ldots, N(p, n)$, and $n = 0, 1, 2, \ldots$. In particular we have the following theorem.

THEOREM. Let $f(\xi)$ be a real continuous function on $|\xi| = 1$. If $(f(\xi), S_{n,j}(\xi)) = 0$ for all spherical harmonics in $\{S_{n,j}(\xi)\}$ then $f(\xi)$ vanishes identically.

Proof. Without loss of generality we can assume that there exists an η such that $f(\eta) > 0$. By the continuity of $f(\xi)$ there exists a neighborhood of η defined by $\tau \le (\xi, \eta) \le 1$ such that $f(\xi) \ge a > 0$ for all ξ in that neighborhood.

Consider the function

$$\phi(t) = 1 - \frac{(1-t)^2}{(1-\tau)^2}. \qquad \tau \le t \le 1$$

$$= 0 \qquad -1 \le t \le \tau.$$

There exists then a positive constant b such that

$$\int_{|\xi|=1} f(\xi)\phi((\xi, \eta))\, d\omega_p \geq b > 0.$$

Using the Weierstrass approximation theorem we can construct a polynomial $p(t)$ so that

$$|\phi(t) - p(t)| \leq \varepsilon, \qquad -1 \leq t \leq 1.$$

It follows that if $|f(\xi)| \leq M$, then

$$\left| \int_{|\xi| \leq 1} f(\xi)[\phi((\xi, \eta)) - p((\xi, \eta))]\, d\omega_p \right| \leq M\omega_p \varepsilon$$

and

$$\int_{|\xi|=1} f(\xi)p((\xi, \eta))\, d\omega_p \geq b - M\omega_p \varepsilon > 0 \tag{1}$$

for sufficiently small ε. But using the completeness properties of the Legendre polynomials we have

$$p(t) = \sum_{k=0}^{n} A_k P_k(t)$$

so that

$$\int_{|\xi|=1} f(\xi) \sum_{k=0}^{n} A_k P_k((\xi, \eta))\, d\omega_p = 0 \tag{2}$$

by hypothesis. But (1) and (2) contradict one another. Hence $f(\xi) = 0$. ■

Using this theorem we can develop expansions. Then if

$$f(\xi) = \sum_{n,j} C_{n,j} S_{n,j}(\xi) \tag{3}$$

the coefficients are obtained from

$$C_{n,j} = \int_{|\xi|=1} f(\xi) S_{n,j}(\xi)\, d\omega_p. \tag{4}$$

The above theorem can be extended to all functions for which

$$\int_{|\xi|=1} |f(\xi)|^2\, d\omega_p < \infty.$$

But to prove such a result one needs certain facts about square-integrable functions. In particular for every ε one can find a continuous function $g(\xi)$ so that

$$\int_{|\xi|=1} |f(\xi) - g(\xi)|^2\, d\omega_p \leq \varepsilon.$$

With this one can show that

$$\lim_{K\to\infty} \int_{|\xi|=1} \left| f(\xi) - \sum_{n=0}^{K} \sum_{j=1}^{N(p,n)} C_{n,j} S_{n,j}(\xi) \right|^2 d\omega_p = 0,$$

where the $C_{n,j}$ are given by (4).

We seek a harmonic function V satisfying a boundary condition on $r = [\sum_1^p x_i^2]^{1/2} = 1$. That is,

$$\begin{aligned} \Delta_p V &= 0 \\ V &= f(\xi) \qquad \text{on} \quad r = 1. \end{aligned} \tag{5}$$

Using (3) we see that

$$V = \sum_{n,j} r^n C_{n,j} S_{n,j}(\xi) \tag{6}$$

must satisfy the boundary condition and by construction must also be harmonic. Hence (6) is the solution of (5).

We can also solve (5) using a Green's function technique. As in Chapter 5 we can construct a delta function, defined as follows. Let

$$\delta(P, P_0) = 0 \qquad P \neq P_0$$

$$\int_{|P-P_0|\leq\varepsilon} \delta(P, P_0)\, dv = 1 \qquad \text{for all} \quad \varepsilon > 0.$$

The Green's function is now defined as the solution of the problem

$$\begin{aligned} \Delta_p G &= \delta(P, P_0) \qquad && |P| < 1 \\ G &= 0 && \text{on} \quad |P| = 1. \end{aligned}$$

To find G we shall first find the so-called free space Green's function. It satisfies

$$\Delta_p G = \delta(P, P_0)$$

but not the boundary condition. By the symmetry of the situation we can assume that it is a function of the single variable $R = |P - P_0|$. Then we consider the following ordinary differential equation.

$$\Delta_p h(R) = h''(R) + \frac{p-1}{R} h'(R) = 0, \qquad R \neq 0.$$

so that a solution, that is also singular at $R = 0$ is given by

$$h(R) = \frac{c}{R^{p-2}}, \qquad p \neq 2$$

The case $p = 2$ must be considered separately; in that case $h(R)$ is logarithmic. To evaluate c we write

$$\int_{R \le \varepsilon} \Delta_p h(R)\, dv = \int_{R \le \varepsilon} \nabla \cdot \nabla h(R)\, dv = \int_{R = \varepsilon} \frac{\partial h(R)}{\partial \nu}\, ds$$

where ν is the outward normal on $R = \varepsilon$. Then

$$\frac{\partial h(R)}{\partial \nu} = \frac{d}{dR} h(R) = \frac{-(p-2)c}{R^{p-1}}$$

and $ds = R^{p-1}\, d\omega_p$, so that

$$\int_{R \le \varepsilon} \Delta_p h(R)\, dv = -c(p-2) \int_{R \le \varepsilon} \frac{R^{p-1} d\omega_p}{R^{p-1}} = -c(p-2)\omega_p.$$

In order for $\Delta_p h(R) = \delta(P, P_0)$ we require that

$$\int_{R \le \varepsilon} \Delta_p h(R)\, dv = \int_{R \le \varepsilon} \delta(P, P_0)\, dv = 1$$

so that $c = (-1/(p-2)\omega_p)$.

Now we can write

$$G = \frac{-1}{(p-2)\omega_p R^{p-2}} + g$$

where g must be so constructed that

$$\Delta_p g = 0$$

$$g = \frac{1}{(p-2)\omega_p R^{p-2}} \qquad \text{on} \qquad r = 1.$$

As in the three dimensional case we can construct g by the method of images. Let P_0' denote a point outside the unit sphere on the ray $(0, P_0)$ whose distance from the origin is $1/r_0$. Let $R' = |P - P_0'|$ and it follows (as in Eq. 5.6.7) that

$$G = -\frac{1}{(p-2)\omega_p}\left[\frac{1}{R^{p-2}} - \frac{1}{(r_0 R')^{p-2}}\right]. \tag{7}$$

If θ denotes the angle between the rays $(0, P_0)$ and $(0, P_0')$, then

$$R = \sqrt{r^2 - 2rr_0 \cos\theta + r_0^2}$$

$$R' = \sqrt{r^2 - \frac{2r}{r_0}\cos\theta + \frac{1}{r_0^2}}.$$

To solve the boundary value problem (5) we now proceed as follows. We use Green's theorem in p dimensions

$$\int_{r\le 1} (V\Delta_p G - G\Delta_p V)\,dv = \int_{r=1} \left(V\frac{\partial G}{\partial \nu} - G\frac{\partial V}{\partial \nu}\right) d\omega_p$$

Since $\Delta_p V = 0$, $\Delta_p G = \delta(P, P_0)$ and $G = 0$ on $r = 1$, and ν is the exterior normal we have

$$\int_{r\le 1} V\delta(P, P_0)\,dv = V(P_0) = \int_{r=1} f(\xi)\frac{\partial G}{\partial r}\bigg|_{r=1} d\omega_p.$$

But

$$\frac{\partial G}{\partial r}\bigg|_{r=1} = \frac{1 - {r_0}^2}{\omega_p[1 - 2r_0\cos\theta + {r_0}^2]^{p/2}}$$

so that we have the alternative representation to (6)

$$V(P_0) = \frac{1}{\omega_p}\int_{|\xi|=1} f(\xi)\frac{1 - {r_0}^2}{[1 - 2r_0\cos\theta + {r_0}^2]^{p/2}}\,d\omega_p. \tag{8}$$

We can rewrite (6) in the form

$$\begin{aligned} V &= \sum_{n=0}^{\infty} {r_0}^n \int_{|\xi|=1} f(\xi) \sum_{j=1}^{N(p,n)} S_{n,j}(\eta)S_{n,j}(\xi)\,d\omega_p \\ &= \sum_{n=0}^{\infty} {r_0}^n \int_{|\xi|=1} f(\xi)\frac{N(p,n)}{\omega_p} P_n(\cos\theta)\,d\omega_p \\ &= \frac{1}{\omega_p}\int_{|\xi|=1} f(\xi)\sum_{n=0}^{\infty} {r_0}^n N(p,n)P_n(\cos\theta)\,d\omega_p. \end{aligned} \tag{9}$$

A comparison of (8) and (9) yields the following generating function

$$\sum_{n=0} r^n N(p,n)P_n(t) = \frac{1 - r_2}{[1 - 2rt + r^2]^{p/2}} \tag{10}$$

Exercises

1. Show that

$$\frac{1}{[1 - 2tr + r^2]^{(p-2/2)}} = \sum_{n=0}^{\infty}\binom{n+p-3}{n}P_n(t)\,r^n.$$

For $p = 3$ the above reduces to the generating functions of the three dimensional Legendre polynomials.

2. Develop the Green's function in section 4 for the case $p = 2$.

3. Show that if $F(X, Y)$ is a polynomial in the components of X and Y, and if for an orthogonal transformation A

$$F(AX, AY) = F(X, Y)$$

then F is a polynomial in the three variables (X, X), (Y, Y), (X, Y).

4. Show that, if $f(t)$ has n continuous derivatives, then

$$\int_{-1}^{1} f(t)P_n(t)(1-t^2)^{(p-3/2)}\,dt = \left(\frac{1}{2}\right)^n \frac{\Gamma((p-1)/2)}{\Gamma(n+(p-1/2))} \int_{-1}^{1} (1-t^2)^{(n+(p-3)/2)} f^{(n)}(t)\,dt.$$

5. Let $r = \sqrt{\sum_1^p x_i^2}$. Show that

$$[\eta.\Delta]^n r^{2-p} = (-1)^n \frac{\Gamma(n+p-2)}{\Gamma(p-2)} \frac{P_n((\xi, \eta))}{r^{n+p-2}}$$

6. Show that if $H_m(X)$ is a harmonic homogeneous polynomial of degree m, such that

$$\frac{\partial H}{\partial x_1} = 0$$

then $\Delta[(1 - 2x_1 t + r^2 t^2)^{-m-1/2(p-2)} H_m(X)] = 0$.

$$r = [\textstyle\sum_1^p x_i^2]^{1/2}$$

This implies that

$$r^{n-m} P_{n-mp+2m}\left(\frac{x_1}{r}\right) H_m(X)$$

is harmonic.

CHAPTER 7

Confluent Hypergeometric Functions

1. Relationship to the Hypergeometric Functions

In Chapter IV it was shown that the hypergeometric differential equation was the only second order linear differential equation with precisely three regular singular points. By means of bilinear changes of the independent variable these could be placed at any three preassigned points. In the canonical form of the equations these are 0, 1 and ∞. The confluent hypergeometric equation results when two of these are allowed to merge into one singular point, which will no longer represent a regular singular point. Consider now

$$z(1-z)y'' + [c-(a+b+1)z]y' - aby = 0$$

and replace z by the new independent variable ζ/b. Then the equation takes the form

$$\zeta\left(1-\frac{\zeta}{b}\right)\frac{d^2y}{d\zeta^2} + \left[c-\frac{a+b+1}{b}\zeta\right]\frac{dy}{d\zeta} - ay = 0.$$

The above equation has regular singular points at 0, b, ∞. If we now allow b to become infinite in the limit we obtain (using z again to symbolize our independent variable)

$$zy'' + [c-z]y' - ay = 0. \tag{1}$$

(1) is a confluent hypergeometric equation. It has two singular points, namely $z = 0$ and $z = \infty$. The former is still a regular singular point, the latter is not. A solution of (1) can be obtained by standard techniques or else by performing a similar limiting process on hypergeometric functions. Consider the function

$$F\left(a, b; c; \frac{\zeta}{b}\right) = \sum_{n=0}^{\infty} \frac{(a)_n(b)_n\zeta^n}{(c)_n n! b^n}$$

Clearly

$$\lim_{b\to\infty} \frac{(b)_n}{b^n} = 1$$

so that, if we define the function

$${}_1F_1(a; c; z) = \lim_{b\to\infty} F\left(a, b; c; \frac{z}{b}\right)$$

we obtain

$${}_1F_1(a; c; z) = \sum_{n=0}^{\infty} \frac{(a)_n z^n}{(c)_n n!}. \tag{2}$$

Note that (2) converges for all finite z. This fact may be confirmed using the ratio test. It also satisfies (1), as may be confirmed by direct substitution. From standard theorems it is known that such a series solution must converge up to the next singularity of the differential equation. In this case the nearest singularity to $z = 0$ is at $z = \infty$. Hence we see from this viewpoint again that (2) converges for all finite values of z.

To obtain a second solution of (1) we consider the following solution of the hypergeometric equation

$$z^{1-c}F(a - c + 1, b - c + 1; 2 - c; z).$$

By a similar limiting process we obtain

$$y_2 = z^{1-c}\, {}_1F_1(a - c + 1; 2 - c; z). \tag{3}$$

(3) will be a solution of (1) not dependent on (2) unless c is an integer. When c is a non-positive integer (2) breaks down. For $c = 1$(2) and (3) coincide and if c is positive integer other than $c = 1$(3) breaks down. It follows that when c is an integer between (2) and (3) we have only one independent solution. We can however, consider the following linear combination of (2) and (3)

$$y_3 = \frac{1}{\sin \pi c}\left[\frac{{}_1F_1(a; c; z)}{\Gamma(c)\Gamma(1 + a - c)} - \frac{z^{1-c}\, {}_1F_1(a - c + 1; 2 - c; z)}{\Gamma(a)\Gamma(2 - c)}\right]. \tag{4}$$

For c nonintegral (4) is clearly independent of (2) or (3). For $c = 1$(4) is indeterminate, but one can easily show that the limit of (4) as c approaches unity exists and yields a solution of (1) independent of (2).

The differential equation (1) is not the only confluent case of the hypergeometric equation. Another can be constructed as follows. With the Riemann Papperitz symbol

$$P\begin{Bmatrix} 0 & c & \infty & \\ \frac{1}{2} + m & c - k & 0 & z \\ \frac{1}{2} - m & k & -c & \end{Bmatrix}$$

we can associate the differential equation

$$zy'' - \frac{1-c}{c-z}zy' + \left[\frac{-c(m^2-\frac{1}{4})}{z(c-z)} + \frac{ck(c-k)}{(c-z)^2}\right]y = 0.$$

If we now allow c to become infinite we obtain

$$zy'' + zy' + \left[\frac{\frac{1}{4}-m^2}{z} + k\right]y = 0 \tag{5}$$

The substitution of $y = z^{1/2+m}\, e^{-z}u$ in (5) yields

$$zu'' + [1 + 2m - z]u' - (\tfrac{1}{2} + m - k)u = 0 \tag{6}$$

and the substitution of $y = e^{-1/2 z}\, V$ yields

$$V'' + \left[-\frac{1}{4} + \frac{k}{z} + \frac{\frac{1}{4}-m^2}{z^2}\right]V = 0. \tag{7}$$

(7) is a confluent hypergeometric equation with one regular singular point at $z = 0$ and another singular point at $z = \infty$. This equation is customarily known as Whittaker's equation. (6) is another form of (1). A comparison of (6) and (7) shows that a solution of (7) can be expressed in the form

$$M_{k,m}(z) = z^{1/2+m}e^{-1/2 z}{}_1F_1(m-k+\tfrac{1}{2};\ 2m+1;\, z) \qquad 2m+1 \neq \text{integer}. \tag{8}$$

Since (7) is invariant under the substitution $m \to -m$, $M_{k,-m}(z)$ will also be a solution, and for general values of m will be independent of $M_{k,m}(z)$. The Wronskian of these two solutions will be given by

$$W(M_{k,m}(z), M_{k,-m}(z)) = -2\,m.$$

Another function that plays a significant role is the classical Whittaker function defined by

$$W_{k,m}(z) = \frac{\Gamma(-2m)M_{k,m}(z)}{\Gamma(\frac{1}{2}-k-m)} + \frac{\Gamma(2m)M_{k,-m}(z)}{\Gamma(\frac{1}{2}-k+m)}. \tag{9}$$

The above is well defined for all values of m, if certain exceptional cases are treated as limiting cases.

2. Applications of These Functions in Mathematical Physics

An equation that arises in many problems concerning wave propagation is

$$\Delta W = \frac{\partial^2 W}{\partial t^2}. \tag{1}$$

In many instances the dependence of W on time is taken to be harmonic, that is we assume that

$$W(x, y, z, t) = e^{ikt}\, u(x, y, z). \tag{2}$$

The insertion of (3) in (1) and cancellation of the terms e^{ikt} leads to

$$\Delta u + k^2 u = 0. \tag{3}$$

Suppose we now consider (3), known as the reduced wave equation in paraboloidal coordinates. These are defined by

$$\begin{aligned} x &= \rho \cos \phi = 2\sqrt{\xi\eta} \cos \phi \\ y &= \rho \sin \phi = 2\sqrt{\xi\eta} \sin \phi \\ z &= \xi - \eta \end{aligned}$$

The surface defined by $\xi = c$ are confocal paraboloids of revolution; hence, the name paraboloidal coordinates. The Laplace operator in these coordinates takes the form

$$\Delta \equiv \frac{1}{2(\xi + \eta)} \left\{ \frac{\partial}{\partial \xi} 2\xi \frac{\partial}{\partial \xi} + \frac{\partial}{\partial \eta} 2\eta \frac{\partial}{\partial \eta} + \frac{\xi + \eta}{2\xi\eta} \frac{\partial^2}{\partial \phi^2} \right\}.$$

If we wish to apply the method of separation of variables to (3) we let

$$u = f_1(\xi) f_2(\eta) e^{ip\Phi}.$$

Then by standard operations we are led to the ordinary differential equations

$$\begin{aligned} (\xi f'_1)' + \left(k^2\xi - \frac{p^2}{4\xi} - c \right) f_1 &= 0 \\ (\eta f'_2)' + \left(k^2\eta - \frac{p^2}{4\eta} + c \right) f_2 &= 0 \end{aligned} \tag{4}$$

where c is a separation constant.

We can reduce both equations in (4) to Whittaker's equation if we make the following substitutions. Let

$$z_1 = 2ik\xi, \qquad z_2 = -2ik\eta, \qquad c = -2ikk'$$

Then both equations in (4) become

$$(zf')' - \left(\frac{z}{4} - k' + \frac{p^2}{4z} \right) f = 0. \tag{5}$$

A simple computation shows that

$$f(z) = z^{-1/2} W_{k', 1/2 p}(z). \tag{6}$$

If we now introduce the parabolic cylinder coordinates

$$x = \xi - \eta$$

$$y = 2\sqrt{\xi\eta}$$

$$z = z$$

the Laplace operator takes the form

$$\Delta \equiv \frac{1}{4(\xi + \eta)} \left\{ \frac{\partial^2}{\partial(\sqrt{\xi})^2} + \frac{\partial^2}{\partial(\sqrt{\eta})^2} \right\} + \frac{\partial^2}{\partial z^2}.$$

If we let

$$u = f_1(\sqrt{\xi}) f_2(\sqrt{\eta}) e^{i\alpha z}$$

in (3) and separate the variables we obtain two ordinary differential equations for f_1 and f_2. If we then introduce the new variables

$$z_1 = 2\sqrt{i\zeta\xi},\ z_2 = 2\sqrt{-i\zeta\eta},\ \zeta = \sqrt{k^2 - \alpha^2}$$

we find that f_1 and f_2 as functions of z_1 and z_2 respectively satisfy

$$f'' + \left[\nu + \frac{1}{2} - \frac{1}{4} z^2\right] f = 0, \qquad \nu + \frac{1}{2} = \frac{c}{4i\zeta} \tag{7}$$

where c is the separation constant.

The solution of (7) is customarily written in the form

$$f = D_\nu(z) = 2^{\frac{1}{2}\nu} \left(\frac{z^2}{2}\right)^{-\frac{1}{4}} W_{\frac{1}{2}\nu + \frac{1}{4},\, \frac{1}{4}} \left(\frac{z^2}{4}\right). \tag{8}$$

The function $D_\nu(z)$ defined in (8) is known as the Weber function.

These equations arise in connection with many problems of mathematical physics. For example, the Schrödinger equation associated with the hydrogen atom is

$$\left[\Delta + \frac{8\pi^2 m}{h^2} \left(W + \frac{ze^2}{r}\right)\right] \psi = 0. \tag{9}$$

Here r is a radial coordinate. We seek a solution, using spherical coordinates in the form

$$\psi = R(r)\Theta(\theta)\Phi(\phi)$$

and denote the separation constants by μ and l. The functions Φ and Θ are found to satisfy

$$\Phi'' + \mu^2\Phi = 0 \tag{10}$$

$$\Theta'' + \cot\theta\Theta' + \left\{l(l+1) - \frac{\mu^2}{\sin^2\theta}\right\}\Theta = 0 \tag{11}$$

(10) clearly has harmonic solutions and for periodicity in the ϕ space we require that μ be an integer. (11) is satisfied by the spherical harmonics in three dimensions. The functions $\Theta(\theta)$ will be bounded, as well as square integrable only if the parameter l is an integer.

To study the function $R(r)$ we introduce the parameters

$$\alpha^2 = -\frac{8\pi^2 m}{h^2}W, \qquad \gamma = \frac{4\pi^2 m}{h^2\alpha}ze^2, \qquad x = 2\alpha r$$

so that

$$R'' + \frac{2}{x}R' - \left\{\frac{1}{4} - \frac{\gamma}{x} + \frac{l(l+1)}{x^2}\right\}R = 0. \tag{12}$$

(12) is satisfied by

$$R = x^l e^{-\frac{1}{2}x} D^l\, {}_1F_1(-\gamma - l;\, 1;\, x). \tag{13}$$

In quantum mechanics one is often interested in solutions of (9) for which

$$\int |\psi|^2 dv < \infty \tag{14}$$

When $\gamma + l$ is an integer in (13) the corresponding function ψ will satisfy (14). When $\gamma + l$ is not integral the function ${}_1F_1(-\gamma - l;\, 1;\, x)$ is not a polynomial and one can show that (14) is no longer satisfied.

Another approach to the same problem is the following. We rewrite (9) in paraboloidal coordinates so that

$$\left(\frac{\partial}{\partial\xi}\xi\frac{\partial}{\partial\xi} + \frac{\partial}{\partial\eta}\eta\frac{\partial}{\partial\eta}\right)\psi + \frac{1}{4\xi\eta}\frac{\partial^2\psi}{2\phi^2} - [\alpha^2(\xi+\eta) - 2v]\psi = 0. \tag{15}$$

We will assume that ψ is independent of ϕ and that

$$\psi(\xi, \eta) = \psi_1(\xi)\psi_2(\eta).$$

Then we can separate our variables and we obtain the following two ordinary differential equations.

$$(\xi\psi_1')' - [\alpha^2\xi - v + c]\psi_1 = 0$$

$$(\eta\psi_2')' - [\alpha^2\eta - v - c]\psi_2 = 0$$

where c is the separation constant. If we select c so that $c = -\alpha + \nu$ we find that a solution for ψ_1 is given by

$$\psi_1 = e^{-\alpha\xi}.$$

For ψ_2 we find that

$$\eta\psi_2'' + \psi_2' - [\alpha^2\eta - 2\nu + \alpha]\psi_2 = 0$$

Letting $\psi_2 = e^{\alpha\eta}\phi(\eta)$ we find

$$\eta\phi'' + (1 + 2\alpha\eta)\phi' + (2\nu)\phi = 0.$$

A simple computation shows that

$$\phi(\eta) = {}_1F_1\left(\frac{\nu}{\alpha}; 1; -2\alpha\eta\right)$$

so that finally

$$\psi = e^{-\alpha(\xi-\eta)}\,{}_1F_1\left(\frac{\nu}{\alpha}; 1; -2\alpha\eta\right).$$

Reverting to rectangular coordinates we have

$$\psi = e^{-\alpha z}\,{}_1F_1\left(\frac{\nu}{\alpha}; 1; -\alpha(r - z)\right). \tag{16}$$

The confluent hypergeometric function in (16) will be a polynomial if ν/α is a negative integer.

3. Integral Representations

Using the fact that the confluent hypergeometric functions are limiting cases of hypergeometric functions we can deduce many of the properties of the former as limiting cases of the latter. A simple example is the following.

$$\left(\frac{d}{dz}\right)^l {}_1F_1(a; c; z) = \frac{(a)_l}{(c)_l}\,{}_1F_1(a + l; c + l; z). \tag{1}$$

Similarly we can deduce integral representations. Suppose we return to the Mellin-Barnes Integral.

$$F(a, b; c; z/b) = \frac{\Gamma(c)}{\Gamma(a)}\frac{1}{2\pi i}\int_{-i\infty}^{i\infty}\frac{\Gamma(a+s)\Gamma(-s)\Gamma(b+s)}{\Gamma(c+s)\Gamma(b)b^s}(-z)^s\,ds.$$

In view of the fact that

$$\lim_{b\to\infty}\frac{\Gamma(b+s)}{\Gamma(b)b^s} = 1$$

we find that

$$ {}_1F_1(a; c; z) = \frac{\Gamma(c)}{\Gamma(a)} \frac{1}{2\pi i} \int_{-i\infty}^{i\infty} \frac{\Gamma(a+s)\Gamma(-s)}{\Gamma(c+s)} (-z)^s \, ds. \tag{2} $$

(2) can of course be derived independently of the above limiting procedure.

Another representation derived earlier is

$$ F(a, b; c; z/b) = \frac{\Gamma(c)}{\Gamma(a)\Gamma(c-a)} \int_0^1 F\left(a, b; a; \frac{zt}{b}\right) t^{a-1}(1-t)^{c-a-1} \, dt. $$

Note that

$$ F\left(a, b; a; \frac{zt}{b}\right) = \left(1 - \frac{zt}{b}\right)^{-b} \to e^{zt}. $$

Then we find that

$$ {}_1F_1(a; c; z) = \frac{\Gamma(c)}{\Gamma(a)\Gamma(c-a)} \int_0^1 e^{zt} t^{a-1}(1-t)^{c-a-1} \, dt. \tag{3} $$

Another approach to derive representations is to consider the differential operator

$$ L_z y = zy'' + (c-z)y' - ay $$

and to seek solutions of

$$ L_z y = 0 \tag{4} $$

in the form

$$ y = \int_\alpha^\beta e^{-zt} V(t) \, dt. $$

Using standard operations we find that $V(t)$ must satisfy

$$ M_t^* V = t(1-t)V' + [(2-c)t + 1 - a]V = 0 \tag{5} $$

and α and β must be such that

$$ P(z, t)\Big|_\alpha^\beta = -e^{-zt} t(1+t)V(t)\Big|_\alpha^\beta = 0. \tag{6} $$

From (5) we find that

$$ V = t^{a-1}(1+t)^{c-a-1} $$

We shall assume that Re $a > 0$ and Re $z > 0$; then (6) will be satisfied for $\alpha = 0$ and $\beta = \infty$. It follows that

$$ \psi(a; c; z) = \frac{1}{\Gamma(a)} \int_0^\infty e^{-zt} t^{a-1}(1+t)^{c-a-1} \, dt \tag{7} $$

satisfies (4). The choice of multiplicative constant is simply dictated by convention.

Had we chosen $\beta = 1$ in (6) and then replaced t by $-t$ in the integral we would have rederived (3).

By replacing z by a new variable of integration τ defined by $zt = \tau$ we obtain from (7)

$$\psi(a; c; z) = \frac{z^{-a}}{\Gamma(a)} \int_0^\infty e^{-\tau}\tau^{a-1}\left(1 + \frac{\tau}{z}\right)^{c-a-1} d\tau. \tag{8}$$

We can use (8) to obtain a Mellin-Barnes type integral for the function $\psi(a; c; z)$. The following can be derived very easily

$$(1+t)^{c-a-1} = \frac{1}{2\pi i} \int_{-i\infty}^{i\infty} \frac{\Gamma(a-c+1+s)\Gamma(-s)}{\Gamma(1+a-c)} t^s \, ds.$$

If in the above we replace t by τ/z and replace the corresponding term in (8) by the above integral we obtain

$$\begin{aligned} \psi(a; c; z) &= \frac{z^{-a}}{2\pi i} \int_{-i\infty}^{i\infty} \frac{\Gamma(a-c+1+s)\Gamma(-s)z^{-s}}{\Gamma(a)\Gamma(1+a-c)} \int_0^\infty e^{-\tau}\tau^{s+a-1} \, d\tau \, ds \\ &= \frac{z^{-a}}{2\pi i} \int_{-i\infty}^{i\infty} \frac{\Gamma(a+s)\Gamma(a-c+1+s)\Gamma(-s)z^{-s}}{\Gamma(a)\Gamma(1+a-c)} ds. \end{aligned}$$

If we replace s by the variable $-s-a$ we obtain

$$\psi(a; c; z) = \frac{1}{2\pi i} \int_{-i\infty}^{i\infty} \frac{\Gamma(a+s)\Gamma(1-c-s)\Gamma(-s)z^s}{\Gamma(a)\Gamma(1+a-c)} ds \tag{9}$$

From (9) we can easily verify that

$$\psi(a-c+1; 2-c; z) = \frac{1}{2\pi i} \int_{-i\infty}^{i\infty} \frac{\Gamma(a-c+1-s)\Gamma(-1+c-s)\Gamma(-s)z^s}{\Gamma(a)\Gamma(1+a-c)} ds.$$

If in the above we make the substitution $s \to s - 1 + c$ and compare the result to (9) we find that

$$\psi(a; c; z) = z^{1-c}\psi(a-c+1; 2-c; z). \tag{10}$$

To relate this function to the functions defined earlier we close the contour on the right in (9). There will be two sets of residues, from $\Gamma(1-c-s)$ and $\Gamma(-s)$. By comparing the resultant power series we obtain

$$\begin{aligned} \psi(a; c; z) &= \frac{\Gamma(1-c)}{\Gamma(1+a-c)} {}_1F_1(a; c; z) \\ &\quad + \frac{\Gamma(c-1)}{\Gamma(a)} z^{1-c} {}_1F_1(1+a-c; 2-c; z). \end{aligned} \tag{11}$$

4. Asymptotic Representations

To study the asymptotic behavior of these functions for large z we return to

$$zy'' + (c - z)y' - ay = 0 \tag{1}$$

and let $z = 1/t$. Then

$$\frac{d^2y}{dt^2} + \left[\frac{2-c}{t} + \frac{1}{t^2}\right]\frac{dy}{dt} - \frac{a}{t^3}\, y = 0 \tag{2}$$

(2) has an irregular singularity at $t = 0$. Nevertheless the series

$$y_1 = \sum_{n=0}^{\infty} \frac{(a)_n(a-c+1)_n(-1)^n}{n!}\, t^{n+a}$$

$$= z^{-a} \sum_{n=0}^{\infty} \frac{(a)_n(a-c+1)_n}{n!}\,(-z)^{-n} \tag{3}$$

satisfies (1). It is not convergent, but can be shown to be asymptotic.

To find a second solution we recall Kummer's relation

$$F(a, b; c; z) = (1-z)^{-b} F\left(c-a, b; c; \frac{z}{z-1}\right).$$

Its limiting confluent form yields

$${}_1F_1(a; c; z) = e^z\, {}_1F_1(c-a; c; -z).$$

This suggests that if $y(a; c; z)$ is any solution of (1) then $e^z y(c-a; c; -z)$ is also a solution. Applying this to y_1 we find that

$$y_2 = e^z\, z^{a-c} \sum \frac{(c-a)_n(1-a)_n}{n!}\, z^{-n}$$

Also satisfies (1), but is clearly independent of (3).

Using (3) and 3.8 we see immediately that

$$\psi(a; c; z) \sim y_1 .$$

Similarly we have

$$e^z \psi(c-a; c; -z) \sim c y_2 ,$$

for a suitable constant c.

By use of the Mellin-Barnes integral 3.9 we find, in analogy to 3.11, that

$$e^z\psi(c-a;c;-z)=\frac{\Gamma(1-c)}{\Gamma(1-a)}\,{}_1F_1(a;c;z)$$

$$+\frac{\Gamma(c-1)}{\Gamma(c-a)}(-z)^{1-c}\,{}_1F_1(1+a-c;2-c;z) \quad (4)$$

In (4), as well as in 3.11 we must restrict the domain of z to a suitable region in order that the terms z^{1-c} and $(-z)^{1-c}$ be well defined. It is convenient to select z so that $-\frac{3}{2}\pi < \arg z < \frac{1}{2}\pi$. By inverting 3.11 and (4) we obtain

$$_1F_1(a;c;z)=\frac{e^{i\pi a}\Gamma(c)}{\Gamma(c-a)}\psi(a;c;z)+\frac{e^{i\pi(a-c)}\Gamma(c)}{\Gamma(a)}e^z\psi(c-a;c;-z)$$

$$z^{1-c}\,{}_1F_1(a-c+1;2-c;z)=-\frac{e^{i\pi(a-c)}\Gamma(2-c)}{\Gamma(1-a)}\psi(a;c;z)$$

$$+\frac{e^{i\pi(a-c)}\Gamma(2-c)}{\Gamma(a-c+1)}e^z\psi(c-a;c;-z) \quad (5)$$

Formula pairs of the type 3.11 and (4), and (5) are often referred to as connection formulas. They connect a pair of linearly independent solutions defined near the origin with a second pair which have known asymptotic properties. By means of these one can immediately express any solution defined near the origin in terms of its asymptotic properties and also any solution defined in terms of some asymptotic property can be expressed in terms of a series convergent near the origin.

Exercises

1. Verify explicitly that

$$_1F_1(a;c;z)=\sum_{n=0}^{\infty}\frac{(a)_nz^n}{(c)_nn!}$$

satisfies

$$zy''+(c-z)y'-ay=0.$$

2. Determine

$$\lim_{c\to 1}\frac{1}{\sin\pi c}\left[\frac{{}_1F_1(a;c;z)}{\Gamma(c)\Gamma(1+a-c)}-\frac{z^{1-c}\,{}_1F_1(a-c+1;2-c;z)}{\Gamma(a)\Gamma(2-c)}\right].$$

3. Show that the Weber function $D_n(z)$, defined in 2.8 is related to the Hermite polynomials via

$$H_n(z)=e^{z^2/4}D_n(z).$$

4. Verify explicitly 3.11. Justify the residue integration.

Chapter 8

Bessel Functions

1. Basic Definitions

We consider the operator Δ_p, that was discussed in chapter VI in connection with the spherical harmonics. It will be necessary for us to express this operator in spherical coordinates in p dimensions. To do so we make use of Green's theorem. We have, for any sufficiently differentiable function V

$$\int_D \Delta_p V \, dv = \int_{\partial D} \frac{\partial V}{\partial \nu} \, ds \tag{1}$$

where D is a suitable simply connected domain, ∂D its bounding surface and ν the exterior normal. By means of (1) we could define $\Delta_p V$ by the following limiting process.

$$\Delta_p V = \lim_{\int_D dv \to 0} \frac{\int_{\partial D} (\partial V/\partial \nu) \, ds}{\int_D dv} \tag{2}$$

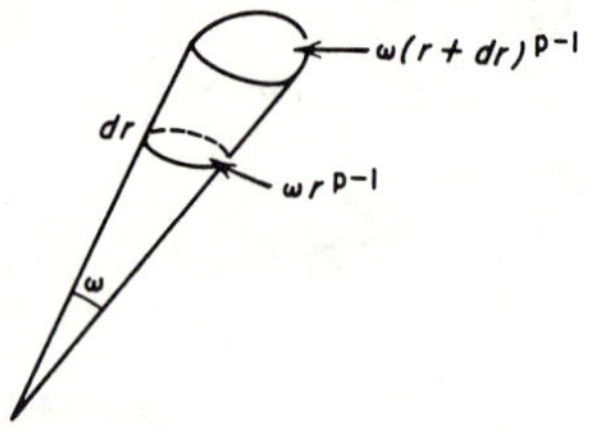

We shall now apply (2) to a suitable domain as shown in the following figure. Essentially D will be a "pill box" bounded by spherical surfaces of radius r and $(r + dr)$ and of angular width ω. Using (2) and noting that

$$\int_D dv = \omega r^{p-1} \, dr$$

we have

$$\Delta_p V = \lim_{\substack{dr \to 0 \\ \omega \to 0}} \frac{\partial V/\partial r\,|_{r+dr}\,\omega(r+dr)^{p-1} - \partial V/\partial r\,|_r\,\omega r^{p-1} + '''}{\omega r^{p-1}\,dr} \tag{3}$$

The first two terms are the contributions to the surface integral from the top and bottom of the pill box and the terms indicated by dots are contributions from the sides. The latter involve derivatives with respect to the angular variables. Taking the limit in (3) we obtain

$$\Delta_p V = \frac{\partial^2 V}{\partial r^2} + \frac{p-1}{r}\frac{\partial V}{\partial r} + \frac{1}{r^2}\Delta_p^* V. \tag{4}$$

The operator Δ_p^* is a differential operator involving only the angular variables and is known as the Beltrami operator. For example for $p = 2$

$$\Delta_2^* V = \frac{\partial^2 V}{\partial \phi^2}$$

and for $p = 3$

$$\Delta_3^* V = \frac{1}{\sin\theta}\frac{\partial}{\partial\theta}\sin\theta\frac{\partial V}{\partial\theta} + \frac{1}{\sin^2\theta}\frac{\partial^2 V}{\partial\phi^2}.$$

Suppose that $H_n(x)$ is a homogeneous, harmonic polynomial. Then $H_n(x) = r^n S_n(\xi)$ and

$$\Delta_p r^n S_n(\xi) = r^{n-2}[\Delta_p^* S_n(\xi) + n(n+p-2)S_n(\xi)] = 0$$

We see therefore that a spherical harmonic of degree n satisfies

$$\Delta_p^* S_n(\xi) + n(n+p-2)S_n(\xi) = 0. \tag{5}$$

In particular for a Legendre polynomial, as was seen earlier, the above reduces to

$$(1-t^2)P_n'' - (p-1)tP_n' + n(n+p-2)P_n = 0. \tag{6}$$

We now seek to determine solutions of the reduced wave equation

$$(\Delta_p + k^2)u = 0$$

in the form $u = f_n(r)S_n(\xi)$ where $S_n(\xi)$ is a spherical harmonic. Using (4) we find

$$\left(f_n'' + \frac{p-1}{r}f_n' + k^2 f_n\right)S_n + \frac{1}{r^2}f_n\,\Delta_p^* S_n = 0$$

and when combined with (5) we find that f_n must satisfy

$$f_n'' + \frac{p-1}{r} f_n' + \left[k^2 - \frac{n(n+p-2)}{r^2}\right] f_n = 0. \tag{7}$$

We now introduce a new independent variable $z = kr$ and let $\xi_n(z)$ denote the solution of the resultant equation

$$\xi_n'' + \frac{p-1}{z} \xi_n' + \left[1 - \frac{n(n+p-2)}{z^2}\right] \xi_n = 0. \tag{8}$$

The substitution $\xi_n(z) = z^{1-p/2} I(z)$ reduces (8) to

$$I'' + \frac{1}{z} I' + \left[1 - \frac{(n + (\frac{1}{2})p - 1)^2}{z^2}\right] I = 0. \tag{9}$$

We define Bessel's equation by

$$I'' + \frac{1}{z} I' + \left[1 - \frac{\nu^2}{z^2}\right] I = 0. \tag{10}$$

For $\nu = n + \frac{1}{2}p - 1$ (9) and (10) coincide. (10) has two singular points, a regular singularity at $z = 0$ and an irregular singularity at $z = \infty$. One solution that can be found near the origin is

$$J_\nu(z) = \left(\frac{z}{2}\right)^\nu \sum_{l=0}^{\infty} \frac{(iz/2)^{2l}}{l!\Gamma(\nu + l + 1)}, \qquad \operatorname{Re} \nu > 0. \tag{11}$$

The above is the standard Bessel function. A solution of (8) is

$$\xi_n(z) = z^{1-p/2} J_{n+\frac{1}{2}p-1}(z).$$

To determine the full solution of (10) we note that if we replace ν by $-\nu$ in (10) it remains unchanged. It follows that $J_{-\nu}(z)$ is also a solution. To determine whether J_ν and $J_{-\nu}$ are linearly independent we proceed as follows. By standard operations we deduce that their Wronskian satisfies

$$W' + \frac{1}{z} W = 0.$$

Then

$$W[J_\nu, J_{-\nu}] = J_\nu J_{-\nu}' - J_{-\nu} J_\nu' = \frac{c}{z}.$$

To deduce c we use the leading terms in (11). Then we obtain

$$\frac{c}{z} = \frac{-2\nu/z}{\nu\Gamma(\nu)\Gamma(1-\nu)} = \frac{-2 \sin \pi\nu}{\pi z}.$$

It follows that when ν is not an integer J_ν and $J_{-\nu}$ are linearly independent.

2. Integral Representations

The wave equation

$$\Delta_p V - \frac{1}{c^2}\frac{\partial^2 V}{\partial t^2} = 0$$

has plane wave solutions of the form

$$V = e^{i[\omega/c\, X \cdot \eta - \omega t]}, \qquad X \cdot \eta = \sum_{i=1}^{p} x_i \eta_i, \qquad |\eta| = 1$$

The latter is a plane wave propagating in the η direction. It follows that, if we let $\omega/c = k$,

$$U = e^{ikX \cdot \eta}$$

satisfies

$$\Delta_p U + k^2 U = 0. \tag{1}$$

By a superposition of such plane waves we can construct more general solutions. For example

$$U = \int_{|\eta|=1} e^{ikr\xi \cdot \eta} S_n(\eta)\, d\omega_p \tag{2}$$

is also a solution of (1). But in view of the fact that (2) is an integral of the Hecke-Funk type we can reduce it to a one-dimensional integral.

$$U = \omega_{p-1}\, S_n(\xi) \int_{-1}^{1} e^{ikrt} P_n(t)(1 - t^2)^{(p-3)/2}\, dt.$$

But we note from our previous discussion that U must be of the form

$$U = S_n(\xi)\xi_n(kr)$$

and it follows that

$$J_{n+\frac{1}{2}p-1}(z) = cz^{p/2-1} \int_{-1}^{1} e^{izt} P_n(t)(1 - t^2)^{(p-3)/2}\, dt \tag{3}$$

for a suitable constant c.

To deduce c in (3) we wish to compare the leading terms in (3) and 1.11. Using Rodrigues' formula

$$P_n(t) = \frac{(-1/2)^n \Gamma((p-1)/2)}{\Gamma(n + (p-1)/2)} (1 - t^2)^{(-p+3)/2} D^n (1 - t^2)^{n+(p-3)/2}$$

in (3) and integrating by parts n times we find that

$$J_{n+\frac{1}{2}p-1}(z) = cz^{\frac{1}{2}p-1}\frac{(\frac{1}{2})^n\Gamma((p-1)/2)}{\Gamma(n+(p-1)/2)}\int_{-1}^{1}(1-t^2)^{n+(p-3)/2}D^n e^{izt}\,dt$$

$$= \frac{cz^{n+\frac{1}{2}p-1}(i/2)^n\Gamma((p-1)/2)}{\Gamma(n+(p-1)/2)}\int_{-1}^{1}(1-t^2)^{n+(p-3)/2}e^{izt}\,dt \quad (4)$$

The integrand in (4) can be expanded in a power series in z and integrated term by term. The resultant integrals can be evaluated in terms of Beta functions, and series 1.11 is thus reconstructed. A comparison of the first two terms shows that

$$c = \frac{(-i)^n}{2^{\frac{1}{2}p-1}\sqrt{\pi}\,\Gamma((p-1)/2)}$$

so that finally

$$J_{n+p/2-1}(z) = \frac{(z/2)^{n+\frac{1}{2}p-1}}{\sqrt{\pi}\,\Gamma(n+(p-1)/2)}\int_{-1}^{1}e^{izt}(1-t^2)^{n+(p-3)/2}\,dt. \quad (5)$$

More generally it can be shown that

$$J_\nu(z) = \frac{(z/2)^\nu}{\sqrt{\pi}\,\Gamma(\nu+1/2)}\int_{-1}^{1}e^{izt}(1-t^2)^{\nu-\frac{1}{2}}\,dt$$

$$= \frac{(z/2)^\nu}{\sqrt{\pi}\,\Gamma(\nu+1/2)}\int_{0}^{\pi}e^{iz\cos\theta}\sin^{2\nu}\theta\,d\theta, \qquad \operatorname{Re}\nu > -\tfrac{1}{2}. \quad (6)$$

In particular for $\nu = \frac{1}{2}$ we find that

$$J_{\frac{1}{2}}(z) = \sqrt{\frac{2}{\pi z}}\sin z$$

and for $\nu = \frac{3}{2}$

$$J_{\frac{3}{2}}(z) = -\sqrt{\frac{2}{\pi z}}\left[\cos z - \frac{\sin z}{z}\right].$$

One can easily deduce from (6) that if $\nu = \frac{1}{2} + m$, where m is an integer, then

$$J_{\frac{1}{2}+m}(z) = \sqrt{\frac{2}{\pi z}}\left[P_m\left(\frac{1}{z}\right)\cos z + Q_m\left(\frac{1}{z}\right)\sin z\right] \quad (7)$$

where P_m and Q_m are suitable polynomials of degree m. We shall rediscover this fact from another point of view at a later stage.

3. Relationship to the Legendre Functions

We shall, in this section, demonstrate that the Bessel functions are confluent cases of the Legendre functions. For simplicity we consider the case $p = 3$. Then

$$\Delta_3 r^n P_n^m (\cos \theta) e^{im\phi} = 0 \tag{1}$$

if we employ the method of separation of variables in spherical coordinates.

Similarly we find in cylindrical coordinates

$$\Delta_3 f(\rho) e^{i\alpha z} e^{im\phi} = 0 \tag{2}$$

where

$$f''(\rho) + \frac{1}{\rho} f'(\rho) - \left[m^2 + \frac{\alpha^2}{\rho^2} \right] f(\rho) = 0. \tag{3}$$

We note that a solution of (3) can be expressed in terms of Bessel functions

$$f(\rho) = J_\alpha(im\rho).$$

The relationship between spherical and cylindrical coordinates can be expressed as follows

$$z = r \cos \theta.$$
$$\rho = r \sin \theta.$$

In particular for small values of θ we find

$$z \approx r$$
$$\rho \approx r\theta.$$

Comparing the θ dependent terms in (1) and (2) we expect that there should be some relationship between $f(\rho)$ and $P_n^m (\cos \theta)$. The function $P_n^m (\cos \theta)$ satisfies

$$y'' + \cot \theta y' + \left[n(n + 1) - \frac{m^2}{\sin^2 \theta} \right] y = 0,$$

and for small values of θ it takes the form

$$y'' + \frac{1}{\theta} y' + \left[n(n + 1) - \frac{m^2}{\theta^2} \right] y = 0 \tag{4}$$

whose solution is $J_m(\sqrt{n(n + 1)}\theta)$. We may suppose therefore that

$$\lim_{\theta \to 0} \frac{P_n^m(\cos \theta)}{J_m(\sqrt{n(n + 1)}\, \theta)}$$

exists and will be a constant depending on n and m. Another way of expressing this is to state that

$$\lim_{n\to\infty} c(n)P_n^m\left(\cos\frac{z}{n}\right) = J_m(z)$$

for a suitable function $c(n)$, which may depend on m. Note that if we replace θ by z/n in (4) and let $n \to \infty$ we obtain

$$\frac{d^2y}{dz^2} + \frac{1}{z}\frac{dy}{dz} + \left[1 - \frac{m^2}{z^2}\right]y = 0$$

which is satisfied by $J_m(z)$.

To make the preceding discussion more precise we return to a familiar integral representation

$$P_n^m(\cos\theta) = \frac{i^m(n+m)!}{2\pi n!}\int_{-\pi}^{\pi}[\cos\theta + i\sin\theta\cos\Psi]^n e^{-im\Psi}\,d\Psi.$$

We let $\theta = z/n$ in the above so that

$$P_n^m\left(\cos\frac{z}{n}\right) = \frac{i^m(n+m)!}{2\pi n!}\int_{-\pi}^{\pi}\left[\cos\frac{z}{n} + i\sin\frac{z}{n}\cos\Psi\right]^n e^{-im\Psi}\,d\Psi \tag{5}$$

For large n we have

$$\cos\frac{z}{n} \approx 1, \qquad \sin\frac{z}{n} \approx \frac{z}{n}$$

and one can easily show that

$$\lim_{n\to\infty}\left[\cos\frac{z}{n} + i\sin\frac{z}{n}\cos\Psi\right]^n = \lim_{n\to\infty}\left[1 + \frac{iz\cos\Psi}{n}\right]^n = e^i\,z\cos\Psi.$$

we also have

$$\lim_{n\to\infty}\frac{(n+m)!}{n!\,n^m} = 1.$$

From (5) we find that

$$\lim_{n\to\infty}\frac{1}{n^m}P_n^m\left(\cos\frac{z}{n}\right) = \frac{i^m}{2\pi}\int_{-\pi}^{\pi} e^{i[z\cos\Psi - m\Psi]}\,d\Psi \tag{6}$$

To determine the relationship between the integral representation in (6) and the Bessel functions, we expand (6) in a power series.

$$\frac{i^m}{2\pi}\int_{-\pi}^{\pi} e^{i[z\cos\Psi - m\Psi]}\,d\Psi = \frac{i^m}{2\pi}\int_{-\pi}^{\pi}\sum_{k=0}^{\infty}\frac{(iz)^k\cos^k\Psi}{k!}e^{-im\Psi}\,d\Psi \tag{7}$$

To evaluate each integral in (7) we let $e^{i\Psi} = t$ and evaluate the integral by means of residue integration over the unit circle in the t domain

$$\int_{-\pi}^{\pi} \cos^k \Psi e^{-im\Psi}\, d\Psi = \frac{1}{i}\int \frac{((t+1/t)/2)^k\, dt}{t^{m+1}} = \text{Res}\ \frac{2\pi}{2^k}\frac{(t+1/t)^k}{t^{m+1}}$$

By the binomial theorem

$$\left(t+\frac{1}{t}\right)^k = \sum_{r=0}^{k}\binom{k}{r} t^{k-2r}$$

and a term of the type t^m can occur in the above only if $k \geq m$ and k has the same parity as m. In that case $k - 2r = m$ for $r = (k - m)/2$. Then

$$\int_{-\pi}^{\pi} \cos^k \Psi e^{-im\Psi}\, d\Psi = \frac{2\pi}{2k}\binom{k}{(k-m)/2}$$

and the right side of (7) reduces to

$$\frac{i^m}{2\pi}\sum \frac{(iz)^k}{k!}\frac{2\pi}{2^k}\binom{k}{(k-m)/2} = \left(\frac{-z}{2}\right)^m \sum_{l=0}^{\infty}\frac{(iz/2)^{2l}}{l!\,(m+l)!}$$

since the sum on the left runs over all indices of the form $m + 2l$. A comparison with 1.11 shows that

$$\lim_{n\to\infty}\frac{1}{n^m}P_n^{\ m}\left(\cos\frac{z}{n}\right) = \frac{i^m}{2\pi}\int_{-\pi}^{\pi} e^{i[z\cos\Psi - m\Psi]}\, d\Psi = (-1)^m J_m(z). \tag{8}$$

If in (8) we let $\Psi = \alpha - \pi/2$, and use the fact that the integration is over a full period of the integrand we obtain

$$J_m(z) = \frac{1^1}{2\pi}\int_{-\pi}^{\pi} e^{i[z\sin\alpha - m\alpha]}\, d\alpha = \frac{1}{\pi}\int_0^{\pi}\cos[z\sin\alpha - m\alpha]\, d\alpha. \tag{9}$$

The last step is a consequence of the fact that the first integral can be decomposed into an even and an odd part, the latter having a zero contribution. Note that in the derivation of (8) and (9) the fact that m is a non-negative integer entered in an essential manner. We shall investigate generalizations of these representations to non-integral indices in a later section.

4. The Generating Function of the Bessel Function

Formula 3.9, namely

$$J_m(z) = \frac{1}{2\pi}\int_{-\pi}^{\pi} e^{i[z\sin\alpha]}\, e^{-im\alpha}\, d\alpha$$

may be given the following interpretation. If the periodic function $g(\alpha)$ is expanded in a Fourier series we have

$$g(\alpha) = \sum_{m=-\infty}^{\infty} a_m e^{im\alpha}$$

where

$$a_m = \frac{1}{2\pi} \int_{-\pi}^{\pi} g(\alpha) e^{-im\alpha} d\alpha.$$

It follows that

$$e^{iz \sin \alpha} = \sum_{m=-\infty}^{\infty} J_m(z) e^{im\alpha} \tag{1}$$

where z must be treated as a parameter.

If we replace $e^{i\alpha}$ by t in (1) we obtain

$$e^{z/2(t-1/t)} = \sum_{m=-\infty}^{\infty} J_m(z) t^m \tag{2}$$

which is known as the generating function of the Bessel Function. The right side of (2) is the Laurent expansion in the domain $0 < |t| < \infty$ in which the left side is regular and single valued. The left has only two essential singularities at $t = 0$ and $t = \infty$ respectively. Hence (2) converges for all $0 < |t| < \infty$.

(2) can actually be used as the definition of the Bessel Function. Replacing z by $-z$ and t by $-t$ in (2) we have

$$e^{z/2(t-1/t)} = \sum_{m=-\infty}^{\infty} (-1)^m J_m(-z) t^m$$

so that

$$J_m(-z) = (-1)^m J_m(z). \tag{3}$$

Similarly if we replace z by $-z$ and t by $1/t$ we have

$$e^{z/2(t-1/t)} = \sum_{m=-\infty}^{\infty} J_{-m}(-z) t^m$$

and comparison with (3) and (2) shows that

$$J_{-m}(-z) = J_m(z) = (-1)^{-m} J_{-m}(z)$$

and

$$J_{-m}(z) = (-1)^m J_m(z). \tag{4}$$

Differentiation of (2) with respect to t and some simple rearrangements leads to

$$\frac{z}{2}\left(t+\frac{1}{t}\right)\sum_{m=-\infty}^{\infty} J_m(z)t^m = \sum_{m=-\infty}^{\infty} mJ_m(z)t^m.$$

A comparison of coefficients of t^m on both sides now shows that

$$\frac{2m}{z} J_m(z) = J_{m+1}(z) + J_{m-1}(z), \tag{5}$$

which is one of the standard recursion formulas for the Bessel functions. A similar approach allows us to find a differential-difference equation. Differentiation of (2) with respect to z leads to

$$\frac{1}{2}\left(t-\frac{1}{t}\right)\sum_{m=-\infty}^{\infty} J_m(z)t^m = \sum_{m=-\infty}^{\infty} J'_m(z)t^m$$

so that

$$J'_m(z) = \tfrac{1}{2}[J_{m+1}(z) - J_{m-1}(z)]. \tag{6}$$

By adding and subtracting (5) and (6) we can obtain

$$J_{m-1}(z) = \frac{m}{z} J_m(z) + J'_m(z)$$

$$J_{m+1}(z) = \frac{m}{z} J_m(z) - J'_m(z).$$

In the top equation we replace m by $m+1$ and thus obtain

$$J_m(z) = \frac{m+1}{z} J_{m+1}(z) + J'_{m+1}(z)$$

$$J_{m+1}(z) = \frac{m}{z} J_m(z) - J'_m(z).$$

Elimination of $J_{m+1}(z)$ finally leads to

$$J''_m(z) + \frac{1}{z} J'_m(z) + \left(1 - \frac{m^2}{z^2}\right) J_m(z) = 0 \tag{7}$$

the differential equation for the Bessel function.

(2) may also be used to obtain the series expansion for $J_m(z)$. Clearly we have

$$\sum_{m=-\infty}^{\infty} J_m(z)t^m = e^{z/2t}e^{-z/2t} = \sum_{n=0}^{\infty} \frac{(z/2)^n t^n}{n!} \sum_{n=0}^{\infty} \frac{(-z/2)^n t^{-n}}{n!}.$$

We now assume that $m \geq 0$, and seek the coefficient of t^m on both sides. Then

$$J_m(z) = \sum_{n-k=m} \frac{(z/2)^n}{n!} \frac{(-z/2)^k}{k!} = \sum_{k=0}^{\infty} \frac{(z/2)^{m+k}(-z/2)^k}{(m+k)!\,k!}$$

$$= \left(\frac{z}{2}\right)^m \sum_{k=0}^{\infty} \frac{(iz/2)^{2k}}{k!\,(m+k)!} \tag{8}$$

An explicit calculation shows easily that (8) satisfies (7).

5. More Integral Representations

From the generating function

$$e^{z/2(t-1/t)} = \sum_{-\infty}^{\infty} J_m(z)t^m$$

we deduce immediately, by Cauchy's integral theorem

$$J_m(z) = \frac{1}{2\pi i} \int_c t^{-m-1}\, e^{z/2(t-1/t)}\, dt. \tag{1}$$

where c is a closed path enclosing the origin. We can generalize the above representation as follows. We suppose that $\operatorname{Re} z > 0$, and select c as shown in the following figure.

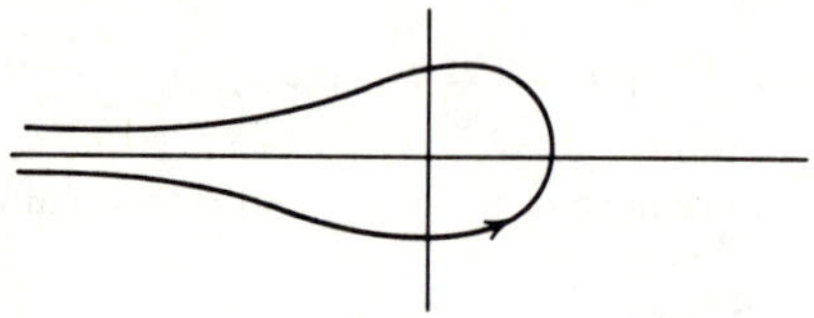

The integral

$$J_\nu(z) = \frac{1}{2\pi i} \int_{-\infty}^{(0+)} t^{-\nu-1} e^{z/2(t-1/t)}\, dt \tag{2}$$

is well defined on this path, and converges exponentially, regardless of ν. For ν an integer it reduces to (1). One can easily verify that it satisfies the differential equation for the Bessel function.

Another way of deriving (2) is to use the series

$$J_\nu(z) = \left(\frac{z}{2}\right)^\nu \sum_{k=0}^{\infty} \frac{(iz/2)^{2k}}{k!\,\Gamma(\nu+k+1)}$$

and the representation

$$\frac{\pi}{\Gamma(\nu+k+1)} = \frac{1}{2\pi i} \int_{-\infty}^{(0+)} e^t t^{-\nu-k-1}\, dt$$

Using the latter in the series we have

$$J_\nu(z) = \frac{(z/2)^\nu}{2\pi i} \int_{-\infty}^{(0+)} t^{-\nu-1} e^t \sum_{k=0}^{\infty} \frac{(-z^2/4t)^k}{k!} \, dt$$

$$= \frac{(z/2)^\nu}{2\pi i} \int_{-\infty}^{(0+)} t^{-\nu-1} e^{t - z^2/4t} \, dt$$

Finally replacing t by $(z/2)t$ we obtain (2) again. For the case Re $z < 0$ we can replace the contour in (2) by a similar one opening to the positive side.

We now take c in (2) to consist of the unit circle and the upper and lower part of Im $t = 0$, Re $t \leqslant -1$, as shown below.

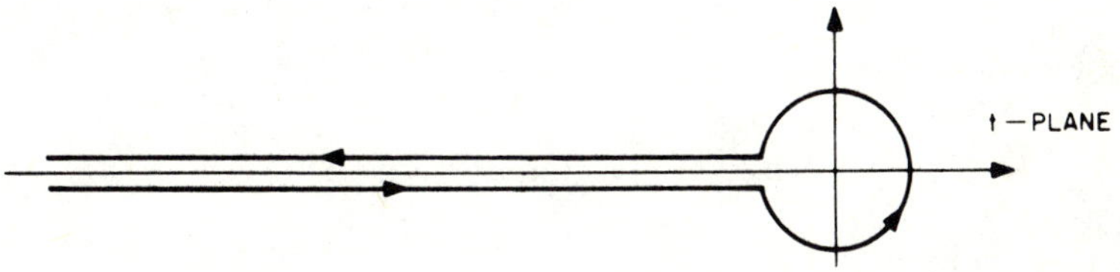

Under the substitution $t = e^\omega$, $z/2(t - 1/t) = z \sinh \omega$, (2) becomes

$$J_\nu(z) = \frac{1}{2\pi i} \int_\Gamma e^{-\nu\omega} e^{z \sinh \omega} \, d\omega, \qquad \text{Re } z > 0 \tag{3}$$

where Γ is as shown in the following figure. By introducing explicit

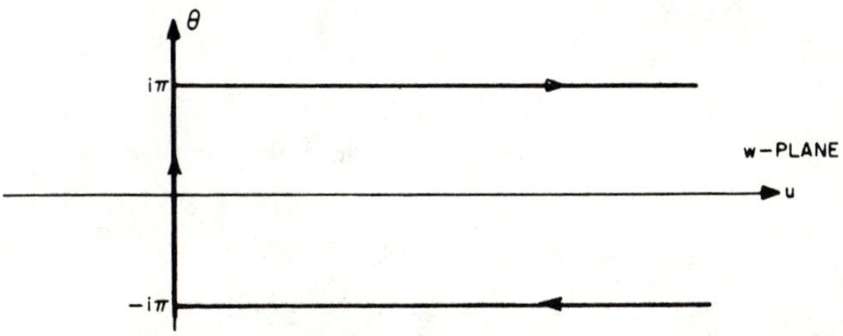

coordinates one can rewrite (3) in the form

$$J_\nu(z) = \frac{1}{\pi} \int_0^\pi \cos(\nu\theta - z \sin \theta) \, d\theta - \frac{\sin \nu\pi}{\pi} \int_0^\infty e^{-\nu u - z \sinh u} \, du \qquad \text{Re } z > 0 \quad (4)$$

(4) is a generalization of 3.9 for non-integer ν. When ν is an integer (4) reduces to 3.9.

In studying Legendre functions we had encountered Maxwell's theory of poles. Similar representations for Bessel functions can be derived. For $t = \cos z/n$ we have

$$\frac{d}{dt} = -\frac{n}{\sin z/n}\frac{d}{dz} \approx -\frac{n^2}{z}\frac{d}{dz}$$

for large n. We then have, using

$$P_n^m(t) = (-1)^m(1-t^2)^{m/2}\left(\frac{d}{dt}\right)^m P_n(t)$$

$$\frac{1}{n^m} P_n^m\left(\cos\frac{z}{n}\right) \approx (-1)^m \frac{(\sin^m z/n)(-1)^m n^{2m}}{n^m}\left(\frac{d}{zdz}\right)^m P_n\left(\cos\frac{z}{n}\right).$$

In view of the fact that

$$\lim_{n\to\infty}\frac{1}{n^m} P_n^m\left(\cos\frac{z}{n}\right) = (-1)^m J_m(z)$$

we find formally

$$J_m(z) = (-1)^m z^m\left(\frac{d}{zdz}\right)^m J_0(z). \tag{5}$$

(5) can be proved independently using the recurrence formulas 4.5 and 4.6. Using (4) we have

$$J_0(z) = \frac{1}{\pi}\int_0^\pi \cos(z\sin\theta)\,d\theta = \frac{2}{\pi}\int_0^{\pi/2}\cos(z\sin\theta)\,d\theta$$

From (5) we have

$$J_1(z) = -\frac{d}{dz}J_0(z) = \frac{2}{\pi}\int_0^{\pi/2}\sin\theta\sin(z\sin\theta)\,d\theta$$

$$= \frac{2z}{\pi}\int_0^{\pi/2}\cos[z\sin\theta]\cos^2\theta\,d\theta$$

by integrating by parts. Similarly

$$J_2(z) = z^2\frac{d}{zdz} x \frac{d}{zdz}J_0(z) = -z\frac{d}{dz}\frac{1}{z}J_1(z)$$

$$= \frac{2z}{\pi}\int_0^{2/\pi}\sin\theta\sin[z\sin\theta]\cos^2\theta\,d\theta$$

$$= \frac{2}{\pi}\frac{z^2}{3}\int_0^{\pi/2}\cos[z\sin\theta]\cos^4\theta\,d\theta.$$

More generally, using a proof by induction we find

$$J_m(z) = \frac{2^m}{\pi} \frac{(m-1)!\, z^m}{(2m-1)!} \int_0^{\pi/2} \cos[z \sin \theta] \cos^{2m} \theta \, d\theta. \tag{6}$$

The latter can be generalized to arbitrary indices by writing

$$J_\nu(z) = \frac{2^\nu}{\pi} \frac{\Gamma(\nu) z^\nu}{\Gamma(2\nu)} \int_0^{\pi/2} \cos[z \sin \theta] \cos^{2\nu} \theta \, d\theta. \qquad \operatorname{Re} 2\nu > -1.$$

Use of the duplication formula

$$\Gamma(\nu)\Gamma(\nu + \tfrac{1}{2}) = 2\sqrt{\pi}\, 2^{-2\nu}\Gamma(2\nu)$$

allows us to write it in the standard form

$$J_\nu(z) = \frac{2(z/2)^\nu}{\sqrt{\pi}\Gamma(\nu + \frac{1}{2})} \int_0^{\pi/2} \cos[z \sin \theta] \cos^{2\nu} \theta \, d\theta. \qquad \operatorname{Re} \nu > -\tfrac{1}{2}. \tag{7}$$

(7) is often referred to as Poisson's integral. It can be verified independently by expanding the integrand in a power series and integrating term by term.

(7) could have been derived by seeking a representation of the form

$$J_m(z) = \int_\alpha^\beta e^{izt} V(t)\, dt.$$

This may also be compared to the method in deriving 2.5 and 2.6.

Finally we return to (5) and replace m by $-m$. Then

$$J_m(z) = z^{-m} \left(\frac{d}{z\, dz} \right)^{-m} J_0(z). \tag{8}$$

In particular for $m = 1$ we have

$$J_1(z) = \frac{1}{z} \left(\frac{d}{z\, dz} \right)^{-1} J_0(z).$$

To interpret $\left(\frac{1}{z} \frac{d}{dz} \right)^{-1}$ we note that if A and B are non-commutative operators then

$$(AB)^{-1} = B^{-1}A^{-1}.$$

Hence

$$J_1(z) = \frac{1}{z} \left(\frac{d}{dz} \right)^{-1} \left(\frac{1}{z} \right)^{-1} J_0(z)$$

$$= \frac{1}{z} \int_0^z \zeta J_0(\zeta)\, d\zeta$$

since the inverse of differentiation is integration. The lower limit is so chosen as to satisfy the initial condition $\lim_{z \to 0} J_1(z) = 0$.

For $m = 2$ we find

$$J_2(z) = \frac{1}{z^2}\left(\frac{d}{z\,dz}\right)^{-1}\left(\frac{d}{z\,dz}\right)^{-1} J_0(z)$$

$$= \frac{1}{z^2}\left(\frac{d}{z\,dz}\right)^{-1} zJ_1(z) = \frac{1}{z^2}\int_0^z \zeta^2 J_1(\zeta)\,d\zeta$$

$$= \frac{1}{z^2}\int_0^z \zeta\left[\int_0^\zeta \tau J_0(\tau)\,d\tau\right] d\zeta$$

$$= \frac{1}{2z^2}\int_0^z \zeta(z^2 - \zeta^2)J_0(\zeta)\,d\zeta.$$

By induction we find

$$J_m(z) = \frac{1}{2^{m-1}\Gamma(m)z^m}\int_0^z (z^2 - \zeta^2)^{m-1} J_0(\zeta)\,d\zeta$$

and letting $\xi = z \sin\theta$

$$J_m(z) = \frac{z^m}{2^{m-1}\Gamma(m)}\int_0^{\pi/2} \sin\theta \cos^{2m-1}\theta J_0(z\sin\theta)\,d\theta.$$

Again the above has a generalization to arbitrary indices

$$J_\nu(z) = \frac{z^\nu}{2^{\nu-1}\Gamma(\nu)}\int_0^{\pi/2} \sin\theta\cos^{2\nu-1}\theta J_0(z\sin\theta)\,d\theta \qquad \text{Re}\,\nu > 0. \quad (9)$$

(9) is known as Sonine's integral.

Sonine's integral may be arrived at in still another fashion. Consider the following integral.

$$I_1 = \int_{\substack{\zeta\geq 0\\ \eta\geq 0}} e^{iz\xi}\zeta^{2m-1}\,ds$$

where the integration is to be carried out over that part of the surface

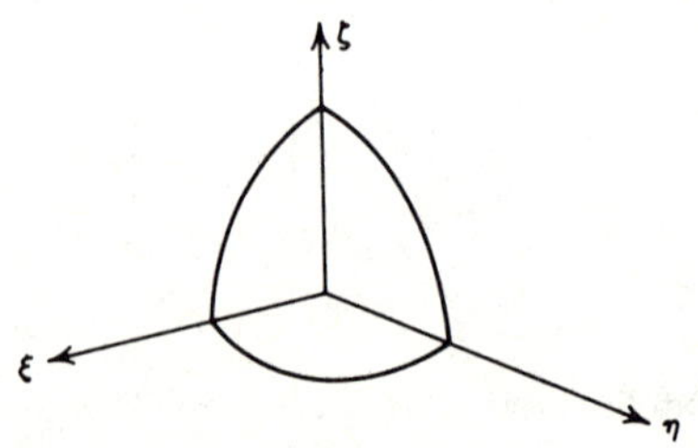

$\xi^2 + \eta^2 + \zeta^2 = 1$ on which $\zeta \geq 0$, $\eta \geq 0$, in other words a certain quarter sphere. A similar integral is given by

$$I_2 = \int_{\substack{\eta \geq 0 \\ \xi \geq 0}} e^{iz\zeta} \eta^{2m-1}\, ds$$

By symmetry we note that

$$I_1 = I_2,$$

since under a suitable rotation of the coordinates we effect the substitution $\xi \to \zeta$, $\zeta \to \eta$, $\eta \to \xi$. Next we introduce the spherical coordinates

$$\xi = \sin\theta\cos\phi, \qquad \eta = \sin\theta\sin\phi, \qquad \zeta = \cos\theta.$$

Then

$$I_1 = \int_0^{\pi/2} \sin\theta(\cos\theta)^{2m-1} \int_0^{\pi} e^{iz\sin\theta\cos\phi}\, d\phi\, d\theta$$

$$= \pi \int_0^{\pi/2} \sin\theta(\cos\theta)^{2m-1} J_0(z\sin\theta)\, d\theta$$

by the use of (4). We also find that

$$I_2 = \int_0^{\pi} e^{iz\cos\theta}(\sin\theta)^{2m} \int_0^{\pi/2} (\sin\phi)^{2m-1}\, d\phi\, d\theta.$$

The inner integral can be evaluated in terms of beta functions

$$\int_0^{\pi/2} (\sin\phi)^{2m-1}\, d\phi = \frac{1}{2} B\left(m, \frac{1}{2}\right) = \frac{\Gamma(m)\Gamma(\frac{1}{2})}{2\Gamma(m+\frac{1}{2})}.$$

For the outer integral we find

$$\int_0^{\pi} e^{iz\cos\theta}(\sin\theta)^{2m}\, d\theta = 2\int_0^{\pi/2} \cos[z\sin\theta](\cos\theta)^{2m}\, d\theta$$

$$= \frac{\sqrt{\pi}\,\Gamma(m+\frac{1}{2})}{(z/2)^m} J_m(z)$$

so that

$$I_2 = \frac{\pi\Gamma(m)J_m(z)}{2(z/2)^m}.$$

By equating I_1 and I_2 we again obtain Sonine's integral.

6. Addition Theorems

In many applications we encounter the need for expansions of the type

$$J_\nu(f(z)) = \sum_{m=-\infty}^{\infty} a_m J_{\nu+m}(z),$$

where $f(z)$ is a prescribed function of z. We shall discuss a number of such cases in the present section.

Taylor's theorem for an analytic function may be written out in one of the following forms

$$f(z+h) = \sum_{m=0}^{\infty} \frac{h^m}{m!} f^{(m)}(z) = \frac{1}{2\pi i} \int_c \frac{f(\zeta)}{\zeta - z - h}\, d\zeta \tag{1}$$

where c is a contour enclosing the point $z + h$. From the expansion

$$J_0(z) = \sum_{n=0}^{\infty} \frac{(-z^2/4)^n}{n!^2}$$

we see that

$$f(z) = J_0(\sqrt{z}),$$

for a suitable determination of $\sqrt{z}$, is an analytic function of z. Applying Taylor's theorem we have

$$J_0(\sqrt{z+h}) = \sum_{m=0}^{\infty} \frac{h^m}{m!} \left(\frac{d}{dz}\right)^m J_0(\sqrt{z}). \tag{2}$$

Replacing z by z^2 in (2) yields

$$J_0(\sqrt{z^2+h}) = \sum_{m=0}^{\infty} \frac{h^m}{2^m m!} \left(\frac{d}{z\,dz}\right)^m J_0(z)$$

and recalling

$$\left(\frac{d}{z\,dz}\right)^m J_0(z) = (-1)^m z^{-m} J_m(z)$$

we have

$$J_0(\sqrt{z^2+h}) = \sum_{m=0}^{\infty} \frac{(-h/2z)^m}{m!} J_m(z). \tag{3}$$

If we let $h = \lambda^2 z^2 - z^2$ we obtain

$$J_0(\lambda z) = \sum_{m=0}^{\infty} \frac{(((1-\lambda^2)/2)z)^m}{m!} J_m(z). \tag{4}$$

(4) is known as the multiplication theorem for $J_0(z)$. More generally one can show that

$$J_n(\lambda z) = \lambda^n \sum_{m=0}^{\infty} \frac{(((1-\lambda^2/)2)z)^m}{m!} J_{n+m}(z). \tag{5}$$

In connection with our study of Legendre functions we had encountered the following addition theorem

$$P_n(\cos\theta\cos\theta' + \sin\theta\sin\theta'\cos(\phi-\phi'))$$
$$= \sum_0^n \varepsilon_m \frac{(n-m)!}{(n+m)!} P_n{}^m(\cos\theta) P_n{}^m(\cos\theta') \cos m(\phi-\phi') \tag{6}$$

resulting from a rotation of coordinates. We expect, therefore, that a suitable limiting form of (6) should lead to a comparable theorem for Bessel functions.

We shall construct a plane tangent to the unit sphere at the north pole. We let

$$\theta = \frac{r}{n}, \qquad \theta' = \frac{\rho}{n}, \qquad \alpha = \frac{R}{n}$$

where $\cos\alpha = \cos\theta\cos\theta' + \sin\theta\sin\theta'\cos(\phi-\phi')$. For small θ, θ', or equivalently large n we find

$$R^2 \approx r^2 + \rho^2 - 2r\rho\cos(\phi-\phi').$$

The above is merely a statement of the law of cosines. Interpreted geometrically, the rotation of coordinates is, for small rotations, a translation on the

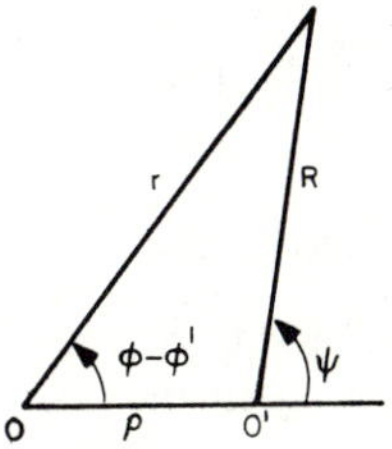

tangent plane. (6) becomes the following

$$P_n\left(\cos\frac{R}{n}\right) = \sum_{m=0}^{n} \varepsilon_m \left[\frac{(n-m)!n^{2m}}{(n+m)!}\right] \frac{P_n{}^m(\cos r/n)}{n^m} \frac{P_n{}^m(\cos\rho/n)}{n^m} \cos m(\phi-\phi')$$

Use of the limiting procedure 3.8 finally yields

$$J_0(R) = \sum_{m=0}^{\infty} \varepsilon_m J_m(r) J_m(\rho) \cos m(\phi-\phi') \tag{7}$$

(7) is known as Gegenbauer's addition theorem. The derivation given is not quite rigorous, but shows the relationship between the addition theorems (6) and (7). We shall discuss a more general result rigorously.

We let

$$Re^{\pm i\Psi} = re^{\pm i\phi} - \rho \tag{8}$$

(see preceding figure with $\phi' = 0$.) From 5.2 we have

$$J_\nu(R)e^{i\nu\Psi} = \frac{1}{2\pi i}\int_{-\infty}^{(0+)} \frac{e^{R/2(t-1/t)+i\nu\Psi}}{t^{\nu+1}}\,dt.$$

If we replace t by $\tau e^{i\Psi}$ and rotate the contour to preserve the convergence of the integral we obtain

$$J_\nu(R)e^{i\nu\Psi} = \frac{1}{2\pi i}\int_c \frac{e^{R/2(\tau e^{i\Psi}-e^{-i\Psi}/\tau)}}{\tau^{\nu+1}}\,d\tau$$

$$= \frac{1}{2\pi i}\int_c \frac{e^{r/2(\tau e^{i\phi}-e^{-i\phi}/\tau)}xe^{-\rho/2(\tau-1/\tau)}}{\tau^{\nu+1}}\,d\tau$$

by the use of (8). By use of the generating function

$$e^{-\rho/2(\tau-1/\tau)} = \sum_{m=-\infty}^{\infty}(-1)^m J_m(\rho)\tau^m$$

we obtain

$$J_\nu(R)e^{i\nu\Psi} = \sum_{-\infty}^{\infty}(-1)^m J_m(\rho)\frac{1}{2\pi i}\int_c \frac{e^{r/2(\tau e^{i\phi}-e^{-i\phi}/\tau)}}{\tau^{\nu-m+1}}\,d\tau$$

$$J_\nu(R)e^{i\nu\Psi} = \sum_{-\infty}^{\infty}(-1)^m J_m(\rho)J_{\nu-m}(r)e^{i(\nu-m)\phi}$$

$$= \sum_{-\infty}^{\infty} J_m(\rho)J_{\nu+m}(r)e^{i(\nu+m)\phi}. \tag{9}$$

(9) is known as Neumann's addition theorem. For $\nu = 0$ it reduces to Gegenbauer's addition theorem. A simple calculation shows that all manipulations performed are legitimate if $|\phi - \Psi| < \pi/2$ or equivalently $\rho < r$. If ν is integral this requirement may be dropped.

(9) can be given the following interpretation. The functions $J_{\nu+m}(r)e^{i(\nu+m)\phi}$ are all solutions of the equation

$$\Delta_2 U + U = 0.$$

The function $J_\nu(R)\,e^{i\nu\Psi}$ satisfies the same equation, but the coordinates (R, Ψ) are referred to a coordinate system obtained by translating the (r, ϕ) coordinate system through a distance ρ. In view of the fact that the differential equation is invariant under all rigid transformations we expect that an expansion of the type

$$J_\nu(R)e^{i\nu\Psi} = \sum_{-\infty}^{\infty} f_m(\rho)J_{\nu+m}(r)e^{i(\nu+m)\phi}$$

should exist. (9) is that expansion. More generally we had seen in section 1, that $R^{1-p/2}J_{(p-2)/2}(R)$ is a solution of

$$\Delta_p U + U = 0.$$

But so are also the functions $r^{1-p/2}J_{n+(p-)/2}(r)S_n(\xi)$, where $S_n(\xi)$ is a suitable spherical harmonic. Here again, the differential equation is invariant under all rigid transformations and R is referred to a translated coordinate system. We may expect an expansion of the following type to exist therefore.

$$\frac{J_{(p-2)/2}(R)}{R^{(p-2)/2}} = \sum_{m=0}^{\infty} f_m(\rho)\frac{J_{m+(p-2)/2}(r)}{r^{(p-2)/2}} S_m(\xi).$$

We first note the following formal identities. Let

$$R = \sqrt{r^2 + \rho^2 - 2r\rho\cos\phi}$$

$$\frac{d}{d(\cos\phi)} f(R) = f'(R)\frac{dR}{d(\cos\phi)} = \rho r\frac{f'(R)}{R}.$$

Then we have

$$\left(\frac{d}{R\,dR}\right)^n \equiv \frac{(-1)^n}{(\rho r)^n} D^n, \qquad D \equiv \frac{d}{d(\cos\phi)}. \tag{10}$$

Secondly we have the known series

$$\ln[1 - 2r\cos\phi + r^2] = -2\sum_{m=1}^{\infty}\frac{\cos m\phi}{m} r^m.$$

Applying D to the above n times we find

$$\frac{(2r)^n(n-1)!}{[1 - 2r\cos\phi + r^2]^n} = 2\sum_{m=1}^{\infty}\frac{D^n\cos m\phi}{m} r^m. \tag{11}$$

In Chapter 6 we had seen that

$$\frac{1}{[1 - 2r\cos\phi + r^2]^n} = \sum_{m=0}^{\infty}\binom{m+2n-1}{m} P_{m,\,2n+2}(\cos\phi)r^m. \tag{12}$$

A comparison of the expansions in (11) and (12) now shows that

$$D^n\cos(m+n)\phi = (m+n)2^{n-1}\Gamma(n)\binom{m+2n-1}{m} P_{m,\,2n+2}(\cos\phi). \tag{13}$$

We had seen earlier that

$$J_0(R) = \sum_{m=0}^{\infty} \varepsilon_m J_m(\rho) J_m(r) \cos m\phi.$$

We now apply the operator defined in (10) to the above and recall that

$$\left(\frac{d}{R\,dR}\right)^n J_0(R) = (-1)^n \frac{J_n(R)}{R^n}.$$

Using (10) and (13) we find that (after replacing n by $(p-2)/2$)

$$\frac{J_{(p-2/2)}(R)}{R^{(p-2/2)}} = 2^{(p-2/2)}\Gamma\left(\frac{p-2}{2}\right) \sum_{m=0}^{\infty} \left(m + \frac{p-2}{2}\right)\binom{m+p-3}{m}$$
$$\times \frac{J_{m+(p-2/2)}(\rho)}{\rho^{(p-2/2)}} \frac{J_{m+(p-2/2)}(r)}{r^{(p-2/2)}} P_{m,p}(\cos\phi). \qquad (14)$$

(14) is often found stated as

$$\frac{J_{(p-2)/2}(R)}{R^{(p-2)/2}} = 2^{((p-2)/2)}\Gamma((p-2)/2) \sum_{m=0}^{\infty}$$
$$\times \left(m + \frac{p-2}{2}\right) \frac{J_{m+((p-2)/2)}(\rho)}{\rho^{(p-2)/2}} \frac{J_{m+((p-2)/2)}(r)}{r^{(p-2)/2}} C_m^{p-2}(\cos\phi). \qquad (15)$$

In terms of Gegenbauer polynomials.

For $p = 2$ we note that (14), or equivalently (15), becomes indeterminate. We note that

$$\Gamma\left(\frac{p-2}{2}\right)\left(m + \frac{p-2}{2}\right)\binom{m+p-3}{m}$$
$$= \frac{\Gamma((p-2)/2)(m+(p-2)/2)(m+p-3)!}{m!\Gamma(p-2)}$$

and

$$\lim_{p\to 2} \frac{\Gamma((p-2)/2)}{\Gamma(p-2)} = 2.$$

It follows that

$$\lim_{p\to 2} \Gamma\left(\frac{p-2}{2}\right)\left(m + \frac{p-2}{2}\right)\binom{m+p-3}{m} = 2 \qquad m \geqslant 1$$
$$= 1 \qquad m = 0.$$

Taking the corresponding limit in (14) and recalling that from Chapter 6 that

$$P_{m,2}(\cos\phi) = \cos m\phi$$

we again obtain

$$J_0(R) = \sum_{m=0}^{\infty} \varepsilon_m J_m(r) J_m(\rho) \cos m\,\phi.$$

For $p = 3$, $P_{m,3}(\cos\phi) = P_m(\cos\phi)$, the classical Legendre polynomial. In this case (14) reduces to

$$\frac{J_{1/2}(R)}{\sqrt{R}} = \sqrt{2\pi}\sum_{m=0}^{\infty}(m + 1/2)\frac{J_{m+1/2}(r)}{\sqrt{r}}\frac{J_{m+1/2}(\rho)}{\sqrt{\rho}} P_m(\cos\phi). \tag{16}$$

Recalling that

$$J_{1/2}(z) = \sqrt{\frac{2}{\pi z}}\sin z$$

(16) reduces to

$$\frac{\sin R}{\pi R} = \sum_{m=0}^{\infty}\left(m + \frac{1}{2}\right)\frac{J_{m+1/2}(r)}{\sqrt{r}}\frac{J_{m+1/2}(\rho)}{\sqrt{\rho}} P_m(\cos\phi).$$

To derive another theorem, often referred to as a singular theorem we will need certain asymptotic formulas for the Bessel functions. We shall defer a precise discussion of these to a later section, but will indicate a heuristic derivation here. It is based on the previously derived formula

$$\frac{1}{n^m} P_n^{\,m}(\cos\theta) = \sqrt{\frac{2}{\pi n \sin\theta}}\cos\left[\left(n + \frac{1}{2}\right)\theta + m\frac{\pi}{2} - \frac{\pi}{4}\right] + \cdots. \tag{17}$$

Setting $\theta = z/n$ and letting $n \to \infty$ we find

$$(-1)^m J_m(z) = \sqrt{\frac{2}{\pi z}}\cos\left[z + m\frac{\pi}{2} - \frac{\pi}{4}\right] + \cdots$$

or equivalently

$$J_m(z) = \sqrt{\frac{2}{\pi z}}\cos\left[z - m\frac{\pi}{2} - \frac{\pi}{4}\right] + \cdots. \tag{18}$$

Since $\theta = z/n$, and (17) is valid for large n, but $0 < \theta < \pi$, (18) must be valid for large z.

Returning to (14) and allowing r to become large, we have first

$$R = \sqrt{r^2 + \rho^2 - 2r\rho\cos\phi} \approx r - \rho\cos\phi,$$

and letting $(p-2)/2 = n$

$$J_n(R) \approx \sqrt{\frac{2}{\pi R}} \cos\left[R - n\frac{\pi}{2} - \frac{\pi}{4}\right]$$

$$\approx \sqrt{\frac{2}{\pi r}} \cos\left[r - \rho\cos\phi - n\frac{\pi}{2} - \frac{\pi}{4}\right]$$

$$J_{m+n}(r) \approx \sqrt{\frac{2}{\pi r}} \cos\left[r - (m+n)\frac{\pi}{2} - \frac{\pi}{4}\right]$$

and (14) reduces to

$$\cos\left[r - \rho\cos\phi - n\frac{\pi}{2} - \frac{\pi}{4}\right]$$

$$= 2^n\Gamma(n)\sum_{m=0}^{\infty}(m+n)\binom{m+2n-1}{m}$$

$$\times \frac{J_{m+n}(\rho)}{\rho}\cos\left[r - (m+n)\frac{\pi}{2} - \frac{\pi}{4}\right]P_{m,\,2n+2}(\cos\phi).$$

By formally applying the operator $i(d/dr)$ to the above and adding the resultant expression to the above we obtain, after cancellation of some terms

$$e^{i\rho\cos\phi} = 2^n\Gamma(n)\sum_{m=0}^{\infty}(m+n)\binom{m+2n-1}{m} i^m \frac{J_{m+n}(\rho)}{\rho^n} P_{m,\,2n+2}(\cos\phi) \quad (19)$$

The latter derivation was purely formal, but (19) can also be established rigorously.

For $p = 2$, or $n = 0$, (19) reduces to

$$e^{i\rho\cos\phi} = \sum_{m=0}^{\infty}\varepsilon_m J_m(\rho) i^m \cos m\phi, \quad (20)$$

which is equivalent to the generating function.

For $p = 3$, or $n = 1/2$, (19) yields

$$e^{i\rho\cos\phi} = \sqrt{\frac{\pi}{2\rho}}\sum_{m=0}^{\infty} i^m(2m+1)J_{m+1/2}(\rho)P_m(\cos\phi). \quad (21)$$

(20) and (21) are expansions of plane waves in terms of spherical functions in two and three dimensions respectively.

7. The Complete Solution of Bessel's Equation

A representation for Bessel's function, encountered earlier is

$$J_\nu(z) = \frac{(z/2)^\nu}{\sqrt{\pi}\,\Gamma(\nu+\frac{1}{2})}\int_{-1}^{1} e^{izt}(1-t^2)^{\nu-\frac{1}{2}}\,dt. \tag{1}$$

More generally it follows by standard techniques that

$$z^\nu \int_\alpha^\beta e^{izt}(t^2-1)^{\nu-\frac{1}{2}}\,dt \tag{2}$$

will satisfy

$$y'' + \frac{1}{z}y' + \left(1 - \frac{\nu^2}{z^2}\right)y = 0 \tag{3}$$

provided α and β are so selected that

$$(t^2-1)^{\nu+\frac{1}{2}}e^{izt}\Big|_\alpha^\beta = 0. \tag{4}$$

For Re $\nu > -\frac{1}{2}$ and $\alpha = -1$, $\beta = 1$ we are led back to (1).

If we assume that Re $z > 0$ we can select for our path of integration either of the contours C_1 or C_2 indicated in the following figure. On each of them

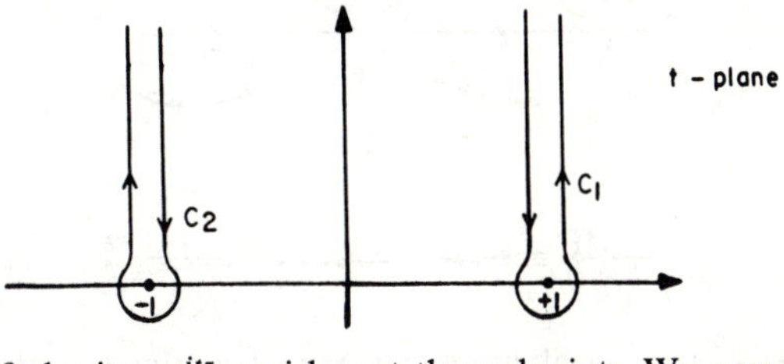

(4) will be satisfied; since e^{itz} vanishes at the endpoints. We now define the following two solutions.

$$H_\nu^{(j)}(z) = \frac{\Gamma(\frac{1}{2}-\nu)(\frac{1}{2}z)^\nu}{\pi i \Gamma(\frac{1}{2})}\int_{C_j} e^{izt}(t^2-1)^{\nu-\frac{1}{2}}\,dt, \qquad j=1,2. \qquad \text{Re } z > 0 \tag{5}$$

Equivalently using $\Gamma(\frac{1}{2}-\nu)\,\Gamma(\frac{1}{2}+\nu) = \pi/\cos \pi\nu$ we have

$$H_\nu^{(j)}(z) = \frac{(\frac{1}{2}z)^\nu}{i\Gamma(\frac{1}{2})\Gamma(\frac{1}{2}+\nu)\cos \pi\nu}\int_{C_j} e^{izt}(t^2-1)^{\nu-\frac{1}{2}}\,dt, \quad j=1,2. \quad \text{Re } z > 0 \tag{6}$$

These are known as the Hankel functions of the first and second kind respectively. It will be shown later that they are indeed linearly independent. In that case it must be possible to express the Bessel function as a linear combination of the Hankel functions.

We shall now demonstrate that

$$J_\nu(z) = \tfrac{1}{2}[H_\nu^{(1)}(z) + H_\nu^{(2)}(z)]. \tag{7}$$

To do so we shall join the two contours C_1 and C_2 as shown in the following figure, and successively deform the path of integration, as shown in the succeeding figures. We can then write

$$\tfrac{1}{2}[H_\nu^{(1)}(z) + H_\nu^{(2)}(z)]$$

$$= \frac{(\frac{1}{2}z)^\nu}{2i\Gamma(\frac{1}{2})\Gamma(\frac{1}{2}+\nu)\cos\pi\nu}\left[\int_{-1}^{1} + \int^{(+1)} + \int_{1}^{-1} + \int^{(-1)}\right] e^{izt}(t^2-1)^{\nu-\frac{1}{2}}\,dt.$$

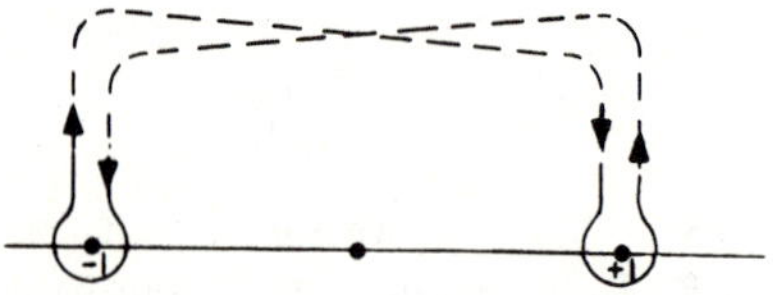

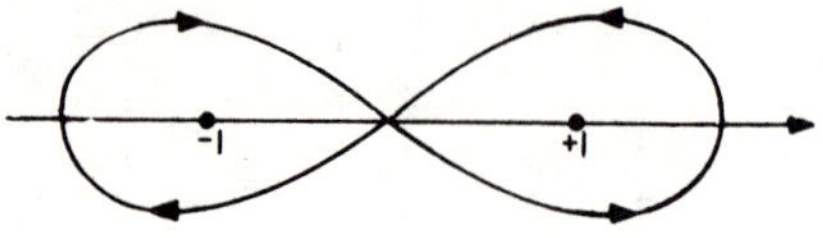

For Re $\nu > -\frac{1}{2}$ the contributions from the integration over the ε circles about ± 1 are on the order of $\varepsilon^{\nu+\frac{1}{2}}$ and as these circles shrink their contribution ultimately vanishes. To evaluate the two remaining integrals we select that branch of our integrand for which

$$\int_{-1}^{1} e^{izt}(t^2-1)^{\nu-\frac{1}{2}}\,dt = e^{-i\pi(\nu-\frac{1}{2})}\int_{-1}^{1} e^{izt}(1-t^2)^{\nu-\frac{1}{2}}\,dt.$$

The second integral then becomes

$$\int_{1}^{-1} e^{izt}(t^2-1)^{\nu-\frac{1}{2}}\,dt = -e^{i\pi(\nu-\frac{1}{2})}\int_{-1}^{1} e^{izt}(1-t^2)^{\nu-\frac{1}{2}}\,dt.$$

Finally

$$\tfrac{1}{2}[H_\nu^{(1)}(z) + H_\nu^{(2)}(z)] = \frac{(\frac{1}{2}z)^\nu}{\sqrt{\pi}\,\Gamma(\nu+\frac{1}{2})}\int_{-1}^{1} e^{izt}(1-t^2)^{\nu-1/2}\,dt = J_\nu(z)$$

by 2.6, thus establishing (7).

8. Asymptotic Expansions for Large Argument

To obtain suitable asymptotic expansions for $H_\nu^{(1)}(z)$, for example, we return to 7.6 and let $t = 1 + iw$. Then

$$H_\nu^{(1)}(z) = \frac{(\frac{1}{2}z)^\nu e^{i[z+\nu\pi/2-\pi/4]}}{\Gamma(\frac{1}{2})\Gamma(\frac{1}{2}+\nu)\cos\pi\nu}\int_C e^{-zw}(2w+iw^2)^{\nu-1/2}\,dw \tag{1}$$

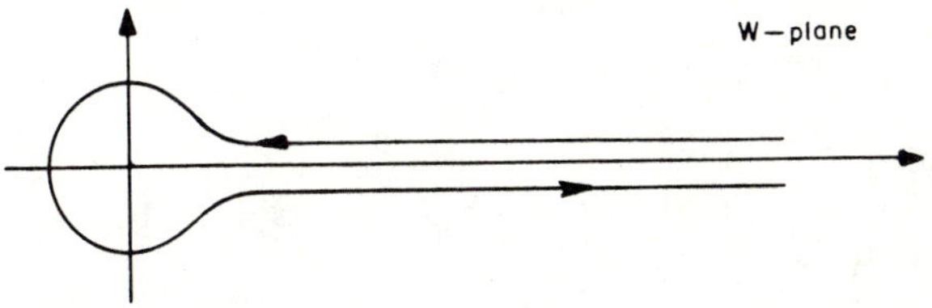

where C is the following contour. The integrand in (1) has two branch points, at $w = 0$ and $w = 2i$, respectively. For $\operatorname{Re} z > 0$ and $\operatorname{Re} \nu > -\frac{1}{2}$ the integral in (1) converges. We can decompose (1) into three integrals one along (∞, ε), one over a circle about the origin C_ε and one along (ε, ∞). The contribution from C_ε can be shown to vanish in the limit and combining the remaining two, by standard operations we obtain

$$H_\nu^{(1)}(z) = \frac{2(\frac{1}{2}z)^\nu e^{i[z-\nu\pi/2-\pi/4]}}{\Gamma(\frac{1}{2})\Gamma(\nu+\frac{1}{2})}\int_0^\infty e^{-zw}(2w+iw^2)^{\nu-1/2}\,dw. \tag{2}$$

(2) is a Laplace transform and we can obtain an asymptotic form by an immediate application of Watson's lemma. We expand the integrand by the binomial theorem.

$$(2w+iw^2)^{\nu-1/2} = (2w)^{\nu-1/2}\left[1+\frac{iw}{2}\right]^{\nu-1/2}$$

$$= (2w)^{\nu-1/2}\sum_{r=0}^{\infty}\binom{\nu-\frac{1}{2}}{r}\left(\frac{iw}{2}\right)^r$$

A term by term integration now yields

$$H_\nu^{(1)}(z) \approx \sqrt{\frac{2}{\pi z}}\, e^{i(z-\nu\pi/2-\pi/4)}\sum_{r=0}^{\infty}\frac{\Gamma(\nu+r+\frac{1}{2})(i/2z)^r}{r!\,\Gamma(\nu-r+\frac{1}{2})}. \tag{3}$$

In view of the fact that $H_\nu^{(1)}(z)$ is a multivalued function it is important to establish the region of validity of (3). In order for (1) to converge we required that Re $zw > 0$. Since the integrand in (1) has a branch point at $2i$ we have to introduce a cut to define a suitable Riemann surface. We will choose a cut along the imaginary axis from $2i$ to infinity. Then the path of integration must be so selected that

$$-\frac{3\pi}{2} < \arg w < \frac{\pi}{2}, \qquad \text{Re } zw > 0. \tag{4}$$

The latter condition is also equivalent to

$$\cos\,[\arg z + \arg w] > 0$$

or

$$-\frac{\pi}{2} < \arg z + \arg w < \frac{\pi}{2}. \tag{5}$$

Both conditions (4) and (5) can be satisfied for $-\pi < \arg z < 2\pi$, as is clear from the following figure. The shaded region shows where both inequalities can be satisfied.

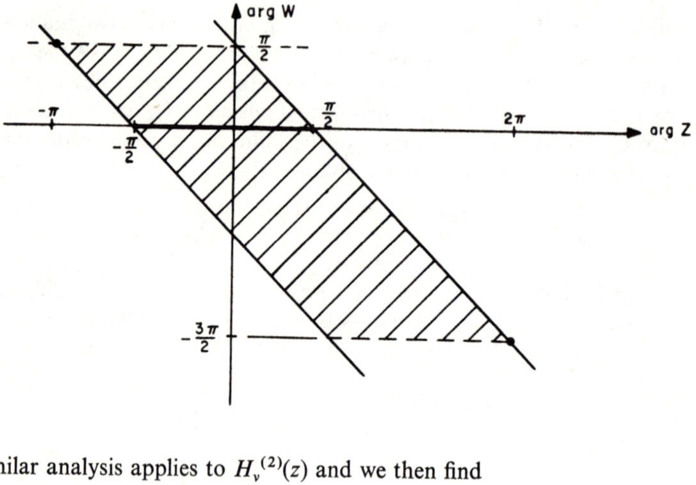

A similar analysis applies to $H_\nu^{(2)}(z)$ and we then find

$$H_\nu^{\binom{1}{2}}(z) \approx \sqrt{\frac{2}{\pi z}}\, e^{\pm i(z - \nu\pi/2 - \pi/4)} \sum_{r=0}^{\infty} \frac{\Gamma(\nu + r + \frac{1}{2})(\pm i/2z)^r}{r!\,\Gamma(\nu - r + \frac{1}{2})}$$

$$\begin{matrix} -\pi \\ -2\pi \end{matrix} < \arg z < \begin{matrix} 2\pi \\ \pi \end{matrix} \tag{6}$$

where all upper symbols refer to $H^{(1)}$ and all lower symbols to $H^{(2)}$. Note that when $\nu = m + \frac{1}{2}$, where m is a non-negative integer, the series in (6) terminate and these are no longer asymptotic, but exact representations. For example for $\nu = \frac{1}{2}$

$$H_{\frac{1}{2}}^{\binom{1}{2}}(z) = \mp i \sqrt{\frac{2}{\pi z}}\, e^{\pm iz}$$

and for $\nu = \frac{3}{2}$

$$H_{\frac{3}{2}}^{\binom{1}{2}}(z) = -\sqrt{\frac{2}{\pi z}}\, e^{\pm iz}\left[1 \pm \frac{i}{z}\right].$$

We now recall that

$$J_\nu(z) = \tfrac{1}{2}[H_\nu^{(1)}(z) + H_\nu^{(2)}(z)]$$

and by combining the two forms in (6) we find that

$$J_\nu(z) \approx \sqrt{\frac{2}{\pi z}} \cos\left[z - \frac{\nu\pi}{2} - \frac{\pi}{4}\right] \sum_{r=0}^{\infty} \frac{\Gamma(\nu + 2r + \frac{1}{2})(-1)^r}{(2r)!\Gamma(\nu - 2r + \frac{1}{2})} (2z)^{-2r}$$

$$- \sqrt{\frac{2}{\pi z}} \sin\left[z - \frac{\nu\pi}{2} - \frac{\pi}{4}\right] \sum_{r=0}^{\infty} \frac{\Gamma(\nu + 2r + \frac{3}{2})(-1)^r}{(2r+1)!\Gamma(\nu - 2r - \frac{1}{2})} (2z)^{-2r-1}$$

$$-\pi < \arg z < \pi. \tag{7}$$

(7) can hold only in the intersections of the region of validity of (6).

To extend the validity of (7) into the domain $-3\pi < \arg z < -\pi$, for example, we proceed as follows. The function $z^{-\nu}J_\nu(z)$, as is evident from the series expansion, is an entire function. If $-3\pi < \arg z < -\pi$ then $-\pi < \arg e^{2\pi i}z < \pi$ so that

$$J_\nu(z) = e^{-2\pi i\nu} J_\nu(e^{2\pi i}z).$$

(7) can be applied to the right of the above so that

$$J_\nu(z) = -e^{-2\pi i\nu} \sqrt{\frac{2}{\pi z}} \cos\left[z - \frac{\nu\pi}{2} - \frac{\pi}{4}\right] \sum_{r=0}^{\infty} \frac{\Gamma(\nu + 2r + \frac{1}{2})(-1)^r}{(2r)!\Gamma(\nu - 2r + \frac{1}{2})} (2z)^{-2r}$$

$$+ e^{-2\pi i\nu} \sqrt{\frac{2}{\pi z}} \sin\left[z - \frac{\nu\pi}{2} - \frac{\pi}{4}\right] \sum_{r=0}^{\infty} \frac{\Gamma(\nu + 2r + \frac{3}{2})(-1)^r}{(2r+1)!\Gamma(\nu - 2r - \frac{1}{2})} (2z)^{-2r-1}$$

$$-3\pi < \arg z < -\pi. \tag{8}$$

When ν is an integer $J_\nu(z)$ is an entire function and we note that the coefficients in (7) and (8) change sign as we pass from one region of validity into the other. Such a discontinuous change in the asymptotic form of a continuous function is often referred to as a Stokes' phenomenon.* We shall return to this topic later.

(6) may be represented more conveniently if the following notation is introduced.

$$ {}_2F_0(a, b; z) = \sum_{r=0}^{\infty} \frac{(a)_r(b)_r z^r}{r!} $$

Then (6) becomes

$$ H_\nu^{\binom{1}{2}}(z) \approx \sqrt{\frac{2}{\pi z}}\, e^{\pm i[z - \nu\pi/2 - \pi/4]}\, {}_2F_0\left(\frac{1}{2} + \nu, \frac{1}{2} - \nu; \frac{\pm 1}{2iz}\right). $$

We can easily compute the Wronskian of the two Hankel functions. We know that

$$ W[H_\nu^{(1)}(z), H_\nu^{(2)}(z)] = \frac{c}{z} $$

for a suitable constant c. By means of (6) we find that $c = {}^{-4i}/\pi$.

A fourth solution, often referred to as the Neumann function $N_\nu(z)$ may be defined as follows

$$ H_\nu^{(1)}(z) = J_\nu(z) + iN_\nu(z) $$

or (9)

$$ H_\nu^{(2)}(z) = J_\nu(z) - iN_\nu(z). $$

It may also be characterized by its asymptotic form

$$ N_\nu(z) \approx \sqrt{\frac{2}{\pi z}} \sin\left(z - \frac{\nu\pi}{2} - \frac{\pi}{4}\right). \tag{10} $$

These functions may be compared to certain corresponding one-dimensional harmonic functions

$$ \begin{aligned} J_\nu(z) &\leftrightarrow \cos z \\ N_\nu(z) &\leftrightarrow \sin z \\ H_\nu^{(1)}(z) &\leftrightarrow e^{iz} \\ H_\nu^{(2)}(z) &\leftrightarrow e^{-iz} \end{aligned} $$

* For an extensive discussion of this and related asymptotic problems the reader is urged to consult K. O. Friedrichs' *Asymptotic Phenomena in Mathematical Physics. Bull. Amer. Math. Soc.* **61** (1955).

When ν is not an integer one can express N_ν in terms of J_ν and $J_{-\nu}$. We will show that

$$N_\nu(z) = \cot \nu\pi J_\nu(z) - \csc \nu\pi J_{-\nu}(z). \tag{11}$$

Using the asymptotic forms (7) and (10) we find that

$$\cot \nu\pi J_\nu(z) - \csc \nu\pi J_{-\nu}(z) - N_\nu(z) = 0\left(\frac{1}{z}\right). \tag{12}$$

The above notation implies that, if $f(z)$ denotes the left side, $zf(z)$ remains bounded for large z. But the general solution of Bessel's equation is of the form

$$\frac{c_1}{\sqrt{z}} e^{iz} P\left(\frac{1}{z}\right) + \frac{c_2}{\sqrt{z}} e^{-iz} Q\left(\frac{1}{z}\right) \tag{13}$$

where P and Q are suitable asymptotic series of the form

$$P\left(\frac{1}{z}\right) = 1 + \frac{a}{z} + \cdots$$

$$Q\left(\frac{1}{z}\right) = 1 + \frac{b}{z} + \cdots.$$

In order for (12) to hold c_1 and c_2 in (13) must vanish. Hence the left side of (12) vanishes and (11) is established.

When ν is integral the right side of (11) is indeterminate, but by suitable limiting processes $N_m(z)$ can be defined, where m is an integer. Thus for example

$$N_0(z)$$

$$= \lim_{\nu\to 0} \left(\cos \nu\pi \left(\frac{z}{2}\right)^\nu \sum_{l=0}^{\infty} \frac{(-z^2/4)^l}{l!\Gamma(\nu + l + 1)} - \left(\frac{z}{2}\right)^{-\nu} \sum_{l=0}^{\infty} \frac{(-z^2/4)^l}{l!\Gamma(-\nu + l + 1)}\right) \Big/ \sin \nu\pi$$

$$= \frac{2}{\pi}\left(\log \frac{z}{2}\right) J_0(z) - \frac{1}{\pi} \sum_{l=0}^{\infty} \frac{(-z^2/4)^l \psi(l+1)}{l!^2}. \tag{14}$$

The addition theorems proved earlier may be extended to more general solutions of Bessel's equation. For example we shall state without proof

$$H_\nu^{(1)\atop(2)}(R)e^{i\nu\psi} = \sum_{m=-\infty}^{\infty} J_m(\rho) H_{\nu+m}^{(1)\atop(2)}(r) e^{i(m+\nu)\phi}, \qquad |r| > |\rho|. \tag{15}$$

9. Airy Functions

We shall now turn to a study of the solutions of the differential equation

$$A''(z) - zA(z) = 0 \tag{1}$$

known as Airy's equation. It is a simple calculation to show that if we let

$$A(z) = \sqrt{z}\, f(\tfrac{2}{3}iz^{3/2})$$

then

$$f''(t) + \frac{1}{t} f'(t) + \left(1 - \frac{1}{9t^2}\right) f = 0.$$

Thus $f(t)$ is a Bessel function of order $\frac{1}{3}$. In other words we can express the general solution of (1) in the form

$$A(z) = \sqrt{z}[c_1 J_{1/3}(\tfrac{2}{3}iz^{3/2}) + c_2 J_{-1/3}(\tfrac{2}{3}iz^{3/2})]. \tag{2}$$

Nevertheless we shall not make use of this relationship to the Bessel functions, but study (1) independently.

Equation (1) is one of the simplest equations in which a so-called turning point occurs. More generally we say that

$$y'' + n(z)y = 0 \tag{3}$$

has a turning point at z_0, if $n(z)/(z - z_0)$ is of constant sign near z_0. In other words $n(z)$ changes sign at z. If $n(z) > 0$ we expect the solutions of (3) to have an oscillatory character. For $n(z) < 0$ the solutions should have an exponential character. Differential equations of this type arise in many problems of mathematical physics. In particular they arise frequently in quantum mechanics and physical optics.

If we seek a solution for (1) in the form of

$$\int_\alpha^\beta e^{tz} v(t)\, dt$$

we can establish that

$$A(z) = \frac{1}{2\pi i} \int_L e^{tz - 1/3 t^3}\, dt \tag{4}$$

where L is any one of the three contours shown in the following diagram.

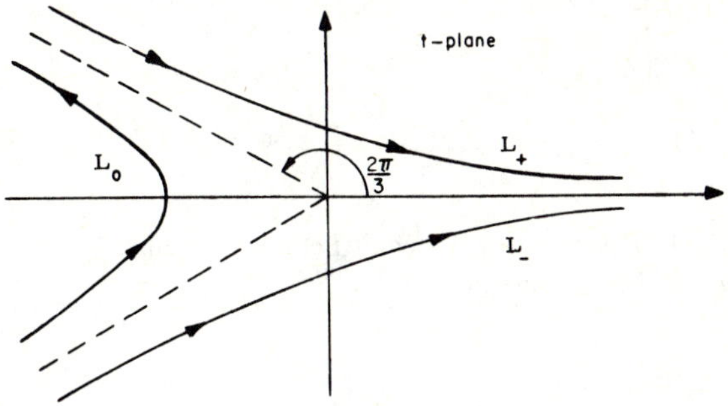

We shall denote the solutions A_+, A_-, A_0 if L in (4) is L_+, L_-, or L_0 respectively. These three solutions cannot be linearly independent and we readily ascertain that

$$A_-(z) - A_+(z) = A_0(z). \tag{5}$$

One of our tasks will be to establish asymptotic forms for each of these solutions. We now define three domains in the z-plane, denoted by S_+, S_-, and S_0, and defined in the following figure. Our subsequent calculations will show that each of our functions A_+, A_-, A_0 has a different asymptotic form in each of the above three domains.

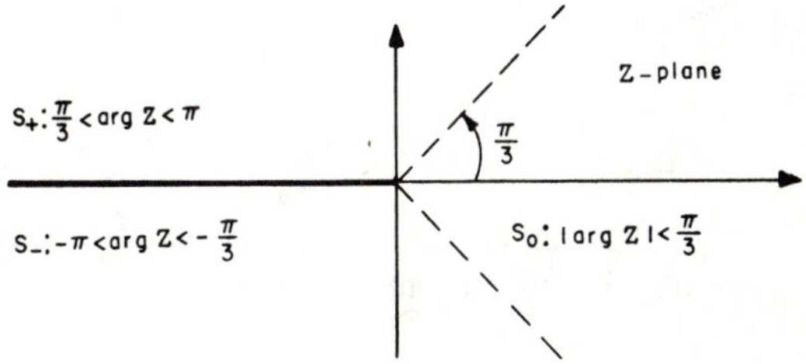

We shall first consider the function $A_0(z)$ and suppose that z is in S_0. Let

$$\begin{aligned} f(t) &= tz - \tfrac{1}{3}t^3 \\ f'(t) &= \ z - t^2 \\ f''(t) &= -2t \\ f'''(t) &= -2. \end{aligned}$$

Then $f'(t)$ will vanish at $t = \pm\sqrt{z}$. But for z in S_0, only $-\sqrt{z}$ will lie on the path L_0. Expanding $f(t)$ in a Taylor series about that point we obtain

$$f(t) = -\tfrac{2}{3}z^{3/2} + \sqrt{z}(t + \sqrt{z})^2 - \tfrac{1}{3}(t + \sqrt{z})^3$$

and

$$A_0(z) = \frac{\exp(-\frac{2}{3}z^{3/2})}{2\pi i} \int_{L_0} e^{\sqrt{z}(t+\sqrt{z})^2 - 1/3(t+\sqrt{z})^3}\, dt. \tag{6}$$

In order to be able to apply Watson's Lemma conveniently to (6) we introduce the new variable of integration

$$w = -iz^{1/4}(t + \sqrt{z}).$$

$$A_0(z) = \frac{z^{-1/4}\exp(-\frac{2}{3}z^{3/2})}{2\pi} \int_{L_0'} e^{-w^2} \exp\left(\frac{i}{3}\, z^{-3/4} w^3\right) dw \tag{7}$$

where L_0' is an appropriately shifted contour of the following type.

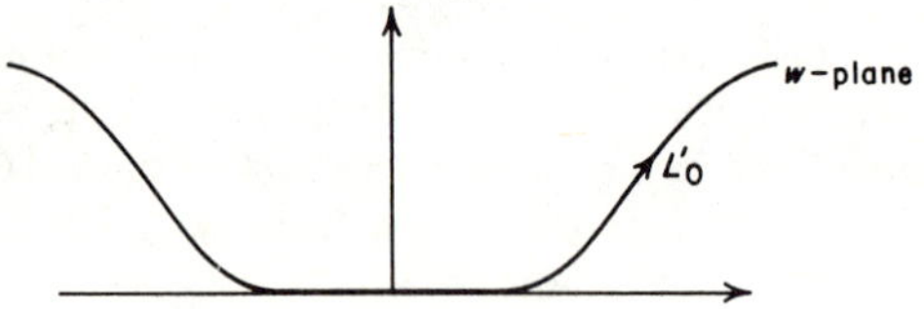

For z large we let

$$e^{(i/3)z^{-3/4}w^3} = \sum_{n=0}^{\infty} \frac{((i/3)w^3)^n}{z^{3n/4} n!}$$

in (6) and interchange summation and integration. The latter step will, by invoking Watson's Lemma, lead to a valid asymptotic solution. We finally obtain

$$A_0(z) \approx \frac{z^{-1/4}\exp(-\frac{2}{3}z^{3/2})}{2\pi} \sum_{n=0}^{\infty} \frac{\Gamma(3n + \frac{1}{2})}{(2n)!} \left(\frac{i}{3z^{3/4}}\right)^{2n}$$

$$= \frac{z^{-1/4}\exp(-\frac{2}{3}z^{3/2})}{2\sqrt{\pi}} + \cdots \qquad z \in S_0. \tag{8}$$

Similar calculations may be applied to A_+ and A_-. It can thus be shown that

$$A_+(z) \approx \frac{z^{-1/4}\exp(\frac{2}{3}z^{3/2})}{2\pi i}\sum_{n=0}^{\infty}\frac{\Gamma(3n+\frac{1}{2})}{(2n)!}\left(\frac{1}{3z^{3/4}}\right)^{2n}$$

$$\approx \frac{z^{-1/4}\exp(\frac{2}{3}z^{3/2})}{2\sqrt{\pi}\, i}$$

$$A_-(z) \approx \frac{z^{-1/4}\exp(\frac{2}{3}z^{3/2})}{2\pi i}\sum_{n=0}^{\infty}\frac{\Gamma(3n+\frac{1}{2})}{(2n)!}\left(\frac{1}{3z^{3/4}}\right)^{2n}$$

$$\approx \frac{z^{-1/4}\exp(\frac{2}{3}z^{3/2})}{2\sqrt{\pi}\, i} \qquad z \in S_0 \tag{9}$$

From (8) and (9) it is evident that $A_0(z) \to 0$ and $A_+(z) \to \infty$ as $z \to \infty$ in S_0.

To deduce the corresponding expansions in $S_\pm$ we proceed as follows, using the results just obtained. Suppose z is in S_+. Then $e^{-2\pi i/3}z$ is in S_0. We then can write

$$A_0(z) = \frac{1}{2\pi i}\int_{L_0} \exp\{(e^{2\pi i/3}t)(e^{-2\pi i/3}z) - \tfrac{1}{3}t^3\}\, dt.$$

If in the above we let $\tau = e^{2\pi i/3}t$ L_0 in the t plane is rotated into L_+ in the τ plane so that

$$A_0(z) = \frac{e^{-2\pi i/3}}{2\pi i}\int_{L^+} \exp\{\tau e^{-2\pi i/3}z - \tfrac{1}{3}\tau^3\}\, d\tau = e^{-2\pi i/3}A_+(e^{-2\pi i/3}z),$$

$$z \in S_+. \tag{10}$$

Using (9) we obtain

$$A_0(z) \approx \frac{z^{-1/4}\exp(-\frac{2}{3}z^{3/2})}{2\pi}\sum_{n=0}^{\infty}\frac{\Gamma(3n+\frac{1}{2})}{(2n)!}\left(\frac{i}{3z^{3/4}}\right)^{2n}, \qquad z \in S_+,$$

$$\approx \frac{z^{-1/4}\exp(-\frac{2}{3}z^{3/2})}{2\sqrt{\pi}}. \tag{11}$$

If we define two asymptotic series

$$V_+ = \frac{z^{-1/4}\exp(\frac{2}{3}z^{3/2})}{2\pi}\sum_{n=0}^{\infty}\frac{\Gamma(3n+\frac{1}{2})}{(2n)!}\left(\frac{i}{3z^{3/4}}\right)^{2n}$$

$$V_- = \frac{z^{-1/4}\exp(-\frac{2}{3}z^{3/2})}{2\pi}\sum_{n=0}^{\infty}\frac{\Gamma(3n+\frac{1}{2})}{(2n)!}\left(\frac{i}{3z^{3/4}}\right)^{2n} \tag{12}$$

we can express every solution in each domain by an expression of the type

$$A = c_- V_- + c_+ V_+$$

where c_- and c_+ depend on the domain in which z lies. The following table shows what the asymptotic behavior is and what the coefficients are for each situation.

Solution		Domain of z: S_0	S_+	S_-
	c_-	1	1	1
A_0		$\to 0$	$\to \infty$	$\to \infty$
	c_+	0	0	0
	c_-	0	0	-1
A_+		$\to \infty$	$\to 0$	$\to -\infty$
	c_+	$-i$	$-i$	0
	c_-	0	1	0
A_-		$\to \infty$	$\to \infty$	$\to 0$
	c_+	$-i$	0	$-i$

Note that, as a check, these must be compatible with (5).

In the aforesaid table the coefficients c_- and c_+ for A_0 remain fixed independent of the domain. In the case of A_+, however, as we move from S_+ to S_- there is a discontinuous change in the coefficients. This phenomenon, as mentioned earlier, is known as a Stokes phenomenon. This discontinuity does not occur in the function itself, but rather in its asymptotic form. The function A_+ is continuous for all z, but it changes rapidly as we move from S_+ to S_-.

Clearly the above results lose their meaning on the boundaries of the domain S_+, S_-, S_0. Suppose we are interested in determining behavior of A_0 on the boundary between S_+ and S_-. Let $z = e^{i\pi}x$, $x > 0$. Then

$$A_0(z) = \frac{1}{2\pi i}\int_{L_0} e^{-tx - \frac{1}{3}t^3}\,dt.$$

If

$$f(t) = -tx - \tfrac{1}{3}t^3$$

then $f'(t) = 0$ at $t = \pm i\sqrt{x}$, and both of these points lie on the path L_0. It follows that the contribution from the integration near $t = i\sqrt{x}$ will be

$$\frac{\exp(-\frac{2}{3}ix^{3/2})}{2\pi i}\int e^{-i\sqrt{x}(t-i\sqrt{x})^2} e^{-\frac{1}{3}(t-i\sqrt{x})^3}\,dt$$

or letting $w = -e^{i\pi/4}x^{1/4}(t - i\sqrt{x})$ we find, using Watson's Lemma

$$\frac{-e^{-i\pi/4}x^{-1/4}\exp(-\frac{2}{3}ix^{3/2})}{2\pi i}\int_{-\infty}^{\infty} e^{-w^2}\exp(-\tfrac{1}{3}e^{i\pi/4}x^{-3/4}w^3)\,dw$$

$$\approx \frac{-e^{-i\pi/4}x^{-1/4}\exp(-\frac{2}{3}ix^{3/2})}{2\pi i}\sum_{n=0}^{\infty}\frac{\Gamma(3n+\frac{1}{2})}{(2n)!}\left(\frac{e^{i\pi/4}}{3x^{3/4}}\right)^{2n}.$$

A similar contribution comes from the neighborhood of $t = -i\sqrt{x}$. Finally

$$A_0(e^{i\pi}x) \approx \frac{x^{-1/4}}{\pi}\sum_{n=0}^{\infty}\frac{\Gamma(3n+\frac{1}{2})\sin[\frac{2}{3}x^{3/2}+\pi/4-n\pi/2]}{(2n)!(3x^{3/4})^{2n}}$$

$$\approx \frac{x^{-1/4}\sin[\frac{2}{3}x^{3/2}+\pi/4]}{\sqrt{\pi}}, \qquad x > 0 \tag{13}$$

For $z = x > 0$ we find in S_0 from (8)

$$A_0(x) \approx \frac{x^{-1/4}\exp(-\frac{2}{3}x^{3/2})}{2\sqrt{\pi}}, \qquad x > 0. \tag{14}$$

A comparison of (13) and (14) shows that the function $A_0(x)$, for real x, oscillates for negative values and decays for positive values. Similar analyses can be performed on all other boundaries.

10. Asymptotic Expansions for Large Indices and Large Arguments

An integral representation that was an immediate consequence of the generating function is the following

$$J_\nu(z) = \frac{1}{2\pi i}\int_{-\infty}^{(0+)}\exp\left[\frac{z}{2}\left(t - \frac{1}{t}\right)\right]t^{-\nu-1}\,dt, \qquad \operatorname{Re} z > 0.$$

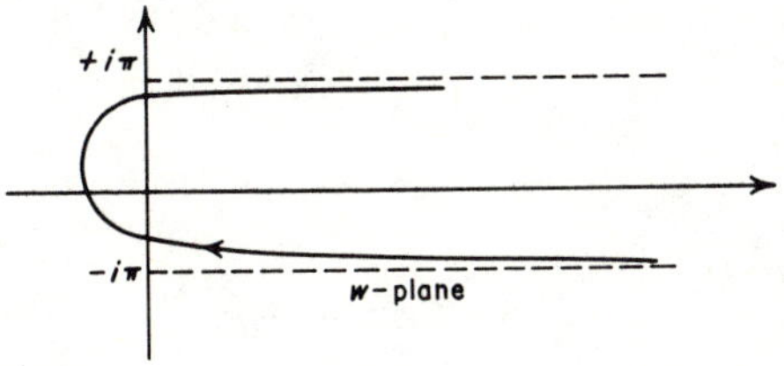

If we let $t = e^w$ we obtain

$$J_\nu(z) = \frac{1}{2\pi i}\int_c e^{z\sinh w - \nu w}\,dw \tag{1}$$

where c is the contour. For Re $z > 0$ the integral is convergent. Two related integrals are the following

$$H_\nu^{(j)}(z) = \frac{(-1)^{j+1}}{\pi i} \int_{c_j} e^{z \sinh w - \nu w}\, dw, \qquad j = 1, 2. \tag{2}$$

where the contours c_j are defined by the following figure. In (2) the integrals

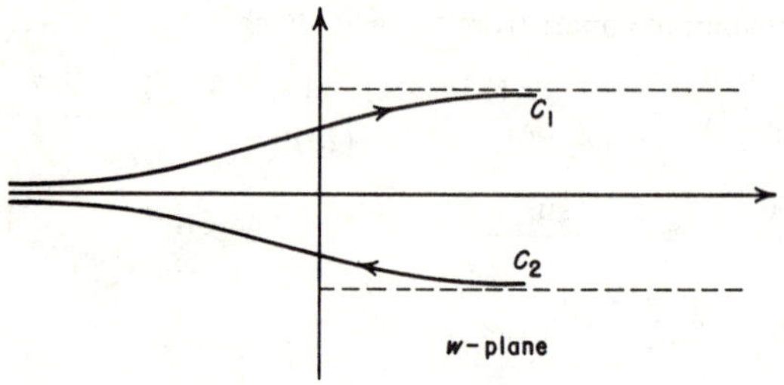

on the right were described by the notation $H_\nu^{(j)}$. This was done in anticipation of some matters that must still be discussed. A direct calculation shows that the integrals in (2) are solutions of Bessel's equation. From the asymptotic results to be discussed shortly it will be evident that they do indeed coincide with the Hankel functions.

Rather than use (2) in our discussion it will be convenient to let $w = -i\tau$. Then

$$H_\nu^{(j)}(z) = \frac{(-1)^j}{\pi} \int_{c_j'} e^{i(\nu\tau - z \sin \tau)}\, d\tau, \qquad j = 1, 2. \tag{3}$$

where c_j' is the image of c_j in the τ plane. (3) will be used to discuss various asymptotic forms of the Hankel functions under a number of hypotheses on ν and z.

1. Suppose $0 < \nu/z < 1$, in which case we let

$$\nu = z \cos \alpha.$$

Let

$$f(\tau) = \tau \cos \alpha - \sin \tau$$

and

$$f'(\tau) = \cos \alpha - \cos \tau.$$

Clearly $f'(\tau)$ will vanish at $\tau = \pm\alpha$. It follows that we can rewrite (3) for $j = 1$ as

$$H_\nu^{(1)}(z) = \frac{-1}{\pi} \int_{c_1'} e^{izf(\tau)}\, d\tau \tag{4}$$

For $0 < \alpha < \pi/2$, the point $\tau = -\alpha$ lies on c, and we will shift our origin so that this point lies on the path of integration. Let

$$\tau = \sigma - \alpha$$

so that

$$\begin{aligned} f(\tau) &= f(\sigma - \alpha) = (\sigma - \alpha) \cos \alpha - \sin (\sigma - \alpha) \\ &= [\sin \alpha - \alpha \cos \alpha] + g(\sigma). \end{aligned}$$

In the above

$$\begin{aligned} g(\sigma) &= (\sigma - \sin \sigma) \cos \alpha + (\cos \sigma - 1) \sin \alpha \\ &= -\frac{\sin \alpha}{2} \sigma^2 + \cdots. \end{aligned} \tag{5}$$

(4) now becomes

$$H_\nu^{(1)}(z) = -\frac{1}{\pi} e^{iz[\sin \alpha - \alpha \cos \alpha]} \int_{c_1'} e^{izg(\sigma)}\, d\sigma. \tag{6}$$

Lastly we wish to evaluate the integral in (6) asymptotically. In order to apply Watson's Lemma we let

$$izg(\sigma) = -u^2$$

and in order to convert (6) to an integral in the u domain we use (5) to solve for σ as a function of u. Then

$$\sigma = \sum_{n=1}^{\infty} a_n \left(\frac{u}{\sqrt{z}}\right)^n, \qquad a_1 = \sqrt{\frac{2}{\sin \alpha}}\, e^{3\pi i/4}.$$

It follows that

$$\begin{aligned} \int_{c_1'} e^{izg(\sigma)}\, d\sigma &= \int_{-\infty}^{\infty} e^{-u^2} \frac{d\sigma}{du}\, du = \int_{-\infty}^{\infty} e^{-u^2} \sum_{n=1}^{\infty} \frac{n a_n u^{n-1}}{z^{n/2}}\, du \\ &\approx \sum_{k=0}^{\infty} \frac{(2k+1) a_{2k+1} \Gamma(k+\frac{1}{2})}{z^{k+1/2}} \end{aligned}$$

so that

$$\begin{aligned} H_\nu^{(1)}(z) &\approx -\frac{1}{\pi} e^{i[z \sin \alpha - \alpha \cos \alpha]} \sum_{k=0}^{\infty} \frac{(2k+1) a_{2k+1} \Gamma(k+\frac{1}{2})}{z^{k+1/2}} \\ &\approx \sqrt{\frac{2}{\pi z \sin \alpha}}\, e^{(z \sin \alpha - \nu\alpha - \pi/4)} + \cdots. \end{aligned} \tag{7}$$

For fixed v and z very large $\alpha \approx \pi/2$ and (7) reduces to

$$H_v^{(1)}(z) \approx \sqrt{\frac{2}{\pi z}}\, e^{i[z - v\pi/2 - \pi/4]} + \cdots$$

a result that was obtained earlier.

2. A second case of interest arises if $v > z$. In this case we let

$$v = z \cosh \alpha$$

and (2) reduces to

$$H_v^{(1)}(z) = \frac{1}{\pi i} \int_{c_1} e^{z[\sinh w - w \cosh \alpha]}\, dw. \tag{8}$$

Applying a similar analysis we let

$$f(w) = \sinh w - w \cosh \alpha.$$

Again we see that $f'(-\alpha) = 0$, since the point $w = -\alpha$ lies on c_1, and

$$f(\tau - \alpha) = -\sinh \alpha + \alpha \cosh \alpha + g(\tau)$$

where

$$g(\tau) = \frac{-1}{2} \sinh \alpha \tau^2 + \cdots. \tag{9}$$

If we introduce the new variable of integration u defined by

$$zg(\tau) = -u^2$$

we see from (9) that

$$\tau = \sum_{n=1}^{\infty} a_n \left(\frac{u}{\sqrt{z}}\right)^n, \qquad a_1 = \sqrt{\frac{2}{\sinh \alpha}}.$$

Then (8) can be rewritten as

$$\begin{aligned} H_\gamma^{(1)}(z) &\approx \frac{1}{\pi i} e^{z[\alpha \cosh \alpha - \sinh \alpha]} \int_{-\infty}^{\infty} e^{-u^2} \sum_{n=1}^{\infty} \frac{n a_n u^{n-1}}{z^{n/2}}\, du \\ &\approx \frac{1}{\pi i} e^{z[\alpha \cosh \alpha - \sinh \alpha]} \sum_{k=0}^{\infty} \frac{(2k+1) a_{2k+1} \Gamma(k + \frac{1}{2})}{z^{k+1/2}} \\ &\approx -i \sqrt{\frac{2}{\pi z \sinh \alpha}}\, e^{v\alpha - z \sinh \alpha}. \end{aligned} \tag{10}$$

The differential equation

$$y'' + \frac{1}{z} y' + \left(1 - \frac{v^2}{z^2}\right) y = 0$$

will, for z very large, have oscillatory solutions. This is borne out by (7). For v large relative to z it will behave like

$$y'' + \frac{1}{z} y' - \frac{v^2}{z^2} y = 0$$

which is an equation of Euler type. It will have power solutions like $z^{\pm v}$. This is also evident in (10). The case $v \approx z$ forms a transition region between these two. In other words we have a turning point near $v \approx z$.

3. We shall now study the case where $v \approx z$. It is convenient to make the substitution $y = z^{1/2} Y$ in Bessel's equation. This leads to

$$Y'' + \left(1 - \frac{v^2 - \frac{1}{4}}{z^2}\right) Y = 0$$

which has turning points at $z^2 = v^2 - \frac{1}{4}$. We now let

$$\frac{v}{z} = 1 + \sigma$$

and study the behavior of $H_v^{(1)}(z)$ as a function of σ.

We return to (2) in the form

$$H_v^{(1)}(z) = \frac{1}{\pi i} \int_{-\infty}^{\infty + \pi i} e^{z[\sinh w - (1+\sigma)w]} \, dw$$

and let

$$f(w) = \sinh w - (1 + \sigma)w.$$

$$= -\sigma w + \frac{w^3}{6} + \cdots.$$

In this instance we see, that for z large, or equivalently σ small, $f'(0)$ is small and the first term of significant magnitude is the cubic term, in the Taylor series. We see that

$$f(w) = -\sigma w + \frac{w^3}{6} + w^5 g(w^2)$$

where $g(w^2)$ is an entire function. Lastly we introduce the new variable of integration V by

$$w = -2^{1/3}\frac{V}{z^{1/3}} \tag{11}$$

so that

$$zf(w) = 2^{1/3}z^{2/3}\sigma V - \frac{V^3}{3} - \frac{2^{5/3}}{z^{2/3}}V^5 g\left(\frac{2^{2/3}V^2}{z^{2/3}}\right).$$

The integral now takes the form

$$H_\nu^{(1)}(z) = \frac{(-2/z)^{1/3}}{\pi i}\int_c e^{\zeta V - V^3/3}\sum_{n=0}^{\infty}\left[\frac{-(2^{5/3}/z^{2/3})V^5}{n!}g\left(\frac{2^{2/3}V^2}{z^{2/3}}\right)\right]^n dV \tag{12}$$

$$\zeta = 2^{1/3}z^{2/3}\sigma$$

where c is the image of the original path of integration. To obtain the leading term in our asymptotic form we drop all terms which are small for large z. Then

$$H_\nu^{(1)}(z) \approx \frac{(-2/z)^{1/3}}{\pi i}\int_c e^{\zeta V - V^3/3}\,dV \tag{13}$$

which is reminiscent of the Airy functions. The critical points in our path of integration are given by

$$V^2 = \zeta$$

and depending on the argument of σ different situations arise. A comparison of the path c with the paths L_0, L_+, L_- discussed in section 9 shows that (13) can be rewritten in the form

$$H_\nu^{(1)}(z) \approx \left(\frac{16}{z}\right)^{1/3} A_-(\zeta), \tag{14}$$

provided z is suitably restricted. The latter, as can be seen from (11), is necessary in order that c is similar to L_-. For convenience we shall take z positive.

We can deduce asymptotic forms from (14) using the results of section 9. These will, of course, depend on whether ζ lies in S_0, S_+ or S_-. A case of particular interest, as regards certain applications to be discussed in the next section is the following. First we note that

$$A_-(\zeta) = -e^{2/3\pi i/3}\,A_0(e^{2/3\pi i}\zeta)$$

which can be verified directly, as was discussed in section 9. Then (14) becomes

$$H_\nu^{(1)}(z) \approx -e^{\frac{2}{3}\pi i}\left(\frac{16}{z}\right)^{\frac{1}{3}} A_0(e^{\frac{2}{3}\pi i}\zeta).$$

For $e^{2/3\pi i}\zeta$ real we can apply formulas 9.13 and 9.14. Under the hypothesis that $e^{2/3\pi i}\zeta = e^{i\pi}x$, where $x > 0$ we obtain from 9.13

$$H_\nu^{(1)}(z) \approx \frac{e^{-i\pi/4}2^{-5/4}}{\sqrt{\pi}\,[z(\nu - z)]^{1/4}} \sin\left[z \times \frac{2^{3/2}}{3}\left[e^{-i\pi/3}\left(\frac{\nu - z}{z}\right)\right]^{3/2} + \frac{\pi}{4}\right]. \quad (15)$$

From (15) we can deduce the following important result. $H_\nu^{(1)}(z)$, for fixed $z > 0$, has as function of the parameter ν zeros and these lie asymptotically for large ν and z near

$$\nu_m \approx z\left\{1 + \frac{e^{i\pi/3}}{2}\left[\frac{3\pi}{4z}(4m + 3)\right]^{2/3}\right\}$$

where m is an integer.

11. Some Applications of Bessel Functions in Physical Optics

Suppose we consider a concave reflecting surface S and a point source of light A, as shown in the following figure. K is the envelope of reflected rays from S and is known as the caustic. These may often be observed if the interior of a ring is used as a reflecting surface. They may be observed in coffee cups and many other concave surfaces. We wish to examine the intensity of light

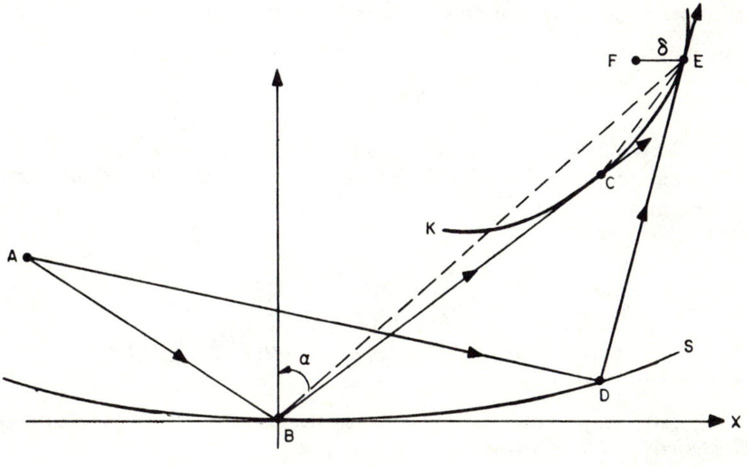

near and on the caustic. This analysis is due to G. B. Airy.* Let $R_1(x) = AD$, the distance from A to a point of S, and let $R_2(x) = DE$ denote the distance from that point along the reflected ray to the caustic, and let

$$R(x) = R_1(x) + R_2(x). \tag{1}$$

Since $R(x)$ is a path of least time it follows that

$$R'(x) = 0. \tag{2}$$

Furthermore the caustic, or envelope of reflected rays must be such that

$$R''(x) = 0. \tag{3}$$

The latter condition can be used to define the caustic. With the possible exception of some special surfaces S, we have

$$R'''(x) \neq 0.$$

We can show that $R'''(x) > 0$. To do so we observe that

$$AB + BC + CE = AD + DE$$

when D is near B. This follows from the fact that $R(x)$ is varying very slowly, as indicated by (2) and (3). But from the triangle inequality

$$BC + CE > BE$$

so that

$$AB + BC + CE = AD + DE > AB + BE.$$

This shows that

$$R(x) > R(0);$$

hence $R(x)$ is an increasing function. If follows that $R'''(x) > 0$.

If we now consider the point F, off the caustic, and compute $R(x)$ measured to F we find

$$\begin{aligned}
R(x + \delta) &= R(x) + \delta R_2'(x) \\
R'(x + \delta) &= R'(x) + \delta R_2''(x) \\
R''(x + \delta) &= R''(x) + \delta R_2'''(x) \\
R'''(x + \delta) &= R'''(x) + \delta R_2^{(IV)}(x)
\end{aligned}$$

* G. B. Airy, *On the Intensity of Light in the Neighborhood of a Caustic*, *Trans. Cambridge Philosophical Soc.*, **6** (1838).

for δ small. If we now expand, as a function of x, we obtain

$$R(x+\delta) = R(\delta) + \delta R_2''(0)x + \tfrac{1}{2}\delta R_2'''(0)x^2 + \tfrac{1}{6}[R'''(0) + \delta R_2{}^{(IV)}(0)]x^3 + '''.$$

By a relabeling of symbols we have

$$R(x+\delta) = R_0 + \delta Ax + \delta Bx^2 + Cx^3 + '''. \tag{4}$$

The symbols indicate that the coefficients of x and x^2 have the same order of magnitude as δ, but the coefficient of x^3 is much larger. By a suitable translation we can eliminate the x^2 term. Then

$$R = R_0 + \delta A(x - x_0) + C(x - x_0)^3 + \cdots \tag{5}$$

where the coefficients in (4) and (5) have slightly different definitions.

Finally the intensity of light at some point F is a superposition of all plane waves emanating at A and being reflected at D. Then

$$I = \int_{-\infty}^{\infty} e^{ikR(x+\delta)}\,dx = e^{ikR_0}\int_{-\infty}^{\infty} e^{ik[\delta A(x-x_0)+C(x-x_0)^3+''']}\,dx \tag{6}$$

and we note that (6) is essentially an Airy function as a function of δ. By the results of section (9) we see that for

1. $\delta > 0$ I will have exponential decay in the shadow zone
2. $\delta < 0$ I will have an oscillatory behavior
3. the point where I has the largest amplitude is not on the caustic, but just off the caustic in the illuminated region.

The qualitative picture is indicated in the following figure.

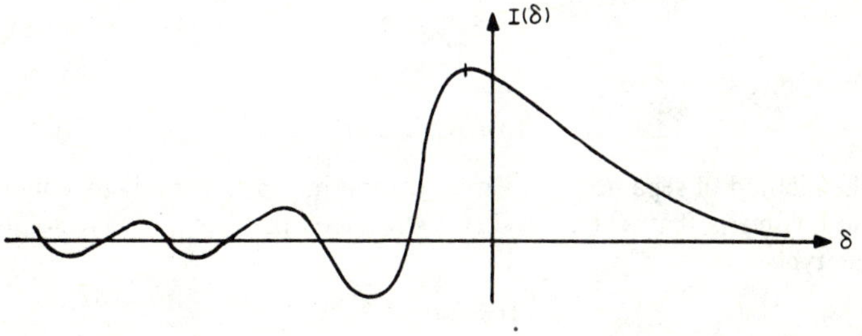

As a second application we consider a problem related to antenna theory. We suppose that a plane wave u_i is incident on a circular cylinder consisting of a perfectly conducting surface. In particular we are interested in investigating the intensity of illumination in the shadow zone. The following figure illustrates the geometry of the situation.

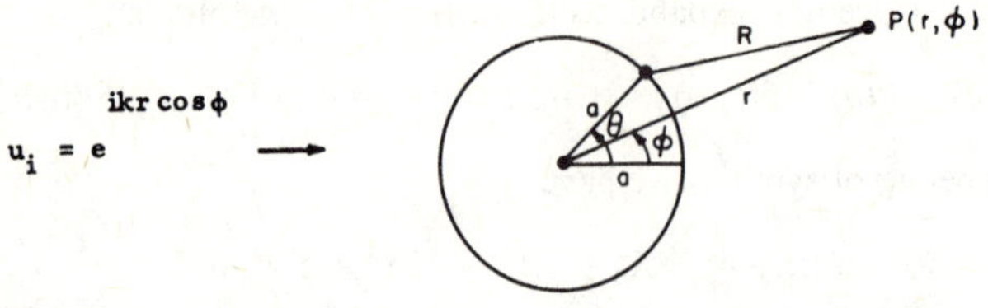

Mathematically we seek the solution of a certain problem. The solution $u(r, \theta)$ should be a function that satisfies the following three conditions.

1. $u(r, \theta)$ should be a solution of the reduced wave equation

$$\Delta u + k^2 u = 0.$$

2. On the surface $r = a$ it is to satisfy the boundary condition

$$u = -u_i = -e^{ika \cos \phi}.$$

3. At infinite distance from the cylinder u has to satisfy the radiation condition. That is

$$\lim_{r \to \infty} r^{1/2} \left[\frac{\partial u}{\partial r} - iku \right] = 0. \tag{7}$$

The latter condition can be deduced on physical grounds. It is equivalent to the statement that the scattered wave u has only finite energy. One can show that Conditions 1 and 2 without 3 would not determine a unique solution. With Condition 3 a unique solution to the stated problem exists. Condition 3 is equivalent to the statement that for large r, u is to have the following asymptotic structure

$$u \approx \frac{e^{kr}}{r^{1/2}} f(\phi). \tag{8}$$

If we solve

$$\Delta u + k^2 u = 0 \tag{9}$$

by the method of separation of variables, and impose a periodicity condition in the ϕ domain, that is $u(r, \phi + 2\pi) = u(r, \phi)$ for all ϕ, we are led to solutions of the type

$$H_n{}^{(1)}(kr)e^{\pm in\phi}. \tag{10}$$

The latter satisfies (9) and from the asymptotic structure of the Hankel function it follows that (10) also satisfies the radiation condition. Finally we seek a solution of the form

$$u = \sum_{n=-\infty}^{\infty} a_n H_n{}^{(1)}(kr)e^{in\phi}. \tag{11}$$

Using the generating function for the Bessel function, or equivalently the singular addition theorem in 2 dimensions, we have

$$u_i = e^{ikr\cos\phi} = \sum_{n=-\infty}^{\infty} i^n J_n(kr)e^{in\phi}. \tag{12}$$

A comparison of (11) and (12) on $r = a$ shows that the boundary condition

$$u = -u_i \quad \text{on} \quad r = a$$

will be satisfied if we select

$$a_n = \frac{-i^n J_n(ka)}{H_n^{(1)}(ka)}$$

so that finally

$$u = -\sum_{n=-\infty}^{\infty} \frac{i^n J_n(ka)}{H_n^{(1)}(ka)} H_n^{(1)}(kr)e^{in\phi} \tag{13}$$

is the solution of our problem. One can easily show that (13) converges for all $r \geq a$ and all ϕ. From the asymptotic formulas for large index we see that $H_n^{(1)}(kr)/H_n^{(1)}(ka)$ is a bounded function of n and

$$J_n(ka) \approx \frac{(ka/2)^n}{n!}$$

The latter is only valid for $n \gg ka$.

In many practical problems we are concerned with values of ka that are large. But the asymptotic formulas for Bessel functions change character near $n = ka$. This leads to great difficulties in obtaining asymptotic estimates for large ka for the function u. In particular for $ka \gg n$ we have

$$\frac{J_n(ka)H_n^{(1)}(kr)}{H_n^{(1)}(ka)} \approx \sqrt{\frac{2}{\pi kr}}\, e^{ik(r-a)} \cos\left[ka - \frac{n\pi}{2} - \frac{\pi}{4}\right]$$

and (13) will converge very slowly. To obtain such asymptotic results we will have to resort to another technique

We seek a solution of the form

$$u(r, \phi) = \int_0^{2\pi} H_0^{(1)}(kR) f(\theta)\, d\theta \tag{14}$$

where

$$R = \sqrt{r^2 + a^2 - 2ra\cos(\phi - \theta)}.$$

(14) satisfies the wave equation and the radiation condition. We shall try to determine $f(\theta)$ so that the boundary condition is also satisfied. To do so we

just expand the Hankel function in (14) by an appropriate addition theorem. For $r > a$ we have

$$H_n^{(1)}(kR) = \sum_{n=-\infty}^{\infty} J_n(ka)H_n^{(1)}(kr)e^{in(\phi-\theta)}.$$

We shall also expand $f(\theta)$ in a Fourier series so that

$$f(\theta) = \frac{1}{2\pi}\sum_{n=-\infty}^{\infty} c_n e^{in\theta}.$$

If we insert the last two expressions in (14) we obtain

$$u(r, \phi) = \sum_{n=-\infty}^{\infty} J_n(ka)H_n^{(1)}(kr)c_n e^{in\phi}.$$

A comparison with (13) shows that

$$c_n = \frac{-i^n}{H_n^{(1)}(ka)}$$

and

$$f(\theta) = \frac{-1}{2\pi}\sum_{n=-\infty}^{\infty}\frac{i^n e^{in\theta}}{H_n^{(1)}(ka)} = \frac{-1}{2\pi}\sum_{n=0}^{\infty}\frac{\varepsilon_n i^n \cos n\theta}{H_n^{(1)}(ka)}. \tag{15}$$

We shall require another representation for $f(\theta)$. We recall that

$$\frac{1}{2}g(0) + \sum_{n=1}^{\infty} g(n) = \frac{1}{2\pi i}\int_C g(z)\frac{\pi \cos \pi z}{\sin \pi z}\,dz \tag{16}$$

for a suitable analytic function $g(z)$. C is shown in the following figure. Note that z goes through the origin and is tangent to the imaginary z axis. Thus only half the residue of $g(z)/z$ is picked up there. This accounts for the term $\frac{1}{2}g(0)$ in (16). Using this result we have

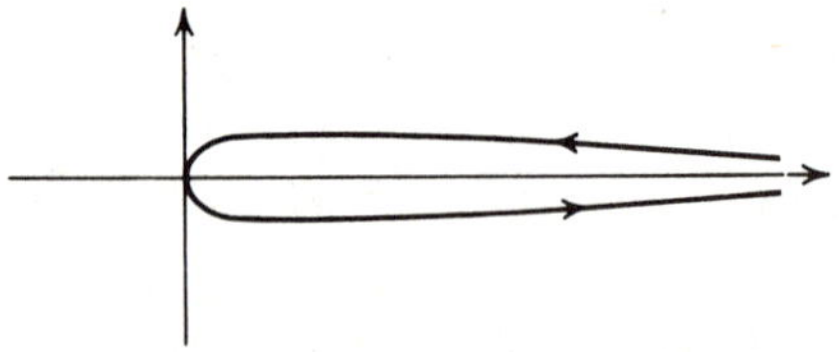

$$f(\theta) = \frac{-1}{2\pi i} \int_C \frac{\cos \nu\theta e^{-i\nu\pi/2}}{H_\nu^{(1)}(ka) \sin \nu\pi} d\nu. \tag{17}$$

The ingenious technique to be used now is in essence due to G. N. Watson.* If we recall from Eq. 8.10.15 that $H_\nu^{(1)}(ka)$ had, as a function of ν, zeros in the complex plane we can deform the contour C so as to pick those residues due to $H_\nu^{(1)}(ka)$, rather than those due to $\sin \nu\pi$. This should lead to another representation of $f(\theta)$ and one that when inserted in (14) will lead readily to asymptotic formulas. The deformed contour is shown in the following figure.

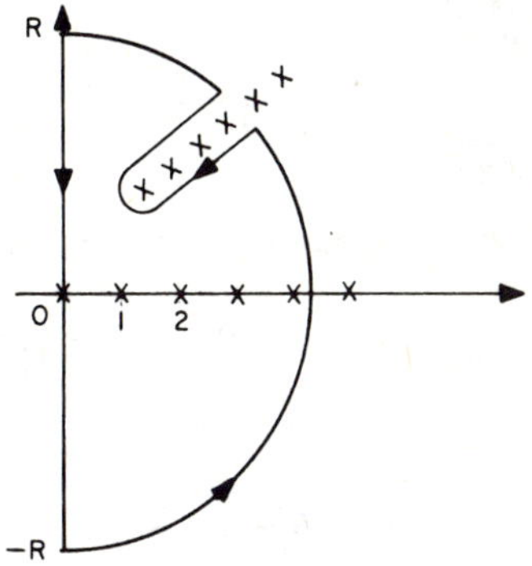

The path selected still includes the residue, due to $\sin \nu\pi$ and excludes those due to $H_\nu^{(1)}(ka)$. The contribution to the integral along the imaginary axis vanishes since the integrand is an odd function of ν. This follows immediately if one uses the fact that $H_{-\nu}^{(1)}(z) = e^{i\pi\nu} H_\nu^{(1)}(z)$. The contributions due to the semicircular arcs will vanish. This is a consequence of the asymptotic formula for the Hankel functions for large $|\nu|$. Finally, the only contribution will be due to the loop around the zeros of $H_\nu^{(1)}(ka)$ so that

$$f(\theta) = \sum_{m=1}^{\infty} \frac{[\cos \nu_m \theta] e^{-i\nu_m\pi/2}}{\sin \nu_m \pi (\partial H_\nu^{(1)}(ka)/\partial \nu)|_{\nu=\nu_m}} \tag{18}$$

where the ν_m are zeros of $H_\nu^{(1)}(ka)$. Insertion of (18) in (14) will lead to another representation of u. To simplify all computations we shall now use the fact that ka is large. Then we have

* See G. N. Watson *The Diffraction of Electrical Waves by the Earth Proc. Roy. Soc.* **A95**, 83–99 (1918). For the details of this analysis see H. Hochstadt, *Diffraction by Some Convex Objects, Arch. Rat. Mech. and Anal.* **3**, 422–438 (1959).

$$\nu_m \approx ka\left[1 + \frac{1}{2}\left(\frac{3\pi}{4}\frac{4m+3}{ka}\right)^{2/3} e^{i\pi/3}\right]$$

$$\left.\frac{\partial H_\nu^{(1)}(ka)}{\partial \nu}\right|_{\nu=\nu_m} \approx (-1)^m e^{-i\pi/6}\, 2i\sqrt{\frac{2}{\pi}}\,(ka)^{-2/3}\left[\frac{3\pi}{4}(4m+3)\right]$$

To evaluate the integrals

$$\int_0^{2\pi} H_0^{(1)}(kR)\cos \nu_m\theta\, d\theta$$

we introduce the asymptotic formula for the Hankel function and obtain integrals of the form

$$\sqrt{\frac{2}{\pi kr}}\, e^{ikr}\int_0^{2\pi} e^{-i(ka\cos(\theta-\phi)\pm\nu_m\theta+\pi/4)}\, d\theta. \tag{19}$$

In order to use Watson's lemma we seek those points θ where

$$\frac{d}{d\theta}\left[ka\cos(\theta-\phi)\pm\nu_m\theta+\frac{\pi}{4}\right] = 0.$$

These are approximately

$$\theta = \phi \pm \frac{\pi}{2}.$$

Geometrically these are the angles corresponding to the points on the cylinder

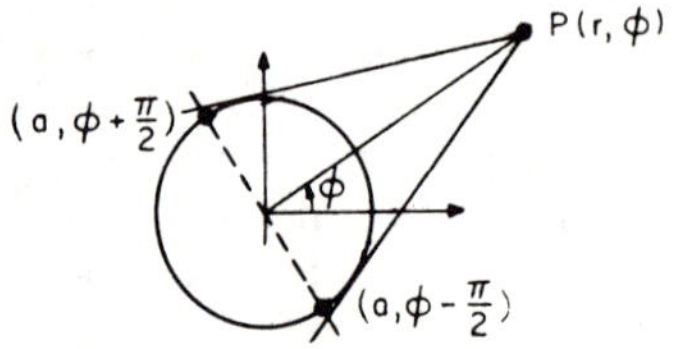

where tangent rays from the point (r, ϕ) touch the cylinder for large values of r. To evaluate an integral of the type

$$\int_0^{2\pi} e^{i[ka\cos(\theta-\phi)+\nu_m\theta]}\, d\theta$$

we let

$$\theta = \phi + \frac{\pi}{2} + u.$$

Note that, since ν_m is complex, $e^{ika\nu_m\theta}$ is no longer periodic in θ for θ real.

Then we have

$$e^{i\nu_m[\phi+\pi/2]}\int_{-\phi-\pi/2}^{-\phi+3\pi/2} e^{i[-ka\sin u+\nu_m u]}\,du.$$

The above integral can be approximated by

$$\int_{-\infty}^{\infty}\exp i\left[\frac{1}{2}\,ka\left(\frac{3\pi}{4}\frac{4m+3}{ka}\right)^{2/3}e^{i\pi/3}u+\frac{ka\,u^3}{6}\right]du$$

$$\approx\left(\frac{2}{ka}\right)^{1/3}e^{-i\pi/3}\int_0^{\infty}\cos[\alpha_m x+\tfrac{1}{3}x^3]\,dx$$

$$\alpha_m=-\left[\frac{3\pi}{8}(4m+3)\right]^{2/3}$$

We let

$$A(\alpha)=\frac{1}{\pi}\int_0^{\infty}\cos[\alpha x+\tfrac{1}{3}x^3]\,dx.$$

Using all these results in (14) we finally obtain

$$u(r,\phi)=\frac{\pi(16ka)^{1/3}}{\sqrt{kr}}\exp\left(\frac{-5\pi i}{12}+ikr\right).$$

$$\sum_{m=1}^{\infty}(-1)^m\frac{[e^{i\nu_m\phi}+e^{i\nu_m(2\pi-\phi)}]}{[1-e^{2\pi i\nu_m}][3\pi/4(4m+3)]^{1/6}}A(\alpha_m) \tag{20}$$

Note that for $0<\phi<2\pi$ the series converges very rapidly, because ν_m is complex.

12. The Zeros of Bessel Functions

There is an extensive body of knowledge regarding the zeros of Bessel functions. We shall not be able to begin to do justice to this topic in our treatment, but we shall only discuss some of the most basic and important properties. In particular we will be concerned with those that will prove of value in certain problems of mathematical physics. For a fuller treatment the reader should consult the treatise by G. N. Watson.

A rather efficient approach is the following. The function

$$\frac{J_\nu(z)}{(z/2)^\nu}=\sum_{n=0}^{\infty}\frac{(iz/2)^{2n}}{n!\,\Gamma(n+\nu+1)} \tag{1}$$

is an entire function.* If $M(r)$ denotes the maximum of $|f(z)|$ on $|z| = r$, and $f(z)$ is entire, then, by definition the order of $f(z)$ is

$$\rho = \limsup_{r\to\infty} \frac{\log \log M(r)}{\log r}.$$

One can show that if $f(z)$ is of finite order and a_n the general coefficient in its Maclaurin series then

$$\rho = \limsup_{n\to\infty} \frac{n \log n}{\log 1/|a_n|}. \tag{2}$$

If we apply (2) to (1) we see that $z^{-\nu}J_\nu(z)$ is of order $\rho = \frac{1}{2}$.

Any entire function of order less than unity is either a polynomial or has an infinite number of zeros. Clearly, since $\rho = 1/2$, $z^{-\nu}J_\nu(z)$ is not a polynomial and hence has an infinite number of zeros. We have the following theorem.

THEOREM Any zero of a solution of

$$y'' + \frac{1}{z}\, y' + \left(1 - \frac{\nu^2}{z^2}\right) y = 0,$$

which is not zero, is a simple zero.

Proof. If at some point $a \neq 0$ $y(a) = y'(a) = 0$ then $y \equiv 0$, since y is a solution of a homogeneous second order equation. If, at a point of regularity its initial values vanish it vanishes identically. ∎

THEOREM $J_\nu(z)$ has an infinity of simple zeros. The only non-simple zero can occur at the origin.

Proof. That $J_\nu(z)$ has an infinity of zeros follows from the fact that $z^{-\nu}J_\nu(z)$ is entire of order 1/2. That all zeros, except $z = 0$, are simple follows from the preceding theorem. For ν an integer $z = 0$ is a zero of multiplicity $|\nu|$. ∎

THEOREM For ν real and $\nu > -1$, all zeros of $J_\nu(z)$ are real.

Proof. Suppose that $\zeta \neq 0$ and $J_\nu(\zeta) = 0$. Since $J_\nu(z)$ is real for real z it follows that $J_\nu(\bar{\zeta}) = 0$. Hence all complex zeros must occur in complex conjugate pairs. If that case were to arise we let

$$f(z) = J_\nu(z\zeta)$$
$$g(z) = J_\nu(z\bar{\zeta}).$$

* For relevant information on entire functions see R. P. Boas Jr., *Entire Functions*, Academic Press, New York, 1954.

Note that $f(z) = \overline{g(z)}$.
Then $f(1) = g(1) = 0$ and also

$$f'' + \frac{1}{z} f' + \left(\zeta^2 - \frac{v^2}{z^2}\right) f = 0$$

$$g'' + \frac{1}{z} g' + \left(\bar{\zeta}^2 - \frac{v^2}{z^2}\right) g = 0.$$

If we multiply the first of the above by zg and the second by zf and subtract we obtain

$$g(zf')' - f(zg')' + (\zeta^2 - \bar{\zeta}^2) zfg = 0. \tag{3}$$

We have

$$\int_0^1 [g(zf')' - f(zg')'] \, dz = z(gf' - fg')\Big|_0^1 = 0$$

since $f(1) = g(1) = 0$ and also

$$z(gf' - fg') \approx kz^{2(1+v)}$$

near the origin and $v > -1$. By integrating (3) we obtain

$$(\zeta^2 - \bar{\zeta}^2) \int_0^1 zfg dz = (\zeta^2 - \bar{\zeta}^2) \int_0^1 zf\bar{f} \, dz = 0$$

so that $\zeta^2 = \zeta^{-2}$. The latter implies that ζ is either real or pure imaginary. The latter cannot be true. Because if $\zeta = ix$, x real, then by (1)

$$\frac{J_v(ix)}{(ix/2)^v} = \sum_{n=0}^{\infty} \frac{x^{2n}}{4^n n! \, \Gamma(n + v + 1)} > 0$$

for real $v > -1$. The above is clearly false. ■

A theorem that applies to all solutions, not only $J_v(z)$ is the following.

THEOREM Suppose v is real and such that $v^2 \geq \frac{1}{4}$. Then if z_n and z_{n+1} are two consecutive real positive zeros of a real solution $y(z)$ such that $1 - (v^2 - \frac{1}{4})/z_n^{\,2} > 0$, of

$$y'' + \frac{1}{z} y' + \left(1 - \frac{v^2}{z^2}\right) y = 0$$

then

$$z_n + \frac{\pi}{\sqrt{1 - (v^2 - \frac{1}{4})/z_{n+1}^2}} < z_{n+1} < z_n + \frac{\pi}{\sqrt{1 - (v^2 - \frac{1}{4})/z_n^{\,2}}}.$$

Proof. Let $y = z^{-1/2} V$ so that

$$V'' + \left[1 - \frac{\nu^2 - \frac{1}{4}}{z^2}\right] V = 0 \tag{4}$$

and let u satisfy

$$u'' + \left[1 - \frac{\nu^2 - \frac{1}{4}}{z_n^{\,2}}\right] u = 0 \tag{5}$$

Consider the solution of (5)

$$u = \sin \sqrt{1 - \frac{\nu^2 - \frac{1}{4}}{z_n^{\,2}}}\,(z - z_n)$$

for which $u(z_n) = 0$. Let ζ_{n+1} be the next larger zero of u. Then

$$\zeta_{n+1} = z_n + \frac{\pi}{\sqrt{1 - (\nu^2 - \frac{1}{4})/z_n^{\,2}}}.$$

If we prove that $z_{n+1} < \zeta_{n+1}$ half of the assertion will have been proved.

Suppose that $z_{n+1} \geq \zeta_{n+1}$ and, without loss of generality $V'(z_n) > 0$. In that case $V > 0$ in (z_n, z_{n+1}). By multiplying (4) by u and (5) by V and subtracting we are led to

$$uV'' - Vu'' + \left(\nu^2 - \frac{1}{4}\right)\left(\frac{1}{z_n^{\,2}} - \frac{1}{z^2}\right) uV = 0.$$

An integration over (z_n, ζ_{n+1}) leads to

$$\left(\nu^2 - \frac{1}{4}\right) \int_{z_n}^{\zeta_{n+1}} \left(\frac{1}{z_n^{\,2}} - \frac{1}{z^2}\right) uV\, dz = -(uV' - Vu')\Big|_{z_n}^{\zeta_{n+1}}$$

$$= V(\zeta_{n+1}) u'(\zeta_{n+1}).$$

By the hypothesis that $z_{n+1} \geq \zeta_{n+1}$ we conclude that the left side is positive and the right side negative. But this is impossible. Hence

$$z_{n+1} < z_n + \frac{\pi}{\sqrt{1 - (\nu^2 - \frac{1}{4})/z_n^{\,2}}}.$$

The second part of the assertion is proved by a similar argument, but z_n is replaced by z_{n+1} in (5). ■

Some of the preceding conclusions may be derived, in a slightly cruder form by use of the asymptotic estimates. For example we have

$$J_\nu(z) \approx \sqrt{\frac{2}{\pi z}} \cos\left[z - \frac{\nu\pi}{2} - \frac{\pi}{4}\right] \sum_{n=0}^{\infty} \frac{a_n}{z^{2n}} - \sqrt{\frac{2}{\pi z}} \sin\left[z - \frac{\nu\pi}{2} - \frac{\pi}{4}\right] \sum_{n=0}^{\infty} \frac{b_n}{z^{2n+1}}.$$

For the specific form of a_n and b_n we refer to 8.8.7. For a zero of $J_\nu(z)$ we see that

$$\tan\left(z - \frac{\nu\pi}{2} - \frac{\pi}{4}\right) = \sum_{n=0}^{\infty} \frac{c_n}{z^{2n-1}}$$

Stokes used the above to derive asymptotic expansions for the zeros.

For example for $\nu = 0$ we find

$$\tan\left(z - \frac{\pi}{4}\right) = -8z - \frac{33}{8z} + \cdots .$$

Clearly for large values of z one zero will have to lie near the point $n\pi - \pi/4$. For convenience let $\zeta_n = n\pi - \pi/4$ and expand z in the form

$$z = \zeta_n + \frac{a}{\zeta_n} + \frac{b}{\zeta_n^{\,3}} + \cdots .$$

By the choice of ζ_n

$$\tan\left(\zeta_n + u - \frac{\pi}{4}\right) = -\cot u$$

so that (6) can be replaced by

$$\tan\left[\frac{a}{\zeta_n} + \frac{b}{\zeta_n^{\,3}} + \cdots\right] = \frac{1}{8z} + \frac{33}{512z^3} + \cdots$$

or equivalently, using the Maclaurin series for the tangent function

$$\frac{a}{\zeta_n} + \frac{b}{\zeta_n^{\,3}} + \frac{a^3}{3\zeta_n^{\,3}} + \cdots = \frac{1}{8\zeta_n} - \frac{a^2}{8\zeta_n^{\,3}} - \frac{33}{512\zeta_n^{\,3}} + \cdots .$$

Comparison of coefficients shows that

$$a = \frac{1}{8}, \qquad b = -\frac{31}{384}, \qquad \cdots$$

and finally

$$z_n \approx \left(n\pi - \frac{\pi}{4}\right) + \frac{1}{8(n\pi - \pi/4)} - \frac{31}{384(n\pi - \pi/4)^3} + \cdots . \tag{6}$$

It is easy to verify by means of the asymptotic formulas that the functions

$$AJ_\nu(z) + BN_\nu(z) \tag{7}$$

$$CJ_\nu(z) + DzJ'_\nu(z) \tag{8}$$

for ν and A, B, C, D real have an infinity of zeros. Asymptotic estimates for these can be found as in the case for $J_\nu(z)$. But it is possible for complex zeros to occur. One can show that (8) will have two complex zeros if $C/D + \nu < 0$.

In view of the fact that $z^{-\nu}J_\nu(z)$ is an entire function it must be possible to produce a representation in terms of its factors. This can be done immediately via the Weierstrass factorization theorem, but we shall do it independently by a direct calculation.

THEOREM The Bessel function can be represented as the following infinite product.

$$J_\nu(z) = \frac{(z/2)^\nu}{\Gamma(\nu+1)} \prod_{n=1}^{\infty} \left(1 - \frac{z^2}{z_n{}^2}\right).$$

Proof. We note first that all zeros $z_n \neq 0$ are simple. Secondly if z_n is a zero so is $-z_n$ since $z^{-\nu}J^\nu(z)$ is an even function of z. The identity (proved immediately by means of the formulas or from the infinite series representation)

$$\frac{d}{dz} z^{-\nu}J_\nu(z) = -z^{-\nu}J_{\nu+1}(z) \tag{9}$$

shows that

$$J_\nu'(z_n) = -J_{\nu+1}(z_n). \tag{10}$$

If we let $G(z) = z^{-\nu}J^\nu(z)$ (9) also shows that

$$F(z) = \frac{G'(z)}{G(z)} = -\frac{J_{\nu+1}(z)}{J_\nu(z)}. \tag{11}$$

If we consider a sequence of circles about the origin, $\{C_n\}$ such that C_n includes all points $\pm z_1, \pm z_2, \ldots \pm z_n$, and not passing through any zeros then F is uniformly bounded on these. This follows from an application of the asymptotic formulas for J_ν and $J_{\nu+1}$ applied to the right of (11).

The sequence of integrals

$$I_n = \frac{1}{2\pi i}\int_{C_n} \frac{F(\zeta)}{\zeta(\zeta - z)}\, d\zeta$$

must converge to zero, in view of the fact that for fixed $z \neq z_n$ and large ζ

$$\left|\frac{F(\zeta)}{\zeta(\zeta - z)}\right| < \frac{M}{|\zeta|^2}.$$

It follows that, if R_n denotes the radius of C_n that

$$|I_n| \le \frac{M}{R_n}$$

so that

$$\lim_{n\to\infty} I_n = 0.$$

A residue integration shows that

$$I_n = -\frac{F(0)}{z} + \frac{F(z)}{z} + \sum_{k=1}^{n}\left[\frac{J_{\nu+1}(z_k)}{z_k(z - z_k)J_\nu'(z_k)} + \frac{J_{\nu+1}(-z_k)}{-z_k(z + z_k)J_\nu'(-z_k)}\right]$$

We have for convenience labeled the zeros $\pm z_1, \pm z_2, \ldots$. By using (10) and collecting all terms we obtain

$$I_n = -\frac{F(0)}{z} + \frac{F(z)}{z} - \sum_{k=1}^{n} \frac{2}{z^2 - z_k^2}.$$

Since $z_k \approx k\pi$ the above series will converge uniformly and we can take the limit. It follows that, since $F(0) = 0$ from (11),

$$F(z) = \frac{G'(z)}{G(z)} = 2\sum_{k=1}^{\infty} \frac{z}{z^2 - z_k^2}.$$

An integration leads to

$$\log G(z) = c_1 + \sum_{k=1}^{\infty} \log\left(\frac{z^2 - z_k^2}{z_k^2}\right).$$

To evaluate c_1 we have

$$z^{-\nu}J_\nu(z) = G(z) = \frac{(\frac{1}{2})^\nu}{\Gamma(\nu + 1)} + \cdots$$

so that

$$c_1 = \log(\tfrac{1}{2})^\nu/\Gamma(\nu + 1)$$

and finally

$$J_\nu(z) = \frac{(z/2)^\nu}{\Gamma(\nu + 1)} \prod_{n=1}^{\infty}\left(1 - \frac{z^2}{z_n^2}\right). \qquad \blacksquare$$

We can use the last theorem to obtain another result with great ease.

THEOREM For $\nu \geq 0$ $J_\nu'(z)$ has only real simple zeros except $z = 0$ and these interlace with the zeros of $J_\nu(z)$.

Proof. From the preceding theorem we have

$$\frac{J_\nu'(z)}{J_\nu(z)} = \frac{\nu}{z} + \sum_{k=1}^{\infty} \frac{2z}{z^2 - z_k{}^2}.$$

The imaginary part of the above is given by

$$\frac{-\nu y}{x^2 + y^2} + \sum_{k=1}^{\infty} \left[\frac{-y}{(x - z_k)^2 + y^2} - \frac{y}{(x + z_k)^2 + y^2}\right].$$

Clearly it can vanish only for $y = 0$ and all zeros of $J_\nu'(z)$ are real. That they are simple is obvious. At a multiple zero of J_ν', J_ν' would also vanish, and by the differential equation so would J_ν. In that case the solution would vanish identically. Finally

$$\frac{d}{dz}\frac{J_\nu'(z)}{J_\nu(z)} = -\frac{\nu}{z^2} - 2\sum \frac{z^2 + z_k{}^2}{(z^2 - z_k{}^2)^2}$$

and for real z

$$\frac{d}{dz}\frac{J_\nu'(z)}{J_\nu(z)} < 0.$$

We see that the function $J_\nu'(z)/J_\nu(z)$ has an infinity of poles, but is of negative slope. Hence between any two zeros of $J_\nu(z)$ there must be precisely one of $J_\nu'(z)$. ■

A similar result is contained in the following theorem.

THEOREM The real zeros of $J_\nu(z)$ and $J_{\nu+1}(z)$ for real ν interlace.

Proof. We consider the following two identities

$$\frac{d}{dz} z^{-\nu} J_\nu(z) = -z^{-\nu} J_{\nu+1}(z)$$

$$\frac{d}{dz} z^{\nu+1} J_{\nu+1}(z) = z^{\nu+1} J_\nu(z).$$

The use of Rolle's theorem tells us that, by the first statement, between two real zeros of $J_\nu(z)$, $J_{\nu+1}(z)$ has to vanish at least one. The second statement tells us that between any two zeros of $J_{\nu+1}(z)$, $J_\nu(z)$ has to vanish at least once. Combining the two we see the truth of the theorem. ■

The zeros of $zJ_\nu'(z) + \alpha J_\nu(z)$ can be discussed in a similar fashion. For $\nu > -1$ and $\alpha + \nu \geq 0$ there is an infinity of real zeros and no complex zeros. For $\alpha + \nu < 0$ there will be precisely two conjugate pure imaginary zeros.

13. Fourier-Bessel Expansions

We have seen that for $\nu > -1$ the function $J_\nu(z)$ has an infinite number of real and simple zeros. Denote the positive zeros by λ_k. We can show that the functions $\{\sqrt{z}J_\nu(\lambda_k z)\}$ form and orthogonal set over the interval (0, 1).

First we shall prove that

$$\int_0^1 zJ_\nu(\lambda_n z)J_\nu(\lambda_m z)\, dz = 0, \qquad n \neq m. \tag{1}$$

From the differential equation, if we let $y = J_\nu(\lambda_n z)$, $u = J_\nu(\lambda_m z)$ we have

$$(zy')' + \left(\lambda_n^2 z - \frac{\nu^2}{z}\right) y = 0$$

$$(zu')' + \left(\lambda_m^2 z - \frac{\nu^2}{z}\right) u = 0.$$

By virtue of the behavior of these functions at $z = 0$ and $z = 1$ it follows that

$$\int_0^1 [u(zy')' - y(zu')']\, dz = z(uy' - yu')\Big|_0^1 = 0.$$

Multiplying the first differential equation by u, the second by y and integrating we have

$$(\lambda_n^2 - \lambda_m^2) \int_0^1 zJ_\nu(\lambda_n z)J_\nu(\lambda_m z)\, dz = 0, \qquad n \neq m.$$

To evaluate

$$\int_0^1 zJ_\nu^2(\lambda_n z)\, dz$$

we proceed as follows. We multiply the differential equation by zy'.

$$zy'(zy')' + \lambda_n^2 z^2 yy' - \nu^2 yy' = 0$$

By integrating the above and applying an integration by parts to the middle term we find

$$(zy')^2\Big|_0^1 + \lambda_n^2 z^2 y^2\Big|_0^1 - 2\lambda_n^2 \int_0^1 z\, y^2 dz - \nu^2 y^2\Big|_0^1 = 0.$$

The second term vanishes identically. The term $(zy')^2 - \nu^2 y^2$ vanishes at $z = 0$ for $\nu > -1$, as can be seen using the series expansion. Finally

$$\int_0^1 z J_\nu^{\,2}(\lambda_n z)\, dz = \tfrac{1}{2} J_\nu'^{\,2}(\lambda_n) = \tfrac{1}{2} J_{\nu+1}^2(\lambda_n). \tag{2}$$

If we let $c_n = [\tfrac{1}{2} J_{\nu+1}^2(\lambda_n)]^{-1/2}$ the set $\{c_n \sqrt{z} J_\nu(\lambda_n z)\}$ is orthonormal over the interval (0, 1). It is natural to enquire whether the set is complete. It will develop that the cases where $\nu \geq 1$ and $-1 < \nu < 1$ have to be treated separately. We shall now define the functions $\phi_n(z)$ by

$$\phi_n(z) = c_n \sqrt{z} J_\nu(\lambda_n z). \tag{3}$$

If we introduce inner product notation we have

$$(\phi_n, \phi_m) = \int_0^1 \phi_n(z)\phi_m(z)\, dz = 0, \qquad n \neq m$$

$$= 1, \qquad n = m.$$

The functions $\phi_n(z)$ are solutions of the differential equation

$$y'' + \left[\lambda_n^{\,2} - \frac{\nu^2 - \frac{1}{4}}{z^2}\right] y = 0. \tag{4}$$

(4) has two solutions which can be characterized by their expansions. Then

$$y_1 = z^{1/2+\nu} + \cdots$$
$$y_2 = z^{1/2-\nu} + \cdots$$

For $\nu > 1$ we see that

$$\int_0^1 y_1^{\,2}\, dz < \infty$$

but $\int_0^1 y_2^{\,2}\, dz$ does not exist. When ν is an integer y_2 takes a slightly different form. Then for $\nu = n$

$$y_2 = \log z(z^{1/2+n} + \cdots) + k_n z^{1/2-n} + \cdots$$

and $\int_0^1 y_2^{\,2}\, dz$ still does not exist.

For $-1 < \nu < 1$ we see that

$$\int_0^1 (c_1 y_1 + c_2 y_2)^2\, dz < \infty$$

for all solutions of (4).

For $\nu \geq 1$ $\phi_n(z)$ can be characterized as the solutions of the eigenvalue problem

$$y'' + \left[\lambda^2 - \frac{\nu^2 - \frac{1}{4}}{z^2}\right] y = 0 \tag{5}$$

with the boundary condition

$$y(1) = 0 \tag{6}$$

and the regularity condition

$$\int_0^1 y^2 \, dz < \infty. \tag{7}$$

For $\nu \geq 1$, (5), (6) and (7) specify, for each λ_n, a unique $\phi_n(z)$. For $-1 < \nu < 1$ (7) no longer helps in selecting a solution. This explains why the two cases must be dealt with separately.

We shall now consider the case $\nu \geq 1$. We first define the Green's function $G(z, \zeta)$ in terms of its differential equation

$$G''(z, \zeta) - \frac{\nu^2 - \frac{1}{4}}{z^2} G(z, \zeta) = \delta(z - \zeta) \tag{8}$$

$$G(1, \zeta) = 0$$

$$\int_0^1 G^2(z, \zeta) \, dz < \infty.$$

We see that

$$\begin{aligned} G(z, \zeta) &= c_1 z^{\frac{1}{2}+\nu}, & z < \zeta \\ &= c_2 (z^{\frac{1}{2}-\nu} - z^{\frac{1}{2}+\nu}), & z > \zeta \end{aligned}$$

by solving the homogeneous case corresponding to (8) and invoking the boundary conditions. To determine c_1 and c_2 we require that G be continuous at $z = \zeta$. Then

$$\lim_{z \to \zeta -} G(z, \zeta) = \lim_{z \to \zeta +} G(z, \zeta).$$

The derivative $G'(z, \zeta)$ must have a unit jump at $z = \zeta$ so that

$$\lim_{z \to \zeta +} G'(z, \zeta) - \lim_{z \to \zeta -} G'(z, \zeta) = 1.$$

These conditions allow us to find c_1 and c_2 so that

$$\begin{aligned} G(z, \zeta) &= -\frac{1}{2\nu} z^{\frac{1}{2}+\nu}(\zeta^{\frac{1}{2}-\nu} - \zeta^{\frac{1}{2}+\nu}), & z < \zeta \\ &= -\frac{1}{2\nu} \zeta^{\frac{1}{2}+\nu}(z^{\frac{1}{2}-\nu} - z^{\frac{1}{2}+\nu}), & z > \zeta. \end{aligned} \tag{9}$$

Note that $G(z, \zeta)$ is symmetric, that is

$$G(z, \zeta) = G(\zeta, z). \tag{10}$$

By the use of this function we can rewrite (5) as an integral equation

$$y(z) = -\lambda^2 \int_0^1 G(z, \zeta) y(\zeta)\, d\zeta. \tag{11}$$

We shall now define the integral operator

$$\begin{aligned} Ky &= -\int_0^1 G(z, \zeta) y(\zeta)\, d\zeta \\ &= \frac{1}{2\nu} [(z^{1/2-\nu} - z^{1/2+\nu}) \int_0^z \zeta^{1/2+\nu} y(\zeta)\, d\zeta \\ &\quad + z^{1/2+\nu} \int_z^1 (\zeta^{1/2-\nu} - \zeta^{1/2+\nu}) y(\zeta)\, d\zeta]. \end{aligned} \tag{12}$$

A direct calculation with (12) shows that

$$(Ky)'' - \frac{\nu^2 - \frac{1}{4}}{z^2}(Ky) = -y. \tag{13}$$

(13) combined with (12) verifies directly that the solution of (11) satisfies (5) and (11) shows that that solution also satisfies the boundary conditions (6) and (7).

We consider the Hilbert space $L_2(0, 1)$ consisting of all real functions $f(z)$ for which

$$\|f\|^2 = \int_0^1 f^2(z)\, dz < \infty.$$

K defines a self-adjoint operator on that space, in the sense that

$$(Kf, g) = (f, Kg)$$

for all f and g. This is an immediate consequence of (10). We also see that K has an infinite number of eigenvalues and eigenvectors since

$$\lambda_n^2 K\phi_n = \phi_n$$

where the $\phi_n(z)$ are defined in (3).

The eigenvalues can be characterized in terms of an extremal property of K.

LEMMA λ_1 can be characterized by

$$\frac{1}{\lambda_1^2} = \sup_{f,\, \|f\|=1} \|Kf\|. \tag{14}$$

In other words the supremum of $\|Kf\|$ over all functions f of norm $\|f\| = 1$ is $1/\lambda_1^2$.

Proof. We know that

$$K\phi_1 = \frac{1}{\lambda_1^2} \phi_1$$

so that

$$\|K\phi_1\| = \frac{1}{\lambda_1^2} \|\phi_1\| = \frac{1}{\lambda_1^2}.$$

We shall show that if f is not linearly dependent on ϕ_1 and $\|f\| = 1$, then

$$\|Kf\| < \frac{1}{\lambda_1^2}.$$

Suppose*

$$\sup_{\|f\|=1} \|Kf\| = \frac{1}{u^2} > \frac{1}{\lambda_1^2}.$$

In that case it can be shown, using the properties of the operator K, that for some element f

$$Kf = \frac{1}{u^2} f.$$

But in that case

$$f'' + \left(u^2 - \frac{\nu^2 - \frac{1}{4}}{z^2}\right) f = 0$$

$$f(1) = 0, \qquad \int_0^1 f^2(z)\, dz < \infty$$

and we took λ_1^2 to be the smallest eigenvalue corresponding to the above. Hence $u^2 \geq \lambda_1^2$.

* That $\sup_{\|f\|=1} \|Kf\|$ exists is an immediate consequence of Schwarz' inequality.

$$\|Kf\|^2 = \int_0^1 \left[\int_0^1 G(z, \zeta) f(\zeta) d\zeta\right]^2 dz \leq \int_0^1 f^2(\zeta) d\zeta \int_0^1 \int_0^1 G^2(z, \zeta)\, dz\, d\zeta = \frac{1}{16(1+\nu)^2(2+\nu)}.$$

If for some element of $f \neq \phi_1$

$$Kf = \frac{1}{\lambda_1^{\,2}} f$$

we would have a second eigenfunction corresponding to $\lambda_1^{\,2}$, which we know cannot be. Hence for all $f \neq \phi_1$ and $\|f\| = 1$

$$\|Kf\| < \frac{1}{\lambda_1^{\,2}}. \qquad \blacksquare$$

By a repetition of this kind or argument we can prove another lemma.

LEMMA λ_{n+1} can be characterized by

$$\frac{1}{\lambda_{n+1}^2} = \sup_{\|f\|=1} \|Kf\|$$

$$(f, \phi_1) = (f, \phi_2) = \cdots = (f, \phi_n) = 0.$$

In other words to obtain $1/\lambda_{n+1}^2$ we compute the supremum of $\|Kf\|$ over all f, such that $\|f\| = 1$ and f orthogonal to the subspace spanned by ϕ_1, $\phi_2, \ldots, \phi_n$.

THEOREM The set of orthonormal functions $\{\phi_i\}$ is complete in the space $L_2(0, 1)$, for the case $\nu \geq 1$.

Proof. Define the functions $g_n(z)$ by

$$g_{n+1} = f - \sum_{i=1}^{n} (f, \phi_i)\phi_i.$$

These functions form a Cauchy sequence since (if $n > m$)

$$\|g_n - g_m\|^2 = \sum_{i=m+1}^{n} (f, \phi_i)^2$$

and the above can be made arbitrarily small for sufficiently large n and m. This follows from an application of Bessel's inequality. Then there exists a function $g(z)$ such that

$$\lim_{n\to\infty} \|g - g_n\|^2 = \lim_{n\to\infty} \int_0^1 (g - g_n)^2 \, dz = 0.$$

Also

$$(g_{n+1}, \phi_i) = 0, \qquad i = 1, 2, \ldots, n.$$

Then by the preceding lemma

$$\frac{\|Kg_n\|}{\|g_n\|} \leq \frac{1}{\lambda_{n+1}^2}.$$

Clearly for all g_n, $\|g_n\| < \|f\|$, so that

$$\|Kg_n\| \leq \frac{\|f\|}{\lambda_{n+1}^2}$$

and since $\lambda_n \to \infty$ we have

$$\lim_{n\to\infty} \|Kg_n\| = \|Kg\| = 0.$$

In order for $Kg = 0$ we require that

$$0 = (Kg)'' + \frac{\nu^2 - \frac{1}{4}}{z^2}(Kg) = -g$$

which can only hold for $g = 0$. Finally we conclude that

$$\lim_{n\to\infty} \left\| f - \sum_1^n (f, \phi_i)\phi_i \right\| = 0. \qquad \blacksquare$$

We return to the case $-1 < \nu < 1$. Clearly the condition

$$\int_0^1 y^2\, dz < \infty$$

does not help in selecting just one solution of the differential equation. Nevertheless, an inspection of the integral equation (11) shows that the only possible solution must be of the form

$$y = z^{1/2+\nu} + \cdots.$$

If we insert

$$y = \sqrt{z}[c_1 J_\nu(\lambda z) + c_2 J_{-\nu}(\lambda z)]$$

in (11) we find by inspecting the terms resulting from an expansion of the Bessel functions that $c_2 = 0$. In other words the kernel of the integral operator in (12) is such that it selects the solution $\sqrt{z}J_\nu(z)$. The boundary condition

$$y(1) = 0$$

again leads to the zeros of $J_\nu(\lambda)$. These eigenvalues can be characterized by the same extremal conditions as before. Finally we can state the following theorem.

THEOREM The set of orthonormal functions $\{[\frac{1}{2}J_{\nu+1}^2(\lambda_n)]^{-1/2}\sqrt{z}J_\nu(\lambda_n z)\}$ is complete in the Hilbert space $L_2[0, 1]$ for all $\nu > -1$.

Proof. The case $\nu \geq 1$ was taken care of in the last theorem. To dispose of the case $-1 < \nu < 1$ we note that all steps of the preceding theorem apply here as well. Finally

$$\lim_{n\to\infty} \left\| f - \sum_1^n (f, \phi_i)\phi_i \right\| = 0. \quad \blacksquare$$

Similar statements can be made in regard to the orthogonal set $\{\sqrt{z}J_\nu(\lambda_n z)\}$ where the λ_n are the zeros of $zJ_\nu'(z) + \alpha J_\nu(z)$. As before, these functions are complete for $\nu > -1$. The resultant expansions are known as Fourier-Dini expansions. Some applications of these will be discussed in the next section.

The results regarding Fourier-Bessel functions can be restated as follows. Suppose $f(z)$ is defined on the interval $(0, a)$ and is such that

$$\int_0^a zf^2(z)\, dz < \infty.$$

Then we have the expansion

$$f(z) = \sum_{n=1}^\infty c_n J_\nu\left(\lambda_n \frac{z}{a}\right) \tag{15}$$

$$c_n = \frac{\int_0^a zf(z)J_\nu(\lambda_n(z/a))\, dz}{\int_0^a zJ_\nu{}^2(\lambda_n(z/a))\, dz}$$

$$= \frac{\int_0^a zf(z)J_\nu(\lambda_n(z/a))\, dz}{\frac{1}{2}a^2J_{\nu+1}^2(\lambda_n)} \tag{16}$$

We now let

$$\zeta_n = \frac{\lambda_n}{a}$$

and define the following monotonically increasing stepfunction

$$\rho(\zeta) = \sum_{k=1}^n \frac{1}{\frac{1}{2}a^2\zeta_k J_{\nu+1}^2(a\zeta_k)}, \qquad \zeta_n \leq \zeta < \zeta_{n+1}. \tag{17}$$

One readily verifies that with this function we can rewrite (15) in the form of the Stieltjes integral

$$f(z) = \int_0^\infty \zeta J_\nu(\zeta z)\left[\int_0^a zf(z)J_\nu(\zeta z)dz\right] d\rho(\zeta). \tag{18}$$

By means of the asymptotic formulas we note that

$$\tfrac{1}{2}a^2\zeta_k J_{v+1}^2(a\zeta_k) \approx \tfrac{1}{2}a^2\zeta_k x \frac{2}{\pi a\zeta_k} = \frac{a}{\pi}$$

since

$$J_{v+1}(a\,\zeta_k) \approx -\sqrt{\frac{2}{\pi a\zeta_k}}.$$

Then

$$\rho(\zeta_{n+1}) - \rho(\zeta_n) \approx \frac{\pi}{a}$$

and

$$\zeta_{n+1} - \zeta_n = \frac{\lambda_{n+1} - \lambda_n}{a} \approx \frac{\pi}{a}$$

so that

$$\frac{\rho(\zeta_{n+1}) - \rho(\zeta_n)}{\zeta_{n+1} - \zeta_n} \approx 1.$$

In the limit as a becomes infinitely large ρ becomes a continuous function and formally we see that

$$\rho'(\zeta) = 1.$$

This can be shown rigorously, but we shall refrain from doing so.* Finally in the limit (18) becomes

$$f(z) = \int_0^\infty \zeta J_v(\zeta z)\left[\int_0^\infty zf(z)J_v(\zeta z)dz\right] d\zeta. \tag{19}$$

The integral

$$F_v(\zeta) = \int_0^\infty zf(z)J_v(\zeta z)\,dz \tag{20}$$

is known as the Hankel transform of order v of $f(z)$. The corresponding inversion formula can be seen from (19) to be

$$f(z) = \int_0^\infty \zeta F_v(\zeta)J_v(\zeta z)\,d\zeta. \tag{21}$$

(19) is the analogue of the Fourier-Bessel expansion for the interval $(0, \infty)$.

* Details of this prodecure may be found in E. C. Titchmarsh, *Eigenfunction Expansions Associated with Second Order Differential Equations*, 2nd edition, Clarendon Press, Oxford, 1962.

EXAMPLE Let $\nu = 0$, consider the function

$$f(z) = 1, \qquad 0 \le z \le a$$
$$= 0, \qquad a < z < \infty.$$

From (20) we see that

$$F_0(\zeta) = \int_0^a zJ_0(\zeta z)\, dz = \frac{a}{\zeta} J_1(a\zeta).$$

(Here we used the formula $[zJ_1(z)]' = zJ_0(z)$.) (21) yields

$$f(z) = \int_0^\infty aJ_0(\zeta z)J_1(\zeta a)\, d\zeta. \tag{22}$$

Integrals of the above type clearly represent discontinuous functions. We shall examine such integrals more systematically in a later section.

The derivation of (20) and (21) was purely formal. We shall not go further in justifying them. The theory of Hankel transforms is of comparable mathematical depth as the theory of Fourier transforms. One can show that for $\nu \ge -\frac{1}{2}$ Hankel transforms have the same convergence properties as Fourier transforms.

14. Applications in Mathematical Physics

In dealing with the problem of flow of heat in a circular plate of some homogeneous and isotropic material we encounter the following type of problem. We seek a solution of the partial differential equation

$$\Delta u = \frac{\partial u}{\partial t} \tag{1}$$

subject to the boundary conditions

$$u(a, \theta, t) = 0 \tag{2}$$

where a is the radius of the plate. As an initial condition we have

$$u(r, \theta, 0) = f(r, \theta) \tag{3}$$

where $f(r, \theta)$ is some temperature distribution.

In order to solve (1) we introduce polar coordinates

$$\frac{\partial^2 u}{\partial r^2} + \frac{1}{r}\frac{\partial u}{\partial r} + \frac{1}{r^2}\frac{\partial^2 u}{\partial \theta^2} = \frac{\partial u}{\partial t}.$$

The method of separation of variables leads to a solution of the type

$$u = e^{-(\lambda/a)^2 t} e^{i\mu\theta} J_\mu\left(\lambda \frac{r}{a}\right)$$

where λ and μ are separation constants. Clearly we require that

$$u(r, \theta + 2\pi, t) = u(r, \theta, t)$$

from which we deduce that the constant μ must be an integer. From (2) we see that λ must be so chosen that

$$J_n(\lambda) = 0.$$

We shall denote the positive zeros of $J_n(\lambda)$ by $\lambda_{n,k}$. Then

$$u = \sum_{n=-\infty}^{\infty} \sum_{k=1}^{\infty} c_{n,k} \exp\left(-\left(\frac{\lambda_{n,k}}{a}\right)^2 t\right) e^{in\theta} J_n\left(\lambda_{n,k}\frac{r}{a}\right) \tag{4}$$

satisfies the differential equation and the boundary condition.

In order to satisfy the initial condition we require that

$$\sum_{n=-\infty}^{\infty} \sum_{k=1}^{\infty} c_{n,k} e^{in\theta} J_n\left(\lambda_{n,k}\frac{r}{a}\right) = f(r, \theta). \tag{5}$$

Using the orthogonality of the functions we obtain

$$c_{n,k} = \frac{1}{\pi a^2 J_{n+1}^2(\lambda_{n,k})} \int_0^a \int_{-\pi}^{\pi} f(r, \theta) e^{-in\theta} r J_n\left(\lambda_{n,k}\frac{r}{a}\right) d\theta\, dr. \tag{6}$$

As another application we wish to solve the following differential equation

$$\Delta u = 0, \qquad r \neq 0. \tag{7}$$

A solution of this equation is

$$u = \frac{1}{r} = \frac{1}{\sqrt{x^2 + y^2 + z^2}} = \frac{1}{\sqrt{\rho^2 + z^2}} \tag{8}$$

depending on whether we use spherical, rectangular or cylindrical coordinates. Physically this solution represents the potential of a point source. A solution of (7), that is independent of the polar angle ϕ, is given by

$$u = e^{-\lambda z} J_0(\lambda \rho).$$

The above can be obtained by separation of variables of (7) in the form

$$u_{\rho\rho} + \frac{1}{\rho} u_\rho + u_{zz} = 0.$$

More generally we can write

$$u = \int_0^\infty f(\lambda) e^{-\lambda|z|} J_0(\lambda\rho)\, d\lambda, \tag{9}$$

where $f(\lambda)$ is an arbitrary function of λ. To determine it we require that

$$\int_0^\infty f(\lambda) e^{-\lambda|z|} J_0(\lambda\rho) d\lambda = \frac{1}{\sqrt{\rho^2 + z^2}} \tag{10}$$

We write $|z|$ in (9) to indicate the fact that we expect u to become vanishingly small at infinity.

(10) is in effect an integral equation for $f(\lambda)$. In particular for $\rho = 0$ it reduces to

$$\int_0^\infty f(\lambda) e^{-\lambda|z|}\, d\lambda = \frac{1}{|z|}.$$

A solution, as is evident by inspection, is $f(\lambda) = 1$. The left side is the Laplace transform of $f(\lambda)$ and it is known that inverse Laplace transforms are unique in the space of continuous functions. We finally see that

$$\int_0^\infty e^{-\lambda|z|} J_0(\lambda\rho)\, d\lambda = \frac{1}{\sqrt{\rho^2 + z^2}} \tag{11}$$

The integral on the left of (11) is the Laplace transform of $J_0(\lambda\rho)$.

Another problem we wish to consider is the following. We seek a solution of

$$\Delta u = 0, \qquad z > 0 \tag{12}$$

such that u satisfies the boundary condition on $z = 0$

$$u(\rho, 0, \phi) = g(\rho). \tag{13}$$

As before we shall seek a solution in the form

$$u = \int_0^\infty \lambda f(\lambda) e^{-\lambda z} J_0(\lambda\rho)\, d\lambda.$$

To satisfy the boundary condition we require that

$$\int_0^\infty \lambda f(\lambda) J_0(\lambda\rho)\, d\lambda = g(\rho).$$

Using Eq. 8.13.21

$$f(\lambda) = \int_0^\infty u g(u) J_0(u\lambda)\, du$$

and finally

$$u = \int^{\infty} \lambda e^{-\lambda z} J_0(\lambda\rho) \left\{ \int_0^{\infty} u g(u) J_0(u\lambda) \, du \right\} d\lambda. \tag{14}$$

As a particular case we take $g(\rho)$ to be defined as follows

$$\begin{aligned} g(\rho) &= 1 \qquad 0 \le \rho \le a \\ &= 0 \qquad a < \rho < \infty. \end{aligned}$$

By use of (8), (13), (22) we find that

$$u = \int_0^{\infty} a e^{-\lambda z} J_0(\lambda\rho) J_1(\lambda a) \, d\lambda. \tag{15}$$

15. Discontinuous Integrals

Many integrals, whose integrands consist of Bessel functions, represent discontinuous functions. We shall now show that

$$I \equiv \int_0^{\infty} e^{-az} z^{-\lambda} J_\nu(zb) \, dz = \left(\frac{b}{2a}\right)^{\nu} \frac{a^{\lambda-1}\Gamma(\nu-\lambda+1)}{\Gamma(\nu+1)}$$

$$\times F\left(\frac{\nu-\lambda+1}{2}, \frac{\nu-\lambda+2}{2}; \nu+1; -\frac{b^2}{a^2}\right) \tag{1}$$

$$R(a \pm ib) > 0, \qquad R(\nu - \lambda + 1) > 0.$$

The indicated permissible ranges for a, b, ν and λ are dictated by the behavior at the endpoints of the interval of integration. This can be verified easily by examining the asymptotic structure near $z = 0$ and $z = \infty$. Under these conditions the integral converges absolutely.

To evaluate I we replace the Bessel functions by its series representation and integrate term by term. This results in

$$I = \left(\frac{b}{2a}\right)^{\nu} a^{\lambda-1} \sum_{k=0}^{\infty} \left(-\frac{b^2}{a^2}\right)^k \frac{\Gamma(\nu-\lambda+1+2k)}{k!\, 2^{2k}\Gamma(\nu+1+k)}.$$

The gamma function in the numerator can be represented as a product of gamma functions by means of the duplication formula

$$\Gamma\left(2\left[\frac{\nu-\lambda+1}{2}+k\right]\right)$$

$$= \frac{2^{\nu-\lambda+2k}}{\sqrt{\pi}} \Gamma\left(\frac{\nu-\lambda+1}{2}+k\right) \Gamma\left(\frac{\nu-\lambda+2}{2}+k\right)$$

$$= \frac{2^{\nu-\lambda+2k}}{\sqrt{\pi}} \Gamma\left(\frac{\nu-\lambda+1}{2}\right) \Gamma\left(\frac{\nu-\lambda+2}{2}\right) \left(\frac{\nu-\lambda+1}{2}\right)_k \left(\frac{\nu-\lambda+2}{2}\right)_k.$$

If this substitution is made the resultant series reduces to a hypergeometric series. Then

$$I = \left(\frac{b}{2a}\right)^{\nu} \frac{a^{\lambda-1}2^{\nu-\lambda}\Gamma((\nu-\lambda+1)/2)\Gamma((\nu-\lambda+2)/2)}{\sqrt{\pi}\Gamma(\nu+1)}$$

$$\times F\left(\frac{\nu-\lambda+1}{2}, \frac{\nu-\lambda+2}{2}; \nu+1; -\frac{b^2}{a^2}\right).$$

To arrive at (1) we note that the product of gamma functions in the numerator can be represented as a single function by means of the duplication formula.

Some special cases of (1) that are of interest are the following. Suppose we let $\lambda = -\nu$

$$\int_0^\infty e^{-az}z^\nu J_\nu(zb)\,dz = \left(\frac{b}{2a}\right)^{\nu} \frac{\Gamma(2\nu+1)}{a^{\nu+1}\Gamma(\nu+1)} F\left(\nu+\frac{1}{2}; \nu+1; \nu+1; -\frac{b^2}{a^2}\right).$$

The above hypergeometric function can be evaluated in terms of a binomial expression. Finally we obtain

$$\int_0^\infty e^{-az}z^\nu J_\nu(zb)\,dz = \left(\frac{b}{2a}\right)^{\nu} \frac{\Gamma(2\nu+1)}{a^{\nu+1}\Gamma(\nu+1)} \left(1+\frac{b^2}{a^2}\right)^{-\nu-\frac{1}{2}}$$

$$= \left(\frac{b}{2}\right)^{\nu} \frac{\Gamma(2\nu+1)}{\Gamma(\nu+1)(a^2+b^2)^{\nu+\frac{1}{2}}} \tag{2}$$

$$R(a \pm ib) > 0, \qquad (2\nu+1) > 0.$$

In particular for $\nu = 0$ we find

$$\int_0^\infty e^{-az}J_0(zb)\,dz = \frac{1}{(a^2+b^2)^{\frac{1}{2}}} \tag{3}$$

$$R(a \pm ib) > 0.$$

(3) was obtained by another technique in Eq. 8.14.11.

By a formal differentiation of (2) with respect to a we obtain

$$\int_0^\infty e^{-az}z^{\nu+1}J_\nu(zb)\,dz = \left(\frac{b}{2}\right)^{\nu} \frac{\Gamma(2\nu+2)a}{\Gamma(\nu+1)(a^2+b^2)^{\nu+\frac{1}{2}}}. \tag{4}$$

$$R(a \pm ib) > 0, \qquad R(2\nu+2) > 0.$$

(4) could, of course, be derived directly from (1) by letting $\lambda = -\nu - 1$ and expressing the hypergeometric function in closed form.

For $\lambda = 0$ (1) reduces to

$$\int_0^\infty e^{-az} J_\nu(zb)\, dz = \left(\frac{b}{2a}\right)^\nu a^{-1} F\left(\frac{\nu+1}{2}, \frac{\nu+2}{2}; \nu+1; -\frac{b^2}{a^2}\right).$$

Here we use the fact that

$$F(b, b+\tfrac{1}{2}; 2b; z) = \frac{((1+\sqrt{1-z})/2)^{1-2b}}{\sqrt{1-z}}$$

so that

$$\int_0^\infty e^{-az} J_\nu(zb)\, dz = \frac{b^\nu[\sqrt{a^2+b^2}+a]^{-\nu}}{\sqrt{a^2+b^2}}$$

$$= \frac{[\sqrt{a^2+b^2}-a]^\nu}{b^\nu\sqrt{a^2+b^2}} \tag{5}$$

$$R(a \pm ib) > 0, \qquad R\nu > -1.$$

Note that for $\nu = 0$ (5) reduces to (3).

A case related to (5) in which a closed form can be found is $\lambda = 1$. Then

$$\int_0^\infty e^{-az} \frac{J_\nu(zb)}{z}\, dz = \left(\frac{b}{2a}\right)^\nu \frac{\Gamma(\nu)}{\Gamma(\nu+1)} F\left(\frac{\nu}{2}, \frac{\nu+1}{2}; \nu+1; -\frac{b^2}{a^2}\right).$$

The appropriate summation in this case is

$$F(b-\tfrac{1}{2}, b; 2b; z) = \left(\frac{1+\sqrt{1-z}}{2}\right)^{1-2b}$$

so that

$$\int_0^\infty e^{-az} \frac{J_\nu(zb)}{z}\, dz = \frac{1}{\nu}\left[\frac{b}{\sqrt{a^2+b^2}+a}\right]^\nu$$

$$= \frac{1}{\nu}\left[\frac{\sqrt{a^2+b^2}-a}{b}\right]^\nu \tag{6}$$

$$R(a \pm ib) > 0, \qquad R(\nu) > 0.$$

(5) could have been obtained from (6) by a formal differentiation with respect to a.

We now return to (1) to investigate the case $a = 0$. Clearly the integral may no longer be absolutely convergent for real values of b. To find the limit as a tends to zero, we replace the hypergeometric function by a suitable integral representation. Then

$$\int^{\infty} e^{-az}z^{-\lambda}J_\nu(zb)\,dz = \left(\frac{b}{2a}\right)^{\nu} \frac{a^{\lambda-1}\Gamma(\nu-\lambda+1)}{\Gamma((\nu-\lambda+2)/2)\Gamma((\nu+\lambda)/2)}$$

$$\times \int_0^1 t^{(\nu-\lambda)/2}(1-t)^{(\nu+\lambda-2)/2}\left(1+\frac{tb^2}{a^2}\right)^{(\lambda-\nu-1)/2} dt$$

$$= \frac{b^{\lambda-1}\Gamma(\nu-\lambda+1)}{2^{\nu}\Gamma((\nu-\lambda+2)/2)\Gamma((\nu+\lambda)/2)}$$

$$\times \int_0^1 t^{(\nu-\lambda)/2}(1-t)^{(\nu+\lambda-2)/2}\left(t+\frac{a^2}{b^2}\right)^{(\lambda-\nu-1/2)} dt$$

For $a = 0$ we then obtain

$$\int_0^{\infty} \frac{J_\nu(zb)}{z^{\lambda}}\,dz = \frac{b^{\lambda-1}\sqrt{\pi}\,\Gamma(\nu-\lambda+1)}{2^{\nu}\Gamma((\nu-\lambda+2)/2)\Gamma((\nu+\lambda+1)/2)}$$

$$= \frac{b^{\lambda-1}\Gamma((\nu-\lambda+1)/2)}{2^{\lambda}\Gamma((\nu+\lambda+1)/2)} \tag{7}$$

$$R(\nu-\lambda+1) > 0, \qquad R(\lambda+\tfrac{1}{2}) > 0,\ b > 0.$$

To investigate integrals in which products of Bessel functions occur we shall make use of Neumann's addition theorem

$$J_\nu(R)e^{i\nu\Psi} = \sum_{m=-\infty}^{\infty} J_m(\rho)J_{\nu+m}(r)e^{i(\nu+m)\phi}$$

where

$$Re^{\pm i\Psi} = re^{\pm i\phi} - \rho. \tag{8}$$

We replace r by zr and ρ by $z\rho$ and multiply by $z^{-\lambda}$. Then integration and use of (7) leads to

$$\frac{\Gamma((\nu-\lambda+1)/2)}{2^{\lambda}\Gamma((\nu+\lambda+1)/2)} R^{\lambda-1}e^{i\nu\Psi} = \sum_{-\infty}^{\infty} e^{i(\nu+m)\phi} \int_0^{\infty} \frac{J_m(z\rho)J_{\nu+m}(zr)}{z^{\lambda}}\,dz. \tag{9}$$

Solving for R in (8) we obtain

$$R^2 = r^2 + \rho^2 - 2r\rho\cos\phi = (r-\rho e^{i\phi})(r-\rho e^{-i\phi}).$$

It follows that

$$R^{\lambda-1}e^{i\nu\Psi} = R^{\lambda-\nu-1}(Re^{i\Psi})^{\nu} = [(r-\rho e^{i\phi})(r-\rho e^{-i\phi})]^{(\lambda-\nu-1)/2}(re^{i\phi}-\rho)^{\nu}. \tag{10}$$

For $\rho < r$ we can rewrite (10) as

$$R^{\lambda-1}e^{i\nu\Psi} = r^{\lambda-1}e^{i\nu\phi}\left[1 - \frac{\rho}{r}e^{i\phi}\right]^{(\lambda-\nu-1)/2}\left[1 - \frac{\rho}{r}e^{-i\nu\phi}\right]^{(\lambda+\nu-1)/2}.$$

By expanding the terms on the right by the binomial theorem and rearranging terms we obtain

$$R^{\lambda-1}e^{i\nu\Psi} = r^{\lambda-1}e^{i\nu\phi}\sum_{-\infty}^{\infty} e^{im\phi}\sum_{k-l=m}\frac{((1+\nu-\lambda)/2)_k((1-\nu-\lambda)/2)_l(\rho/r)^{k+l}}{k!\,l!}.$$

The inner sum can be evaluated in terms of a hypergeometric function so that

$$R^{\lambda-1}e^{i\nu\Psi} = r^{\lambda-1}e^{i\nu\phi}\sum_{-\infty}^{\infty} e^{im\phi}u_m\left(\frac{\rho}{r}\right), \qquad \rho < r \tag{11}$$

where

$$u_m\left(\frac{\rho}{r}\right)$$

$$= \frac{((1-\lambda+\nu\operatorname{sgn} m)/2)_{|m|}}{|m|!}\left(\frac{\rho}{r}\right)^{|m|} F\left(\frac{1-\lambda-\nu}{2}, \frac{1-\lambda+\nu}{2} + |m|; 1 + |m|; \frac{\rho^2}{r^2}\right)$$

$$\begin{aligned} \operatorname{sgn} m &= 1 \qquad m > 0 \\ &= -1 \qquad m < 0. \end{aligned} \tag{12}$$

A comparison of (11) with (9) shows that

$$\int_0^\infty \frac{J_m(z\rho)J_{\nu+m}(zr)}{z}\,dz = \frac{\Gamma((\nu-\lambda+1)/2)r^{\lambda-1}((1-\lambda+\nu\operatorname{sgn} m)/2)_{|m|}(\rho/r)^{|m|}}{2\,\Gamma((\nu+\lambda+1)/2)\,|m|!}$$

$$\times F\left(\frac{1-\lambda-\nu}{2}, \frac{1-\lambda+\nu}{2} + |m|; 1 + |m|; \frac{\rho^2}{r^2}\right). \tag{13}$$

In order for this to hold we require that

$$0 < \rho < r, \qquad R\lambda > -1, \qquad R(2m + \nu - \lambda + 1) > 0 \qquad \text{if} \quad m \geq 0.$$

Although the latter was derived under the restriction that m was an integer one can use analytic continuation to show that the result is valid for more general values. We then suppose that $m \geq 0$. In that case

$$\frac{((1-\lambda+\nu\operatorname{sgn} m)/2)_{|m|}}{|m|!} = \frac{((1-\lambda+\nu)/2)_m}{m!} = \frac{\Gamma((1-\lambda+\nu)/2) + m)}{\Gamma(m+1)\Gamma((1-\lambda+\nu)/2)}.$$

(13) can be rewritten after replacing m by μ and ν by $\nu - \mu$ as

$$\int_0^\infty \frac{J_\mu(z\rho)J_\nu(zr)}{z}\,dz = \frac{r^{\lambda-1}\Gamma((1-\lambda+\nu+\mu)/2)(\rho/r)^\mu}{2^\lambda\Gamma(\mu+1)\Gamma((1+\lambda+\nu-\mu)/2)}$$

$$\times F\left(\frac{1-\lambda-\nu+\mu}{2}, \frac{1-\lambda+\nu+\mu}{2}; 1+\mu; \frac{\rho^2}{r^2}\right) \quad (14)$$

$$R\lambda > -1, \qquad R(\mu+\nu-\lambda+1) > 0, \qquad 0 < \rho < r.$$

An interchange of μ and ν in (14) effectively alters the roles of r and ρ. Then we obtain

$$\int_0^\infty \frac{J_\mu(z\rho)J_\nu(zr)}{z}\,dz = \frac{\rho^{\lambda-1}\Gamma((1-\lambda+\nu+\mu)/2)(r/\rho)^\nu}{2^\lambda\Gamma(\nu+1)\Gamma((1+\lambda+\mu-\nu)/2)}$$

$$\times F\left(\frac{1-\lambda-\mu+\nu}{2}, \frac{1-\lambda+\nu+\mu}{2}; 1+\nu; \frac{r^2}{\rho^2}\right) \quad (15)$$

$$R\lambda > -1, \qquad R(\mu+\nu-\lambda+1) > 0, \qquad 0 < r < \rho.$$

For $r = \rho$ we can express the hypergeometric function as a product of Gamma functions. Then

$$\int_0^\infty \frac{J_\mu(z\rho)J_\nu(z\rho)}{z^\lambda}\,dz$$

$$= \frac{\rho^{\lambda-1}\Gamma((1-\lambda+\nu+\mu)/2)\Gamma(\lambda)}{2^\lambda\Gamma((1+\lambda+\mu-\nu)/2)\Gamma((1+\nu+\mu+\lambda)/2)\Gamma((1+\lambda+\nu-\mu)/2)}. \quad (16)$$

(14), (15) and (16) are often referred to as the formulas of Sonine and Schafheitlin.

For $\mu = \nu$, $\lambda = 1$ they reduce to

$$\int_0^\infty \frac{J_\mu(z\rho)J_\mu(zr)}{z}\,dz = \begin{cases} \dfrac{1}{2\mu}\left(\dfrac{\rho}{r}\right)^\mu, & \rho < r \\[2ex] \dfrac{1}{2\mu}\left(\dfrac{r}{\rho}\right)^\mu, & \rho > r. \end{cases} \quad (17)$$

Exercises

1. Use the method of Frobenius to solve

$$I'' + \frac{1}{z}I' + \left[1 - \frac{\nu^2}{z^2}\right]I = 0.$$

2. Find an integral representation for $I(z)$, whose differential equation is given in (1). Use as a kernel in

$$L(z) = \int_{\alpha}^{\beta} K(z, t)V(t)\, dt$$

$$K(z, t) = e^{izt}$$

3. Use 8.2.6 to show that

$$\sin z = z\left[J_0(z) + 2\sum_{k=1}^{\infty} \frac{(-1)J_{2k}(z)}{1-4k^2}\right]$$

4. Show that

$$J_0\left(z\left(t+\frac{1}{t}\right)\right) = J_0{}^2(z) + \sum_{n=1}^{\infty} (-1)^n J_n(z)[t^{2n} + t^{-2n}]$$

$$J_0(z \sin \alpha) = J_0{}^2\left(\frac{t}{2}\right) + 2\sum J_n{}^2\left(\frac{z}{2}\right) \cos 2n\alpha.$$

5. Show that

$$J_n(z+\zeta) = \sum_{k=-\infty}^{\infty} (-1)^k J_n(z) J_{k+n}(\zeta).$$

6. Show that

$$\frac{1}{2\pi i}\int_{-\infty}^{(0+)} t^{-\nu-1} \exp\left(\frac{z}{2}\left(t-\frac{1}{t}\right)\right) dt$$

satisfies Bessel's equation, by direct substitution in the differential equation.

7. Use

$$J_m'(z) = \tfrac{1}{2}[J_{m-1}(z) - J_{m+1}(z)]$$

$$\frac{m}{z} J_m(z) = \tfrac{1}{2}[J_{m-1}(z) + J_{m+1}(z)]$$

to prove that

$$J_m(z) = (-1)^m z^m \left(\frac{d}{z\,dz}\right)^m J_0(z).$$

and more generally

$$J_{m+\nu}(z) = (-1)^m z^{m+\nu} \left(\frac{d}{z\,dz}\right)^m J_\nu(z).$$

8. Prove the validity of 8.5.6 by induction.

9. Expand the integrand in 5.7 and integrate term by term to verify that the resultant series agrees with the series for the Bessel function.

10. Show that

$$\int_0^z J_\nu(\zeta)\,d\zeta = 2\sum_{k=0}^{\infty} J_{\nu+2k+1}(z), \qquad \text{Re}\,\nu > -1.$$

11. Show that

$$\left(\frac{d}{zdz}\right)^m z^{-\nu} J_0(z) = (-1)^m z^{-m-\nu} J_{m+\nu}(z)$$

and

$$\left(\frac{d}{zdz}\right)^m z^\nu J_\nu(z) = z^{\nu-m} J_{\nu-m}(z).$$

12. Use the result of exercise 1 to show that

$$\int_0^z \zeta^\mu J_\nu(\zeta)d\zeta = z^\mu J_{\nu+1}(z) - (\mu-\nu-1)\int_0^z \zeta^{\mu-1} J_{\nu+1}(\zeta)d\zeta \quad \text{if} \quad \text{Re}\,(\mu+\nu) > -1.$$

13. Discuss integrals of the form

$$\int_0^z \zeta^m J_\nu(\zeta)d\zeta, \qquad m = 0,1,2,\ldots, \qquad \text{Re}(\nu+m) > -1$$

using the results of the previous exercises.

14. Show that

$$J_n(\lambda z) = \lambda^n \sum_{m=0}^{\infty} \frac{(((1-\lambda^2)/2)z)^m}{m!} J_{n+m}(z).$$

15. Show that

$$P_n(\cos\theta - h\sin\theta) = \sum_{m=0}^{\infty} \frac{h^m}{m!} P_n{}^m \cos\theta).$$

16. Replaces θ by z/n and h by $h/2zn$ in 15 and take the limit as n becomes infinitely large. This will provide another proof for the multiplication theorem for $J_0(z)$.

17. Show that

$$\sum_{n=0}^{\infty} \frac{r^n P_n(\cos\theta)}{n!} = e^{r\cos\theta} J_0\,(r\sin\theta).$$

18. Show that

$$\sum_{n=0}^{\infty} \frac{J_n(z)t^n}{n!} = J_0(\sqrt{z^2 - 2zt}.)$$

19. Show that

$$J_n(z) = \frac{2^n n!}{(2n)!} z^n \left(1 + \frac{d^2}{dz^2}\right)^n J_0(z).$$

20. Show that when $\nu = m + \frac{1}{2}$, m a non-negative integer, the asymptotic formulas for the Hankel functions 8.8.6 reduce to finite sums.

21. In view of the fact that the solutions of Bessel's equation are in general multivalued functions with a single branch point in the finite part of the complex plane they must satisfy certain circuits relation. Show that

$$J_\nu(e^{m\pi i} z) = e^{m\nu\pi i} J_\nu(z)$$

$$N_\nu(e^{m\pi i} z) = e^{m\nu\pi i} N_\nu(z) + 2i \sin m\nu\pi \cot \nu\pi\, J_\nu(z)$$

$$H_\nu^{(1)}(e^{m\pi i} z) = -\frac{\sin (m-1)\nu\pi}{\sin \nu\pi} H_\nu^{(1)}(z) - e^{-\nu\pi i} \frac{\sin m\nu\pi}{\sin \nu\pi} H_\nu^{(2)}(z)$$

$$H_\nu^{(2)}(e^{m\pi i} z) = e^{\nu\pi i} \frac{\sin m\nu\pi}{\sin \nu\pi} H_\nu^{(1)}(z) + \frac{\sin (m+1)\nu\pi}{\sin \nu\pi} H_\nu^{(2)}(z)$$

where m is any integer.

22. Show that

$$H_\nu^{(1)}(e^{i\pi} z) = -H_{-\nu}^{(2)}(z) = -e^{-\nu\pi i} H_\nu^{(2)}(z)$$

$$H_\nu^{(2)}(e^{-i\pi} z) = -H_{-\nu}^{(1)}(z) = -e^{\nu\pi i} H_\nu^{(1)}(z)$$

$$\overline{H_\nu^{(2)}(z)} = H_{\bar{\nu}}^{(1)}(\bar{z})$$

23. Consider the second order differential equation

$$z^2 y'' + (2\alpha - 2\nu\beta + 1) z y' + [\beta^2 \gamma^2 z^{2\beta} + \alpha(\alpha - 2\nu\beta)] y = 0.$$

Let $Z_\nu(z)$ denote any solution of

$$z^2 y'' + z y' + (z^2 - \nu^2) y = 0.$$

Then $z^{\beta\nu-\alpha} Z_\nu(\gamma z^\beta)$ satisfies the former equation.

24. Use the method of exercise 23 to discuss the solutions of

$$y'' - zy = 0.$$

$$y'' - a^2 z^{2q-2} y = 0.$$

25. Show that

$$y'' + (e^{2z} - \nu^2) y = 0$$

is satisfied by $J_\nu(e^z)$.

26. Show that

$$z^4 y'' + (e^{2/z} - \nu^2)\, y = 0$$

is satisfied by $z J_\nu(e^{1/z})$.

27. Consider the function

$$\phi(u) = \frac{1}{\sqrt{z^2 - u^2}}, \qquad 0 < u < z$$

$$= 0 \qquad z < u < \pi.$$

Show that, when expanded as a cosine series we obtain

$$\phi(u) = \frac{1}{2} + \sum_{n=1}^{\infty} J_0(nz) \cos n\, u.$$

28. Show that if $f(z)$ has a continuous derivative in $[0, \pi]$, then

$$f(z) = f(0) + \frac{2}{\pi} \int_0^z \frac{u\, du}{\sqrt{z^2 - u^2}} \int_0 \frac{f'(u\xi)\, d\xi}{\sqrt{1 - \xi^2}}.$$

[*Hint:* Let $\xi = \sin \phi$ and consider the resultant double integral as an integral over a quadrant of a circle or radius z in the plane with polar coordinates u, ϕ.]

29. Use the results of the preceding two exercises to prove Schloemlich's expansion

$$f(z) = f(0) + \tfrac{1}{2} a_0 + \sum_{n=1}^{\infty} a_n J_0(nz)$$

where

$$a_n = \frac{2}{\pi} \int_0^{\pi} u \cos n\, u\, du \int_0^1 \frac{f'(u\xi)\, d\xi}{\sqrt{1 - \xi^2}}.$$

30. Prove the addition theorem 8.815 by using 6.9 and a similar theorem for $J_{-\nu}(R)e^{i\nu\Psi}$ and construct $H_\nu^{(1)}(R)e^{i\nu\Psi}$ as a superposition of such functions.

31. The Airy function (see 8.9.4)

$$A_0(z) = \frac{1}{2\pi i} \int_{L_0} e^{tz - \frac{1}{3}t^3}\, dt$$

can be expanded in a Taylor series

$$A_0(z) = a_0 + a_1 z + \cdots$$

where

$$a_0 = \frac{1}{2\pi i} \int_{L_0} e^{-\frac{1}{3}t^3}\, dt$$

$$a_1 = \frac{1}{2\pi i} \int_{L_0} t e^{-\frac{1}{3}t^3}\, dt.$$

Evaluate the 1st two integrals explicitly. Use these to determine c_1 and c_2 in

$$A_0(z) = z\, [c_1 J_{1/3}(\tfrac{2}{3} i z^{3/2}) + c_2 J_{-1/3}(\tfrac{2}{3} i z^{3/2})].$$

Do the same for $A_+(z)$ and $A_-(z)$.

32. Perform the necessary calculations to deduce the asymptotic formulas 8.9.9.

33. Prove that the Neumann function has the following integral representation

$$\pi N(z) = \int_0^\pi \sin[z \sin t - \nu t]\, dt - \int_0^\infty [e^{\nu t} + e^{-\nu t} \cos \nu\pi] e^{-z \sinh t}\, dt \quad \mathrm{Re}\, z > 0.$$

34. If $\{\lambda_n\}$ denotes the zeros of $J_\nu(z)$, where $\lambda_n \neq 0$, evaluate exactly

$$\sum_{n=1}^{\infty} \frac{1}{\lambda_n^2} \quad \text{and} \quad \sum_{n=1}^{\infty} \frac{1}{\lambda_n^2}.$$

(*Hint:* Compare the infinite series representation with the infinite product representation.)

35. Derive asymptotic forms of the large zeros of $J_\nu(z)$ using Stokes method.

36. Show that $zJ_\nu'(z) + \alpha J_\nu(z)$, with $\nu > -1$ and $\alpha + \nu \geq 0$ has an infinity of real zeros and no complex zeros. For $\alpha + \nu < 0$ precisely two complex conjugate zeros will occur.

37. Show that the set $\{\sqrt{z}\, J_\nu(\lambda_n z)\}$ is complete in $L_2[0, 1]$ if $\nu > -1$ and the λ_n are the zeros of $zJ_\nu'(z) + \alpha J_\nu(z) = 0$.

38. Find $u(\rho, z, \phi)$ such that

$$\Delta u = 0$$

and $u(\rho, 0, \phi) = g(\rho)$.
(*Hint:* Consider solutions of the form $u = \int_0^\infty \lambda f(\lambda) e^{-\lambda z} J_0(\lambda\rho)\, d\lambda$.)

39. Find $u(\rho, z, \phi)$ such that

$$\Delta u = 0$$

and

$$u = f(\rho) \quad \text{on} \quad z = 0$$
$$= g(\rho) \quad \text{on} \quad z = L.$$

(*Hint:* Consider solutions of the form

$$u = \int_0^\infty \lambda J_0(\lambda\rho)\{A(\lambda)\sinh \lambda z + B(\lambda)\sinh \lambda(L - z)\} d\lambda.)$$

40. Consider the differential equation

$$\frac{\partial u}{\partial t} = \Delta u$$

in polar coordinates. Find u so that

$$u(r, \phi, 0) = f(r) \qquad r_1 \leq r \leq r_2.$$
$$u = 0 \quad \text{for} \quad r = r_1$$
$$= 0 \quad \text{for} \quad r = r_2.$$

41. Consider the differential equation

$$\frac{\partial u}{\partial t} = \Delta u$$

in polar coordinates. Find u so that

$$u(r, \phi, 0) = f(r) \qquad 0 \le r \le a$$

$$u = 0 \quad \text{on} \quad \phi = 0, \phi = \theta,$$

$$\frac{\partial u}{\partial r} + hu = 0 \quad \text{on} \quad r = a$$

42. Verify that the integral in 8.15.1 converges for the indicated parameter ranges.

43. Show that

$$\int_0^\infty \cos wz\, J_0(zb)dz = \frac{1/2}{\sqrt{b^2 - w^2}}, \; 0 \le w < b$$

$$= 0 \qquad\qquad b < w.$$

44. Show that

$$\int_{-\infty}^\infty \frac{\sin a(z+\zeta)}{z+\zeta} J_0(b\zeta)d\zeta = \pi J_0(bz), \; b \le a$$

$$= 2\int_0^a \frac{\cos \omega z\, d\omega}{\sqrt{b^2 - \omega^2}}, \; b \ge a.$$

(*Hint:* Recognize the left side as a convolution integral and take the Fourier transform.)

CHAPTER 9

Hill's Equation

1. Mathieu's Equation

We consider the elliptic cylinder coordinate system defined by

$$x = c \cosh \xi \cos \eta$$

$$y = c \sinh \xi \sin \eta.$$

The domains of ξ and η are given by

$$0 \le \xi < \infty, \qquad 0 \le \eta \le 2\pi.$$

The surface $\xi = \xi_0$ reduces to the ellipse

$$\left(\frac{x}{c \cosh \xi_0}\right)^2 + \left(\frac{y}{c \sinh \xi_0}\right)^2 = 1$$

and the surface $\eta = \eta_0$ reduces to the hyperbolas

$$\left(\frac{x}{c \cos \eta_0}\right)^2 - \left(\frac{y}{c \sin \eta_0}\right)^2 = 1.$$

When the reduced wave equation

$$\Delta u + k^2 u = 0$$

is expressed in these coordinates it becomes

$$\frac{\partial^2 u}{\partial \xi^2} + \frac{\partial^2 u}{\partial \eta^2} + \frac{k^2 c^2}{2}(\cosh 2\xi - \cos 2\eta)u = 0.$$

If we now let

$$u = f(\xi)\, g(\eta)$$

and use the technique of separation of variables we obtain two ordinary differential equations for f and g.

$$f'' + \left[-\lambda + \frac{c^2k^2}{2}\cosh 2\xi\right]f = 0$$

$$g'' + \left[\lambda - \frac{c^2k^2}{2}\cos 2\eta\right]g = 0. \tag{1}$$

Here λ denotes a separation constant. The first of the above equations is known as the modified Mathieu equation and the second as the Mathieu equation. Note that formally the substitution of $\xi = i\eta$ converts the modified equation into a Mathieu equation.

This coordinate system is closely related to polar coordinates. As a matter of fact, the variable η plays a role identical to a polar angle and e^{ξ} is similar to a radial variable. In many applications one is interested in those solutions for which

$$g(\eta + 2\pi) = g(\eta).$$

For $k = 0$ it is obvious that this can hold only for special values of λ, namely those for which $\lambda = n^2$, and n an integer. We shall examine the structure of equations of this type to see what can be said about such problems and their eigenvalues.

2. Hill's Equation

We shall now turn our attention to the equation

$$y'' + [\lambda - q(t)]y = 0 \tag{1}$$

where $q(t + \pi) = q(t)$. Note that for $q(t) = (c^2k^2/2)\cos 2t$ (1) reduces to the Mathieu equation. We shall also make the following hypotheses on $q(t)$. First $q(t)$ will be assumed to be integrable on the interval $(0, \pi)$; that is

$$\int_0^{\pi} |q(t)|\, dt < \infty.$$

Without loss of generality we can assume that

$$\mu = \frac{1}{\pi}\int_0^{\pi} q(t)\, dt = 0.$$

Otherwise we rewrite the equation in the form

$$y'' + [\lambda - \mu - (q(t) - \mu)]y = 0$$

and work with the parameter $\lambda - \mu$ instead of λ.

Two linearly independent solutions of (1) can be defined by

$$\begin{aligned} y_1(0) &= 1, \qquad & y_2(0) &= 0 \\ y_1'(0) &= 0, \qquad & y_2'(0) &= 1, \end{aligned} \tag{2}$$

and clearly their Wronskian is given by

$$W[y_1, y_2] = y_1 y_2' - y_1' y_2 = 1. \tag{3}$$

Another quantity that plays a fundamental role in the theory of Hill's equation is the discriminant of Hill's equation. It is defined by

$$\Delta(\lambda) = y_1(\pi) + y_2'(\pi) \tag{4}$$

and note that it is a function of λ since y_1 and y_2 depend on λ. It will be convenient for future purpose to consider the roots of the quadratic equation

$$\rho^2 - \Delta\rho + 1 = 0. \tag{5}$$

Let

$$\rho_1 = \frac{\Delta + \sqrt{\Delta^2 - 4}}{2}$$

$$\rho_2 = \frac{\Delta - \sqrt{\Delta^2 - 4}}{2}.$$

For λ real clearly $\Delta(\lambda)$ will be real and the following statements regarding ρ_1 and ρ_2 can be made. For $|\Delta| > 2$

$$|\rho_1| > 1, \qquad |\rho_2| < 1,$$

for $|\Delta| = 2$

$$\rho_1 = \rho_2$$

and for $|\Delta| < 2$ ρ_1 and ρ_2 will be of the form

$$\rho_1 = e^{i\alpha\pi}, \quad \rho_2 = e^{-i\alpha\pi}$$

for some real α such that $\Delta = 2\cos\alpha\pi$.

In general we cannot expect the solution of (1) to be periodic, merely because the coefficient is. But as will be seen shortly solutions with the translation property

$$y(t + \pi) = \rho y(t) \tag{6}$$

for some suitable ρ exist. In view of the fact that the differential equation is invariant under the change of variable $t \to t + \pi$, if $y(t)$ is a solution so is $y(t + \pi)$. To find solutions satisfying (6) we let

$$\begin{aligned} y &= c_1 y_1 + c_2 y_2 \\ y' &= c_1 y_1' + c_2 y_2' \end{aligned} \tag{7}$$

and it follows that

$$\begin{aligned} c_1 y_1(\pi) + c_2 y_2(\pi) &= \rho[c_1 y_1(0) + c_2 y_2(0)] = \rho c_1 \\ c_1 y_1'(\pi) + c_2 y_2'(\pi) &= \rho[c_1 y_1'(0) + c_2 y_2'(0)] = \rho c_2 \end{aligned}$$

or equivalently

$$\begin{aligned} c_1(y_1(\pi) - \rho) + c_2 y_2(\pi) &= 0 \\ c_1 y_1'(\pi) + c_2(y_2'(\pi) - \rho) &= 0. \end{aligned} \tag{8}$$

In order for non-vanishing c_1 and c_2 to exist and satisfy the above we require that ρ be such that

$$\begin{vmatrix} y_1(\pi) - \rho & y_2(\pi) \\ y_1'(\pi) & y_2'(\pi) - \rho \end{vmatrix} = 0.$$

Using (3) and (4) we see that the above reduces to

$$\rho^2 - \Delta\rho + 1 = 0.$$

We can now state and prove the following theorem.

FLOQUET'S THEOREM. If $\rho_1 \neq \rho_2$ (where ρ_1 and ρ_2 are the solutions of (5)) then we can find two linearly independent solutions of (1), say f_1 and f_2 such that

$$f_1(t) = e^{i\alpha t} p_1(t), \qquad f_2(t) = e^{i\alpha t} p_2(t)$$

where p_1 and p_2 are periodic functions with period π.

If $\rho_1 = \rho_2$ (1) has at least one solution with period π or 2π depending on whether $\rho_1 = \rho_2 = 1$ or $\rho_1 = \rho_2 = -1$ respectively. Let $p(t)$ denote such a periodic solution. Any other independent solution of (1) is such that

$$y(t + \pi) = \rho_1 y(t) + \tau p(t)$$

where τ is a suitable constant. This second solution will be periodic, or equivalently $\tau = 0$, if and only if

$$y_2(\pi) = y_1'(\pi) = 0.$$

Proof. If $|\Delta| \neq 2$, $\rho_1 \neq \rho_2$ and there will be two independent sets of coefficients c_1 and c_2 such that

$$f_1(t) = c_1 y_1(t) + c_2 y_2(t)$$

$$f_2(t) = c_3 y_1(t) + c_4 y_2(t)$$

and

$$f_1(t + \pi) = \rho_1 f_1(t)$$

$$f_2(t + \pi) = \rho_2 f_2(t).$$

We now define α such that

$$\rho_1 = e^{i\alpha\pi}$$

$$\rho_2 = e^{-i\alpha\pi}$$

(in view of the fact that $\rho_1 \rho_2 = 1$). Then the function

$$p_1(t) = e^{-i\alpha t} f_1(t)$$

is clearly periodic with period π. To see this we note that

$$p_1(t + \pi) = e^{-i\alpha(t+\pi)} f_1(t + \pi) = \frac{1}{\rho_1} e^{-i\alpha t} \rho_1 f_1(t) = p_1(t).$$

Then

$$f_1(t) = e^{i\alpha t} p_1(t)$$

and similarly for $f_2(t)$.

To prove the second part of the theorem we see that for $\rho_1 = \rho_2$ there may exist only one set of coefficients c_1 and c_2 leading to a solution $p(t)$ such that

$$p(t + \pi) = \rho_1 p(t).$$

But if $\rho_1 = \rho_2$ $\Delta^2 = 4$ and there are only two possibilities; either $\rho_1 = 1$ or $\rho_1 = -1$. In the first case $p(t)$ has period π and in the second case

$$p(t + 2\pi) = -p(t + \pi) = p(t)$$

so that $p(t)$ has period 2π.

We now wish to examine the second solution. First we assume that $y_2(\pi) \neq 0$. This is equivalent to saying that $p(t)$ and $y_2(t)$ are independent. Since $y_2(t + \pi)$ is also a solution it follows that

$$y_2(t + \pi) = c_1 y_2(t) + c_2 p(t).$$

To determine c_1 and c_2 we consider

$$y_2(\pi) = c_1 y_2(0) + c_2\, p(0) = c_2\, p(0)$$
$$y_2'(\pi) = c_1 y_2'(0) + c_2\, p'(0)$$

so that, after solving for c_1 and c_2

$$y_2(t+\pi) = \frac{p(0)y_2'(\pi) - p'(0)y_2(\pi)}{p(0)}\, y_2(t) + \frac{y_2(\pi)}{p(0)}\, p(t).$$

But

$$\frac{p'(0)}{p(0)} = \frac{p'(\pi)}{p(\pi)}$$

so that

$$\frac{p(0)y_2'(\pi) - p'(0)y_2(\pi)}{p(0)} = \frac{p(\pi)y_2'(\pi) - p'(\pi)y_2(\pi)}{p(\pi)}$$
$$= \frac{p(0)y_2'(0) - p'(0)y_2(0)}{p(\pi)} = \frac{p(0)}{p(\pi)} = \frac{1}{\rho_1} = \rho_1.$$

Here we use the fact that $p(\pi)y_2'(\pi) - p'(\pi)y_2(\pi)$ is the Wronskian of p and y_2. Finally

$$y_2(t+\pi) = \rho_1 y_2(t) + \frac{y_2(\pi)}{p(0)}\, p(t).$$

When $y_2(\pi) = 0$ we see from

$$y_1(\pi)y_2'(\pi) - y_2(\pi)y_1'(\pi) = 1$$

that

$$y_2'(\pi) = \frac{1}{y_1(\pi)}.$$

But for the case under discussion

$$\Delta = y_1(\pi) + y_2'(\pi) = y_1(\pi) + \frac{1}{y_1(\pi)} = \pm 2.$$

Hence if $\rho_1 = 1$ $y_1(\pi) = 1$ and if $\rho_1 = -1$ $y_1(\pi) = -1$. In the former case both y_1 and y_2, and then all solutions will have period π. In the latter case all solutions will have period 2π. ■

From the preceding discussion a number of conclusions may be drawn. For $|\Delta| > 2$ we have $|\rho_1| > 1$. Then

$$f_1(t+\pi) = \rho_1 f_1(t)$$

and more generally

$$f_1(t + k\pi) = \rho_1{}^k f_1(t).$$

it follows that $f_1(t)$ will become unboundedly large for increasing t. We refer to this situation as an unstable case. When $|\Delta| < 2$ we have $|\rho_1| = |\rho_2| = 1$ and all solutions are bounded. This is known as the stable case. For $|\Delta| = 2$ we have $|\rho_1| = |\rho_2| = 1$ and at least one solution is periodic. The second solution is either periodic or unbounded.

3. The Discriminant

The discriminant $\Delta(\lambda)$ is an entire function of λ. Many of the interesting properties of Hill's equations can be studied in terms of the structure of $\Delta(\lambda)$. This section will be devoted to a study of the analytical properties of $\Delta(\lambda)$.

THEOREM. $\Delta(\lambda)$ is an entire function of λ of order $\frac{1}{2}$. That is there exists a constant $c_1 > 0$ such that

$$|\Delta(\lambda)e^{-c_1\sqrt{|\lambda|}}|$$

is bounded for all λ, and another constant $c_2 > 0$ such that

$$|\Delta(\lambda)e^{-c_2\sqrt{|\lambda|}}| \to \infty$$

as $\lambda \to -\infty$.

Proof. Let $\omega^2 = \lambda$, $u_0 = \cos \omega t$ and

$$u_n(t) = \int_0^t \frac{\sin \omega(t - \tau)}{\omega} q(\tau)u_{n-1}(\tau)\, d\tau, \qquad n = 1, 2, \ldots .$$

and we consider $0 \le t \le \pi$.

Clearly

$$|u_0(t)| = |\cos \omega t| \le \cosh |\omega| t \le e^{|\omega| t}$$

so that

$$|u_1(t)| \le \int_0^t \frac{e^{|\omega|(t-\tau)}|q(\tau)|}{|\omega|} e^{|\omega|\tau}\, d\tau$$

$$\le \frac{e^{|\omega|t}}{|\omega|} \int_0^t |q(\tau)|\, d\tau$$

and it follows by induction that

$$|u_n| \le \frac{e^{|\omega|t}}{n!\, |\omega|^n} \left[\int_0^t |q|\, d\tau\right]^n.$$

From these estimates we see that u_n is an entire function of $\lambda = \omega^2$ of precise order $\frac{1}{2}$ for every n. Also the function

$$y(t) = \sum_{n=0}^{\infty} u_n(t)$$

is defined by the above uniformly convergent series. Since

$$u_n(0) = u_n'(0) = 0 \qquad \text{for} \quad n = 1, 2, \ldots$$

$$u_0(0) = 1, \qquad u_0'(0) = 0$$

we have

$$y(0) = 1, \; y'(0) = 0.$$

It also follows by a direct calculation that

$$y'' + \omega^2 y = \sum_{n=0}^{\infty} u_n'' + \omega^2 u_n = \sum_{n-1}^{\infty} q(t)u_{n-1}(t)$$
$$= q(t)y,$$

so that y satisfies Hill's equation and by virtue of the initial conditions y can be identified with $y_1(t)$. It follows that

$$|y_1(t)| \leq e^{|\omega|t} \sum_{n=0}^{\infty} \frac{[1/|\omega| \int_0^t |q| \, d\tau]^n}{n!}$$
$$= \exp\left[|\omega|t + \frac{1}{|\omega|}\int_0^t |q| \, d\tau\right]$$

and

$$|y_1(\pi)| \leq \exp\left[\sqrt{|\lambda|}\, \pi + \frac{1}{\sqrt{|\lambda|}}\int_0^\pi |q| \, d\tau\right]$$

so that $y_1(\pi)$ is an entire function of λ of order at most $\frac{1}{2}$.

Had we selected for $u_0(t) = 1/\omega \sin \omega t$ we would have arrived at similar conclusions for $y_2(t)$ defined by

$$y_2 = \sum_{n=0}^{\infty} u_n(t).$$

Then

$$|\Delta(\lambda)| \leq 2 \exp\left[\sqrt{|\lambda|}\, \pi + \frac{1}{\sqrt{|\lambda|}}\int_0^\pi |q| \, d\tau\right]$$

and $\Delta(\lambda)$ is entire of order at most $\frac{1}{2}$.

We still have to show that the order is precisely so. To do this we need suitable lower estimates on $\Delta(\lambda)$. First we shall assume that q is bounded, $|q| < M$, and we let $\lambda = -\omega^2 - M$ and suppose ω large so that

$$-\lambda + q = \omega^2 + M + q \geq \omega^2 > 0.$$

Then

$$y_1'' = (-\lambda + q)y_1 \geq \omega^2 y_1 .$$

By integrating the above twice and using the initial conditions corresponding to y_1 we obtain

$$y_1 \geq 1 + \omega^2 \int_0^t (t - \tau) y_1 \, d\tau.$$

Combining the last two inequalities we have

$$y_1'' \geq \omega^2 + \omega^4 \int_0^t (t - \tau) y_1 \, d\tau.$$

Again integrating we find

$$y_1 \geq 1 + \frac{\omega^2 t^2}{2} + \omega^4 \int_0^t \frac{(t - \tau)^3}{3!} y_1 \, dt.$$

By repeating this process we find

$$y_1 \geq \cosh \omega t.$$

Similarly

$$y_2 \geq \frac{\sinh \omega t}{\omega}.$$

As before we see that

$$\Delta(\lambda) \geq 2 \cosh \sqrt{-\lambda + M}\pi, \qquad -\lambda + M > 0$$

showing that $\Delta(\lambda)$ is of precise order $\frac{1}{2}$.

To take care of the case where q is not bounded we write

$$q(t) = q^*(t) + \delta(t)$$

where q^* is bounded and $\delta(t)$ is an integrable function such that

$$\int_0^\pi |\delta(t)| \, dt < \varepsilon$$

for some sufficiently small and positive ε. We now define u_1 and u_2 as the solution of

$$u'' + [\lambda - q^*(t)]u = 0.$$

Writing

$$y'' + [\lambda - q^*(t)]y = \delta(t)y$$

we obtain, using the classical method of variation of parameters

$$y_1 = u_1 - \int_0^t \delta(\tau)y_1(\tau)[u_1(t)u_2(\tau) - u_1(\tau)u_2(t)]\,d\tau.$$

If $\lambda = -\omega^2 - M$, where $|q^*| \leq M$, we have

$$u_1 \geq \cosh \omega t$$

$$y_1(t) \leq K_1 e^{\omega t}$$

$$u_1(t)u_2(\tau) - u_1(\tau)u_2(t) \leq K_2 \frac{e^{\omega(t-\tau)}}{\omega}$$

for suitable constants K_1 and K_2. Then

$$y_1 \geq \cosh \omega t - \varepsilon K_1 K_2 \frac{e^{\omega t}}{\omega}$$

and we see that y_1 is of precise order $\frac{1}{2}$. Similar results hold for y_2 and finally for $\Delta(\lambda)$. ■

Using the estimates of this theorem we see that

$$y_1 = \cos\sqrt{\lambda}\,t + \frac{1}{\sqrt{\lambda}}\int_0^t \sin\sqrt{\lambda}(t-\xi)\cos\sqrt{\lambda}\,\xi q(\xi)\,d\xi + 0\left(\frac{1}{\lambda}\right).$$

Similarly

$$y_2' = \cos\sqrt{\lambda}\,t + \frac{1}{\sqrt{\lambda}}\int_0^t \cos\sqrt{\lambda}(t-\xi)\sin\sqrt{\lambda}\,\xi q(\xi)\,d\xi + 0\left(\frac{1}{\lambda}\right).$$

Recall that $\int_0^\pi q(\xi)d\xi = 0$, so that

$$\Delta(\lambda) = y_1(\pi) + y_2'(\pi) = 2\cos\sqrt{\lambda}\,\pi + 0\left(\frac{1}{\lambda}\right). \tag{1}$$

An immediate corollary of the above theorem is the following

COROLLARY. For λ sufficiently small $\Delta > 2$.
As was seen earlier when for some values of λ

$$\Delta(\lambda) = 2 \tag{2}$$

the equation

$$y'' + [\lambda - q(t)]y = 0$$

has a solution of period π. We consider the boundary value problem (3) coupled with

$$y(0) = y(\pi)$$
$$y'(0) = y'(\pi)$$

All eigenvalues of this problem must be solutions of (2).

Similarly the solutions of

$$\Delta(\lambda) = -2$$

are such that (3) has solutions with the periodicity property

$$y(t + \pi) = -y(t).$$

We see then that these values are eigenvalues of (3) coupled with

$$y(0) = -y(\pi)$$
$$y'(0) = -y'(0).$$

We can now state and prove the following theorem.

THEOREM. The functions $\Delta(\lambda) - 2$ and $\Delta(\lambda) + 2$ each have an infinitude of zeros. All of these zeros are real.

Proof. It is well known that all entire functions of order less than 1, which are not polynomials must have an infinite number of zeros. A function of precise order $\frac{1}{2}$ cannot be a polynomial and hence must vanish infinitely often. To show that the zeros of both $\Delta(\lambda) - 2$ and $\Delta(\lambda) + 2$ are real we refer to the corresponding boundary value problem. If $\Delta(\lambda) - 2$ has a complex zero λ then its conjugate $\bar{\lambda}$ is also a zero since the function $\Delta(\lambda)$ is real for real λ. Let y and $\bar{y}$ denote the corresponding solutions of (2). Then

$$y'' + [\lambda - q]y = 0$$
$$\bar{y}'' + [\bar{\lambda} - q]\bar{y} = 0.$$

Multiplying the first by $\bar{y}$ and the second by y and integrating we have

$$\int_0^\pi (\bar{y}y'' - y\bar{y}'')\,dt + (\lambda - \bar{\lambda})\int_0^\pi |y|^2\,dt = 0.$$

But by virtue of the boundary conditions

$$\int_0^\pi (\bar{y}y'' - y\bar{y}'')\,dt = \bar{y}y' - y\bar{y}'\Big|_0^\pi = 0.$$

Since $\int_0^\pi |y|^2 dt > 0$ we see that $\lambda = \bar{\lambda}$ and hence is real. In other words the boundary value problem is selfadjoint. An identical proof applies to the zeros of $\Delta(\lambda) + 2$. ■

Use of (1) allows us to estimate these zeros. Thus from

$$\Delta(\lambda) - 2 = 2\cos\sqrt{\lambda}\,\pi - 2 + 0\left(\frac{1}{\lambda}\right) = 0$$

we see that

$$\sin\sqrt{\lambda}\,\pi/2 = 0\left(\frac{1}{\sqrt{\lambda}}\right). \tag{4}$$

It follows that for large zeros

$$\sqrt{\lambda} \approx 2n$$

and a refinement by use of (4) shows that

$$\sqrt{\lambda} = 2n + 0\left(\frac{1}{n}\right)$$

Similarly from

$$\Delta(\lambda) + 2 = 2\cos\sqrt{\lambda}\,\pi + 2 + 0\left(\frac{1}{n}\right) = 0$$

we see that

$$\sqrt{\lambda} = 2n + 1 + 0\left(\frac{1}{n}\right).$$

We shall show next that the zeros of $\Delta(\lambda) + 2$ and those of $\Delta(\lambda) - 2$ are arranged in a particular way. To do that we require the use of a theorem due to Laguerre, and it will be convenient to state and prove this theorem first.

THEOREM. Let $f(z)$ be an entire function of order less than 2, which is real for real values of z. If $f(z)$ has only real zeros, so does $f'(z)$ and the zeros of $f(z)$ and $f'(z)$ interlace.

Proof. By the Weierstrass factorization theorem we can represent this function by

$$f(z) = Ke^{az}\prod_{n=1}^{\infty}\left(1 - \frac{z}{z_n}\right)e^{z/z_n}$$

where K, u and z_n are all real by hypothesis. The logarithmic derivatives of this function is given by

$$\frac{f'}{f} = a + \sum_{n=1}^{\infty}\left(\frac{1}{z - z_n} + \frac{1}{z_n}\right).$$

A simple computation shows that its imaginary part is given by

$$Im\,\frac{f'}{f} = y \sum_{n=1}^{\infty} \frac{1}{(x - z_n)^2 + y^2}$$

If $f'(z) = 0$ for some z then also $Im\,(f'/f) = 0$. But we see that this can only happen for $y = 0$. Hence all zeros of $f'(z)$ are real.

Next we see that

$$\left(\frac{f'}{f}\right)' = - \sum_{n=1}^{\infty} \frac{1}{(z - z_n)^2}$$

so that, for real z, f'/f is a decreasing function. It also has poles at all zeros of $f(z)$. The following diagram shows the structure of $f(z)$.

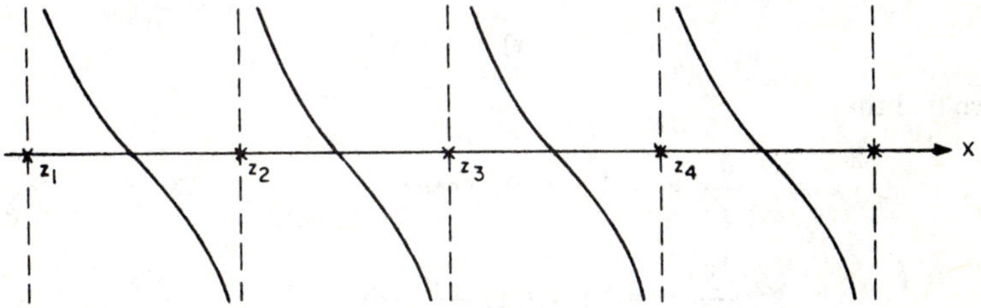

We see therefore that the zeros of $f(z)$ alternate with the zeros of $f'(z)$. ■

We require one other lemma

LEMMA. If $\Delta^2(\lambda) - 4 = 0$ has a multiple zero, then it must be a double zero.

Proof. Suppose for some λ

$$\Delta(\lambda) = 2$$

$$\Delta'(\lambda) = 0.$$

To prove the lemma it is enough to show that at the point $\Delta''(\lambda) \neq 0$. Now

$$\Delta''(\lambda) = \frac{\partial^2}{\partial \lambda^2}\, y_1(\pi) + \frac{\partial^2}{\partial \lambda^2}\, y_2'(\pi).$$

We also know that in this case all solutions of (3) have period π. Then

$$y_1(\pi) = 1, \qquad y_1'(\pi) = 0$$

$$y_2(\pi) = 0, \qquad y_2'(\pi) = 1.$$

We also have from (3)

$$\frac{\partial y_1''}{\partial \lambda} + (\lambda - q(t)) \frac{\partial y_1}{\partial \lambda} = -y_1$$

$$\frac{\partial^2 y_1''}{\partial \lambda} + (\lambda - q(t)) \frac{\partial^2 y_1}{\partial \lambda^2} = -2 \frac{\partial y_1}{\partial \lambda}.$$

Letting

$$K(t, \tau) = y_2(t)y_1(\tau) - y_2(\tau)y_1(t)$$

where

$$K(\pi, \tau) = -y_2(\tau)$$

$$K_t(\pi, \tau) = -y_1(\tau)$$

we find that

$$\frac{\partial y_1}{\partial \lambda} = -\int_0^t K(t, \tau) y_1(\tau)\, d\tau$$

$$\frac{\partial^2 y_1}{\partial \lambda^2} = -2 \int_0^t K(t, \tau) \frac{\partial y_1(\tau)}{\partial \lambda}\, d\tau$$

$$= 2 \int_0^t K(t, \tau)\, d\tau \int_0^\tau K(\tau, \zeta) y_1(\zeta)\, d\zeta.$$

A similar operation applied to y_2 finally shows that

$$\Delta'' = -2 \int_0^\pi d\tau \int_0^\tau K^2(\tau, \zeta)\, d\zeta < 0.$$

Thus for $\Delta(\lambda) = 2$, $\Delta'(\lambda) = 0$ we have a double zero. A similar calculation applies at $\Delta(\lambda) = -2$. ■

We are now in a position to state and prove one of the most important theorems pertaining to the discriminant.

OSCILLATION THEOREM. Let $\lambda_0, \lambda_1, \lambda_2, \lambda_3 \ldots$ denote the zeros of $\Delta(\lambda) - 2$ and $\lambda_1', \lambda_2', \lambda_3', \lambda_4', \ldots$ those of $\Delta(\lambda) + 2$. Then

$$\lambda_0 < \lambda_1' \le \lambda_2' < \lambda_1 \le \lambda_2 < \lambda_3' \le \lambda_4' < \cdots.$$

For all λ in the intervals $(-\infty, \lambda_0)$, (λ_1', λ_2'), $(\lambda_1, \lambda_2), \ldots |\Delta(\lambda)| > 2$. These intervals are known as the instability intervals.

Proof. Let $\mu_1, \mu_2, \ldots$ denote the zeros of $\Delta'(\lambda)$. We apply Laguerre's theorem to the function $\Delta(\lambda) - 2$. Then

$$\lambda_0 < \mu_1 < \lambda_1 \leq \mu_2 \leq \lambda_2 < \mu_3 < \cdots.$$

The reason for $\lambda_1 \leq \mu_2 \leq \lambda_2$ is that a double zero might occur at this point, and similarly at all pairs $\lambda_{2n-1}, \lambda_{2n}$. Similarly

$$\lambda_1' \leq \mu_1 \leq \lambda_2' < \mu_2 < \lambda_3' \leq \mu_3 \leq \lambda_4' < \mu_4 < \cdots.$$

We had seen earlier that for λ small $\Delta(\lambda) > 2$. It follows that $\lambda_0 < \lambda_1$. By combining these inequalities we see that

$$\lambda_0 < \lambda_1' \leq \mu_1 \leq \lambda_2' < \lambda_1 \leq \mu_2 \leq \lambda_2 < \lambda_3' \leq \mu_3 \leq \cdots.$$

from which we can read off the main conclusion of the theorem. The following diagram illustrates the character of $\Delta(\lambda)$. For large λ we have $\Delta(\lambda) = 2 \cos\sqrt{\lambda}\pi + 0\,(1/\lambda)$.

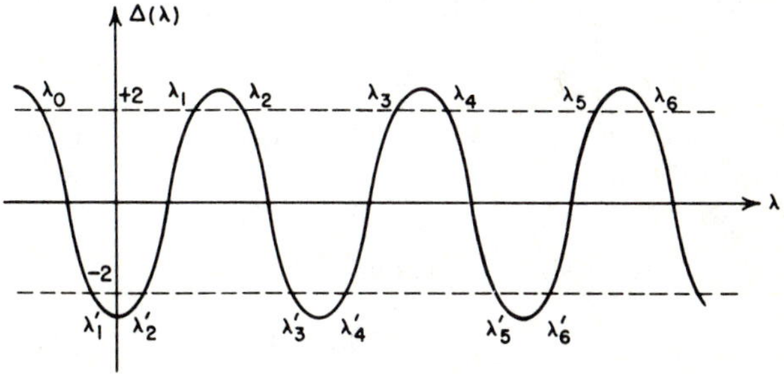

Since in the intervals formed by $\mu_1, \mu_2, \ldots \Delta'(\lambda)$ has constant slope we see that $|\Delta| > 2$ in the instability intervals. ■

Clearly if $|\Delta| > 2$, as seen earlier, one solution of Hill's equation is unbounded. Hence the name instability intervals. Similarly for λ in the remaining intervals, the stability intervals, all solutions of Hill's equation are bounded.

THEOREM. $\lambda_0 < 0$, unless $q = 0$, almost everywhere.

Proof. When $\lambda = \lambda_0$, there exists a solution y_0 such that

$$y_0'(0) - y_0'(\pi) = 0$$
$$y_0(0) - y_0(\pi) = 0.$$

λ_0 is the smallest eigenvalue corresponding to these boundary conditions and hence y_0 has least oscillation of all eigenfunctions. Then $y_0 \neq 0$ for all t. We now write Hill's equation in the form

$$\frac{y_0''}{y_0} + \lambda_0 - q(t) = 0$$

and integrate over $(0, \pi)$. We had assumed that $\int_0^\pi q(t)dt = 0$ so that

$$\int_0^\pi \frac{y_0''}{y_0}\, dt + \lambda_0 \pi = 0.$$

By an integration by parts, and use of the boundary conditions we find

$$\int_0^\pi \left(\frac{y_0'}{y_0}\right)^2 dt + \lambda_0 \pi = 0.$$

Hence either $\lambda_0 < 0$ or $y_0' = 0$. In the latter instance $y_0 =$ constant is a solution and then $q(t) = 0$ a.e. ■

Under suitable hypotheses we can obtain lower bounds on λ_1'. One such result is the following.

THEOREM. Consider the equation

$$y'' + \lambda p(t)\, y = 0 \tag{5}$$

where

$$p(t + \pi) = p(t) \geq 0$$

and

$$\int_0^\pi p(t)\, dt \leq \frac{4}{\pi}.$$

We impose the boundary conditions

$$y(0) + y(\pi) = 0$$

$$y'(0) + y'(\pi) = 0.$$

The smallest eigenvalue corresponding to these conditions, say λ_1', satisfies

$$\lambda_1' > 1.$$

Proof. One can show that (15), although we have $\lambda p(t)$ instead of $\lambda + p(t)$, has a theory associated with it identical to that of Hill's equation. We note that $\lambda_0 = 0$ in this case. λ_1' will be the smallest zero of

$$\Delta(\lambda) + 2 = y_1(\pi) + y_2'(\pi) + 2 = 0,$$

and $\lambda_1' > \lambda_0 = 0$. We have

$$y'' = -\lambda p(t) y$$

so that y'' and y are of opposite sign. In other words y is always concave toward the t axis. From the boundary conditions we see that y has to vanish at least once in $(0, \pi)$. The eigenfunction corresponding to the smallest eigenvalue λ_1 must have minimum oscillation and will therefore vanish precisely once. This shows that y can have at most one local maximum or minimum in $(0, \pi)$. The following figure illustrates a typical case if $y(0) > 0$ and $y'(0) > 0$.

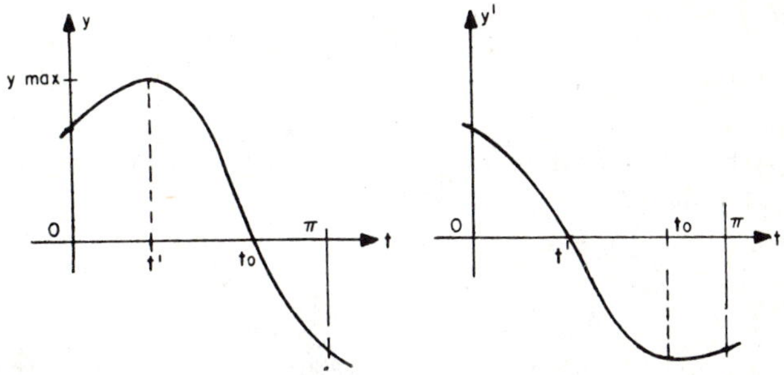

We see that the boundary conditions severely limit the qualitative character of y. If t' denotes the location of the point where $y' = 0$,

$$\begin{aligned}\int_0^\pi |y'|\, dt &= \int_0^{t'} |y'|\, dt + \int_{t'}^\pi |y'|\, dt \\ &= |y|_{\max} - y(0) + |y|_{\max} - y(\pi) \\ &= 2|y|_{\max}\end{aligned}$$

by virtue of the boundary conditions. By Schwarz' inequality we have

$$\int_0^\pi |y'|\, dt \le \sqrt{\pi \int_0^\pi y'^2\, dt}$$

so that

$$2|y|_{\max} \le \sqrt{\pi \int_0^\pi y'^2\, dt}. \tag{6}$$

Again using the boundary conditions we have

$$\int_0^\pi yy''\,dt = -\int_0^\pi y'^2\,dt.$$

If we multiply the differential equation (5) by y and integrate over $(0, \pi)$ we obtain

$$\int_0^\pi y'^2\,dt = \lambda_1' \int_0^\pi p(t)y^2\,dt.$$

From the meanvalue theorem and the hypotheses on $p(t)$, and use of (6), it follows that

$$\int_0^\pi y'^2\,dt < \lambda_1' y_{\max}^2 \int_0^\pi p(t)\,dt \le \lambda_1' \int_0^\pi y'^2\,dt,$$

so that

$$\lambda_1' > 1. \qquad ■$$

It may be added that this estimate is the best possible. In order for $\lambda_1' = 1$ we would require that

$$\int_0^\pi p(t)y^2\,dt = y_{\max}^2 \int_0^\pi p(t)\,dt$$

and the latter can occur only if y is a constant, in which case $p = 0$ or if $p(t)$ is a delta function. A simple computation shows that if

$$p(t) = \frac{4}{\pi}\,\delta(t - \xi), \qquad \pi \int_0^\pi p(t)\,dt = 4$$

$$\lambda = 1$$

$$y = t - \xi + \frac{\pi}{2} \qquad t < \xi$$

$$= -t + \xi + \frac{\pi}{2} \qquad t > \xi.$$

Evidently we must be able to find a $p(t)$ for which $\pi\int_0^\pi p(t)dt > 4$ and such that $\lambda_1' < 1$.

This last result may be used to prove a stability criterion due to Lyapunov.

STABILITY CRITERION. If

$$p(t + \pi) = p(t) \geq 0$$

and

$$\pi \int_0^\pi p(t)\, dt \leq 4$$

the all solutions of

$$y'' + p(t)y = 0$$

are stable.

Proof. For the Hill's equation

$$y'' + \lambda\, p(t)y = 0$$

we have $\lambda_0 = 0$, and under the above hypotheses $\lambda_1' > 1$. Hence for all λ in (λ_0, λ_1') the differential equation is stable since $-2 < \Delta < 2$ in that interval. Clearly $\lambda_0 < 1 < \lambda_1'$ so that

$$y'' + p(t)y = 0$$

is stable. ■

As was pointed out earlier, the above estimate is the best possible.

4. Expansion Theorems

We consider the boundary value problem

$$y'' + [\lambda - q(t)]y = 0$$

where as before

$$q(t + \pi) = q(t)$$

$$\int_0^\pi q(t)\, dt = 0$$

with the boundary conditions

$$\begin{aligned} y(-L\pi) &= y(L\pi) \\ y'(-L\pi) &= y'(L\pi). \end{aligned} \tag{2}$$

First we shall examine the case $q \equiv 0$. Then we see that the general solution of (1) is

$$y = c_1 \cos \sqrt{\lambda}\, t + c_2 \sin \sqrt{\lambda}\, t$$

and (2) shows that the corresponding eigenvalues are given by

$$\lambda_n = \frac{n^2}{L^2} \tag{3}$$

With $\lambda_0 = 0$ we associate the eigenfunction $y_0 = 1$. All other eigenvalues are double eigenvalues and with λ_n we associate

$$y_{2n-1} = \frac{1}{\sqrt{L\pi}} \cos \frac{n}{L}\, t, \qquad y_{2n} = \frac{1}{\sqrt{L\pi}} \sin \frac{n}{L}\, t. \tag{4}$$

From the theory of Fourier series we know that this set of orthonormal functions is complete on the space $L_2[-L\pi, L\pi]$.

We shall show that the boundary value problem (1), (2) also leads to a complete and orthonormal set of functions on $L_2[-L\pi, L\pi]$. To determine the corresponding eigenvalues we see that in order for y to satisfy (2) it must be such that

$$y(t+\pi) = \rho y(t)$$
$$y(t + 2L\pi) = \rho^{2L} y(t) = y(t)$$

so that $\rho^{2L} = 1$. In other words ρ must be a $2L$'th root of unity. Then

$$\rho = e^{ik/L}, \qquad k = 0, 1, \ldots, 2L-1.$$

To determine λ we must solve the equations

$$\Delta(\lambda) = \rho + \frac{1}{\rho} = 2 \cos \frac{\pi k}{L}, \qquad k = 0, 1, \ldots, 2L-1. \tag{5}$$

In 9.3.1 we saw that

$$\Delta(\lambda) = 2 \cos \sqrt{\lambda}\, \pi + 0\left(\frac{1}{\lambda}\right).$$

Combining this with (5) we can show easily that for large λ

$$\sqrt{\lambda} = 2m + \frac{k}{L} + 0\left(\frac{1}{\lambda}\right).$$

In particular we can write n in the form

$$n = s2L + r \qquad 0 \le r \le 2L-1$$

where r is a suitable remainder. For n large we have

$$\sqrt{\lambda_n} = 2s + \frac{r}{L} + 0\left(\frac{1}{n}\right) = \frac{n}{L} + 0\left(\frac{1}{n}\right). \tag{6}$$

(3) and (6) agree asymptotically.

In the previous section we had shown that asymptotically

$$y_1 = \cos\sqrt{\lambda}\,t + \frac{\sin\sqrt{\lambda}\,t}{2\sqrt{\lambda}}\int_0^t q(\xi)\,d\xi + 0\left(\frac{1}{\sqrt{\lambda}}\right).$$

Similarly

$$y_2 = \frac{\sin\sqrt{\lambda}\,t}{\sqrt{\lambda}} - \frac{\cos\sqrt{\lambda}\,t}{2\lambda}\int_0^t q(\xi)\,d\xi + 0\left(\frac{1}{\lambda}\right).$$

Standard arguments show that eigenfunctions of the boundary problem (1), (2) are orthogonal. Using the above asymptotic series we can easily show that

$$y_{2n-1} = \frac{1}{\sqrt{L\pi}}\cos\frac{n}{L}\,t + 0\left(\frac{1}{n}\right) \tag{7}$$

$$y_{2n} = \frac{1}{\sqrt{L\pi}}\sin\frac{n}{L}\,t + 0\left(\frac{1}{n}\right),$$

where the terms $|0(1/n)| < K/n$ for some K independent of n. To prove that the orthonormal set given in (7) is complete on $L_2[-L\pi, L\pi]$ we invoke the following theorem.

THEOREM. Let H be a Hilbert space and $\{\phi_n\}$ a complete orthonormal set. Let $\{\Psi_n\}$ be as second orthonormal set such that

$$\sum_{n=0}^{\infty} \|\phi_n - \Psi_n\|^2 < \infty.$$

Then $\{\Psi_n\}$ is also complete.

We shall prove this theorem in an appendix. To apply it we select for H the space $L_2[-L\pi, L\pi]$. For $\{\phi_n\}$ we select the set of eigenfunctions corresponding to (1), (2) with $q \equiv 0$. These are given in (4). For $\{\Psi_n\}$ we select the set given in (7). Then

$$\|\phi_n - \Psi_n\|^2 = \int_{-L\pi}^{L\pi}\left[0\left(\frac{1}{n}\right)\right]^2 dt \le \frac{K^2}{n^2}$$

for a suitable constant K. Then

$$\sum_{n=0}^{\infty} \|\phi_n - \Psi_n\|^2 < \infty$$

and it follows that the eigenfunctions given by (7) are complete.

Note that all eigenvalues of (1), (2) that correspond to complex values of ρ are double eigenvalues. If ρ_1 is such a value it must satisfy

$$\rho^2 - \Delta(\lambda)\,\rho + 1 = 0.$$

In that case $1/\rho_1$ is a second value. For this λ we find one solution for which

$$y(t+\pi) = \rho_1\, y(t)$$

and a second solution for which

$$y(t+\pi) = \frac{1}{\rho_1}\, y(t).$$

Since $\rho^{2L} = 1$ we have for both

$$y(t + 2L\pi) = y(t).$$

We had seen earlier that $\Delta(\lambda)$ was such that $|\Delta(\lambda)| \le 2$ for all λ in the stability intervals. These are given by $[\lambda_0, \lambda_1']$, $[\lambda_2', \lambda_1]$, $[\lambda_2, \lambda_3']$, ..., $[\lambda_{2n}, \lambda_{2n+1}']$, $[\lambda_{2n+2}', \lambda_{2n+1}]$,.... We now seek all solutions of the equations

$$\Delta(\lambda) = 2\cos\frac{k\pi}{L}, \qquad k = 0, 1, \ldots, 2L-1, \tag{8}$$

such that $\lambda_0 \le \lambda \le \lambda_1'$. Clearly for $k > 0$, k and $2L - k$ furnish the same value of λ. There will be a total of $L + 1$ distinct solutions of (8). We shall denote these by $\mu_{0,1} < \mu_{1,1} < \ldots \mu_{L,1}$.

Then

$$\Delta(\mu_{0,1}) = 2$$

$$\Delta(\mu_{1,1}) = 2\cos\frac{\pi}{L}$$

$$\vdots$$

$$\Delta(\mu_{L,1}) = -2,$$

and

$$\mu_{0,1} = \lambda_0, \qquad \mu_{L,1} = \lambda_1'$$

Similarly if we solve (8) in $\lambda_2' \le \lambda \le \lambda_1$ we obtain $\mu_{0,2} < \mu_{1,2} < \cdots < \mu_{L,2}$ so that

$$\Delta(\mu_{0,2}) = -2$$

$$\Delta(\mu_{1,2}) = -2\cos\frac{\pi}{L}$$

$$\vdots$$

$$\Delta(\mu_{L,2}) = 2$$

and $\mu_{0,2} = \lambda_2'$, $\mu_{L,2} = \lambda_1$. More generally the solutions of (8) in the n'th interval of stability will be denoted by

$$\mu_{i,n},\ i = 0, 1, \ldots, L.$$

With all $\mu_{i,n}$ for which $|\Delta(\mu_{i,n}| \neq 2$ we associate two solutions of (1). According to Floquet's theorem one will be such that

$$y^+(t+\pi) = \rho y^+(t) \tag{9}$$

and a second such that

$$y^-(t+\pi) = \frac{1}{\rho}\, y^-(t). \tag{10}$$

We adopt the convention that y^+ corresponds to ρ such that $\rho = e^{ik(\pi/L)}$ and y^- to $1/\rho = e^{-ik(\pi/L)}$. In general these will be complex solutions. Corresponding to $\mu_{0,n}$ and $\mu_{L,n}$ we obtain in general only a single periodic solution. We shall normalize these solutions so that

$$\int_{-L\pi}^{L\pi} |y^\pm|^2\, dt = 2L\pi.$$

According to our completeness theorem we can expand all square integrable functions as follows.

$$f(t) = \sum_{n=1}^{\infty} \left\{ \sum_{k=0}^{L} y_{n,k}^+(t) \frac{1}{2L\pi} \int_{-L\pi}^{L\pi} f(\tau)\bar{y}_{n,k}^+(\tau)\, d\tau \right.$$
$$\left. + \sum_{k=1}^{L-1} y_{n,k}^-(t) \frac{1}{2L\pi} \int_{-L\pi}^{L\pi} f(\tau)\bar{y}_{n,k}^-(\tau)\, d\tau \right\}. \tag{11}$$

The outer sum is over all stability intervals and the inner sums over all eigenfunctions corresponding to eigenvalues in the n'th interval.

(11) is a general statement that holds for all integral values of L. We shall now consider the case where L becomes infinitely large. We saw that

$$\lambda_0 = \mu_{0,1} < \mu_{1,1} < \cdots < \mu_{L-1,1} < \mu_{L,1} = \lambda_1'$$

and

$$\Delta(\mu_{k,1}) = 2\cos\frac{k\pi}{L}.$$

As L becomes large the eigenvalues $\mu_{k,1}$ fill the interval $[\lambda_0, \lambda_1']$ densely. Clearly

$$\Delta(\mu_{k+1,\,1}) - \Delta(\mu_{k,\,1}) = 2\cos\frac{k+1}{L}\pi - 2\cos\frac{k\pi}{L}. \tag{12}$$

Using the mean-value theorem we have

$$\Delta(\mu_{k+1,\,1}) - \Delta(\mu_{k,\,1}) = \Delta'(\tilde{\mu})[\mu_{k+1,\,1} - \mu_{k,\,1}]$$

$$2\cos\frac{k+1}{L}\pi - 2\cos\frac{k\pi}{L} = -\frac{2\pi}{L}\sin\frac{k+\theta}{L}\pi$$

where $\mu_{k,1} \le \tilde{\mu} \le \mu_{k+1,1}$ and $0 \le \theta \le 1$. But

$$2\sin\frac{k+\theta}{L}\pi = \sqrt{4 - \left(2\cos\frac{k+\theta}{L}\pi\right)^2}$$

$$= \sqrt{4 - \Delta^2(\tilde{\tilde{\mu}})}$$

where $\mu_{k,1} \le \tilde{\tilde{\mu}} \le \mu_{k+1,1}$ since Δ is monotone in $[\lambda_0, \lambda_1]$. (12) can now be replaced by

$$\frac{|\Delta'(\tilde{\mu})|\,[\mu_{k+1,\,1} - \mu_{k,\,1}] \times (L/\pi)}{\sqrt{4 - \Delta^2(\tilde{\tilde{\mu}})}} = 1 \tag{13}$$

We used the fact that $\Delta' \le 0$ in $[\lambda', \lambda_1']$ so that $-\Delta' = |\Delta'|$. (13) shows that as L becomes infinite $\mu_{k+1,1} - \mu_{k,1}$ must become small with $1/L$. It follows that

$$\lim_{L\to\infty} \sum_{k=0}^{L} y_{1,\,k}^{+}(t)\frac{1}{2L\pi}\int_{-L\pi}^{L\pi} f(\tau)\bar{y}_{1,\,k}^{+}(\tau)\,d\tau$$

$$= \lim_{L\to\infty} \sum_{k=0}^{L} y_{1,\,k}^{+}(t)\frac{1}{2L\pi}\int_{-L\pi}^{L\pi} f(\tau)\bar{y}_{1,\,k}^{+}(\tau)\,d\tau$$

$$\times \frac{|\Delta'(\tilde{\mu})|\,[\mu_{k+1,\,1} - \mu_{k,\,1}]L/\pi}{\sqrt{4 - \Delta^2(\tilde{\tilde{\mu}})}}$$

$$= \frac{1}{2\pi^2}\int_{\lambda_0}^{\lambda'_1} \frac{y^+(t,\mu)\,|\Delta'(\mu)|}{\sqrt{4 - \Delta^2(\mu)}}\,d\mu \int_{-\infty}^{\infty} f(\tau)\bar{y}^+(\tau,\mu)\,d\tau.$$

By $y^+(t, \mu)$ we denote that solution for which

$$y^+(t+\pi) = \rho y^+(t)$$

and $Im\ \rho \ge 0$. Then

$$y^-(t+\pi) = \frac{1}{\rho}y^-(t).$$

These are so normalized that

$$\lim_{L\to\infty} \frac{1}{2L\pi} \int_{-L\pi}^{L\pi} |y^{\pm}(t)|^2 \, dt = \frac{1}{\pi} \int_0^{\pi} |y^{\pm}(t)|^2 \, dt = 1.$$

Applying this argument to all terms of (11) we finally obtain

$$\begin{aligned} f(t) = \sum_{n=0}^{\infty} \frac{1}{2\pi^2} \int_{\lambda_{2n}}^{\lambda'_{2n+1}} & \frac{|\Delta'(\mu)| \, d\mu}{\sqrt{4 - \Delta^2(\mu)}} \\ & \times \int_{-\infty}^{\infty} f(\tau)[y^+(t, \mu)\bar{y}^+(\tau, \mu) + y^-(t, \mu)\bar{y}^-(\tau, \mu)] \, d\tau \\ & + \frac{1}{2\pi^2} \int_{\lambda'_{2n+2}}^{\lambda_{2n+1}} \frac{|\Delta'(\mu)| \, d\mu}{\sqrt{4 - \Delta^2(\mu)}} \\ & \times \int_{-\infty}^{\infty} f(\tau)[y^+(t, \mu)\bar{y}^+(\tau, \mu) + y^-(t, \mu)\bar{y}^-(\tau, \mu)] \, d\tau \end{aligned} \tag{14}$$

For $q = 0$ (11) reduces to a standard Fourier series. Similarly (14) reduces to the Fourier integral theorem. We have, in this case

$$\Delta(\mu) = 2 \cos \sqrt{\mu}\pi$$

so that

$$\lambda_0 = 0, \qquad \lambda'_1 = \lambda'_2, \qquad \lambda_1 = \lambda_2, \ldots$$

$$y^+ = e^{i\sqrt{\mu}t}, \qquad y^- = e^{i\sqrt{\mu}t}$$

and (14) reduces to

$$f(t) = \int_0^{\infty} \frac{d\mu}{2\pi\sqrt{\mu}} \int_{-\infty}^{\infty} f(\tau) \cos \sqrt{\mu}(t - \tau) \, d\tau. \tag{15}$$

5. Inverse Problems

We have seen that with the differential equation

$$y'' + [\lambda - q(t)]y = 0 \tag{1}$$

$$q(t + \pi) = q(t)$$

$$\int_0^{\pi} q(t) \, dt = 0$$

we can associate the discriminant $\Delta(\lambda)$. The latter depends, of course, on $q(t)$. The question can then be raised as to what information regarding $q(t)$ can be deduced from the knowledge of $\Delta(\lambda)$. We shall state and prove two such theorems.

THEOREM. If all zeros of $\Delta(\lambda) + 2$ are double zeros, or equivalently if all odd instability intervals vanish, then the function $q(t)$ in (1) has period $\pi/2$.

Proof. With (1) we associate a related equation

$$u'' + [\lambda - q(t + \tau)]u = 0 \tag{2}$$

where τ is an arbitrary translation. If y_1 and y_2 are solutions of (1) with initial values (1, 0) and (0, 1) respectively, the corresponding solutions of (2) are given by

$$\begin{aligned} u_1(t) &= y_2'(\tau)y_1(t + \tau) - y_1'(\tau)y_2(t + \tau) \\ u_2(t) &= y_1(\tau)y_2(t + \tau) - y_2(\tau)y_1(t + \tau). \end{aligned} \tag{3}$$

The verification of (3) is straightforward.

The discriminants of (1) and (2) are determined by all eigenvalues corresponding to the boundary conditions

$$y(t + \pi) \pm y(t) = 0$$
$$u(t + \pi) \pm u(t) = 0.$$

respectively. But the latter can be rewritten as

$$y(t + \tau + \pi) \pm y(t + \tau),$$

showing that both sets of eigenvalues are the same.

From this we conclude that the discriminant of (2) is independent of τ and the discriminants of (1) and (2) are identical.

In order to accomplish the proof we note that for every double zero of $\Delta(\lambda) + 2$ all solutions of (1) satisfy

$$y(t + \pi) + y(t) = 0.$$

The eigenvalue may in this case be related to the boundary condition

$$y_2\left(\frac{\pi}{2}\right) + y_2\left(-\frac{\pi}{2}\right) = 0$$

and also

$$u_2\left(\frac{\pi}{2}\right) + u_2\left(-\frac{\pi}{2}\right) = 0.$$

Use of the previously discussed asymptotic formula

$$u_2 = \frac{\sin\sqrt{\lambda}\,t}{\sqrt{\lambda}} - \frac{\cos\sqrt{\lambda}\,t}{2\lambda}\int_0^t q(\xi + \tau)\,d\xi + o\left(\frac{1}{\lambda}\right)$$

allows us to rewrite the above in the form

$$u_2\left(\frac{\pi}{2}\right) + u_2\left(-\frac{\pi}{2}\right) = -\frac{\cos\sqrt{\lambda}\,\pi/2}{\lambda}\int_0^{\pi/2} q(\xi + \tau)\,d\xi + o\left(\frac{1}{\lambda}\right) = 0.$$

The above, as a function of λ, is entire of order 1/2. Its zeros must be independent of τ. It follows that $\int_0^{\pi/2} q(\xi + \tau)\,d\xi$ either reduces to a constant independent of τ or else factors out of all terms. The latter eventuality cannot arise, since the higher order terms are nonlinear functionals of q. In order for the aforesaid integral to be independent of τ, q must indeed have period $\pi/2$, as was to be shown. ■

A second theorem to be proved is the following.

THEOREM. If all finite instability intervals vanish, then $q(t) = 0$ almost everywhere.

To illustrate different techniques we shall furnish two proofs.

Proof 1. If all instability finite intervals vanish we have

$$\lambda_1' = \lambda_2',\ \lambda_1 = \lambda_2,\ \lambda_3' = \lambda_4', \ldots.$$

In other words with the exception of λ_0, all zeros of

$$\Delta^2(\lambda) - 4 = 0$$

are double zeros. It follows that $\Delta'^2(\lambda)$ and $((4 - \Delta^2(\lambda))/(\lambda - \lambda_0))$ are both entire functions of order $\frac{1}{2}$ and they vanish at $\lambda_1', \lambda_1, \lambda_2', \lambda_2, \ldots$ to the same order. Hence they differ by at most a constant. Then

$$\Delta'^2 = K\,\frac{4 - \Delta^2}{\lambda - \lambda_0}.$$

The above is a differential equation for Δ. Letting

$$\Delta = 2\cos\theta$$

we obtain

$$\theta'^2 = \frac{K}{\lambda - \lambda_0}.$$

Solving the latter we find that

$$\Delta = 2\cos\left[2\sqrt{K(\lambda - \lambda_0)} + \theta_0\right].$$

But $\Delta(\lambda_0) = 2$ so that $\theta_0 = 0$. In 9.3.1, we saw that

$$\Delta = 2\cos\sqrt{\lambda}\,\pi + 0\left(\frac{1}{\lambda}\right).$$

Comparing the two forms for Δ we see that

$$2\sqrt{K} = \pi$$

and

$$\cos\pi\sqrt{\lambda - \lambda_0} = \cos\sqrt{\lambda}\,\pi + 0\left(\frac{1}{\lambda}\right).$$

Expanding the left side we obtain

$$\cos\pi\sqrt{\lambda} + \sin\pi\sqrt{\lambda}x\,\frac{\pi\lambda_0}{2\sqrt{\lambda}} = \cos\sqrt{\lambda}\pi + 0\left(\frac{1}{\lambda}\right).$$

The above can hold for large λ if and only if $\lambda_0 = 0$. But by the theorem proved in 9.3 this implies that $q(t) = 0$ almost everywhere. ■

Proof 2. If all instability intervals vanish, then all zeros of $\Delta(\lambda) + 2$ are certainly double zeros. Hence, by the previous theorem, $q(t)$ has period $\pi/2$. We now consider equation (1) as a Hill's equation over the interval $\pi/2$. The eigenvalues corresponding to the boundary conditions

$$y\left(t + \frac{\pi}{2}\right) \pm y(t) = 0$$

are the same as those corresponding to

$$y(t + \pi) - y(t) = 0.$$

The latter, with the exception of λ_0, has double eigenvalues and so those corresponding to

$$y\left(t + \frac{\pi}{2}\right) + y(t) = 0$$

are also double eigenvalues. It now follows that $q(t)$ has period $\pi/4$, by again invoking the previous theorem. By repeating this argument q has every period of the form $\pi/2^n$. Hence q is a constant almost everywhere. But $\int_0^\pi q(t)dt = 0$ so that $q = 0$ almost everywhere. ■

Another theorem that will be stated without proof, is the following.

THEOREM. If precisely one of the finite instability intervals does not vanish, then $q(t)$ is necessarily an elliptic function, almost everywhere.

The proof* is too involved to reproduce here. It is similar to the one given for the first theorem of this section. Using the asymptotics one can show that $q(t)$ satisfies a differential equation of the form

$$q'' = 3q^2 + aq + b.$$

It is well known that solutions of the latter are elliptic functions.

It has been known for a long time that when $q(t)$ is elliptic it may happen that all but a finite number of the finite instability intervals vanish. The previous theorem is a converse to these results.

6. Hill's Equations with Even Coefficients

We now consider equations of the type

$$y'' + [\lambda - q(t)]y = 0 \tag{1}$$

$$q(t + \pi) = q(t)$$

$$\int_0^\pi q(t)\, dt = 0$$

where we impose the additional condition

$$q(-t) = q(t). \tag{2}$$

With the restriction (2) we can make some additional statements. If $y(t)$ satisfies (1) so does $y(-t)$, since the equation is invariant under the change of variable $t \to -t$. Then $y(t) + y(-t)$ is even and $y(t) - y(-t)$ is odd. It follows that if λ is a zero of $\Delta(\lambda) - 2$ and $y(t)$ has period π then we can find a periodic solution that is purely even or purely odd. Suppose $y_1(t)$, which in this case is even, also has period π. Then

$$y_1(t + \pi) = y_1(t)$$

$$y_1'(t + \pi) = y_1'(t)$$

and $y_1'(t)$ is odd. Letting $t = -\pi/2$ we have

$$y_1\left(\frac{\pi}{2}\right) = y_1\left(-\frac{\pi}{2}\right)$$

$$y_1'\left(\frac{\pi}{2}\right) = y'\left(-\frac{\pi}{2}\right) = -y'\left(\frac{\pi}{2}\right)$$

and necessarily $y_1'(\pi/2) = 0$.

* See H. Hochstadt., *On the Determination of a Hill's equation from its Spectrum, Arch. Rat. Mech. and Anal.* **19**, 353–362 (1965).

We can also show that if $y_1'(\pi/2) = 0$ then $y_1(t)$ has period π. Clearly

$$y_1(t+\pi) = c_1 y_1(t) + c_2 y_2(t)$$
$$y_1'(t+\pi) = c_1 y_1'(t) + c_2 y_2'(t).$$

To determine c_1 and c_2 we let $t = -\pi/2$. Then

$$y_1\left(\frac{\pi}{2}\right) = c_1 y_1\left(\frac{\pi}{2}\right) - c_2 y_2\left(\frac{\pi}{2}\right)$$
$$y_1'\left(\frac{\pi}{2}\right) = -c_1 y_1'\left(\frac{\pi}{2}\right) + c_2 y_2'\left(\frac{\pi}{2}\right).$$

If $y_1'(\pi/2) = 0$, $y_2(\pi/2) \neq 0$, otherwise y_1 and y_2 would be linearly dependent. Then $c_2 = 0$ from the second equation. Since $y_1(\pi/2) \neq 0$ the first shows that $c_1 = 1$ and

$$y_1(t+\pi) = y_1(t)$$

so that y_1 has period π. Such reasoning leads to the following theorem.

THEOREM. If $q(t)$ is even in (1) then there exists a nontrivial periodic solution that is:

1. even and of period π if and only if $y_1'(\pi/2) = 0$
2. odd and of period π if and only if $y_2(\pi/2) = 0$
3. even and of period 2π if and only if $y_1(\pi/2) = 0$
4. odd and of period 2π if and only if $y_2'(\pi/2) = 0$.

The first of these cases was proved above. The rest follows by similar reasoning.

7. Mathieu's Equation Revisited

In 9.1 we introduced Mathieu's equation and subsequently treated Hill's equation as a generalization. Consider

$$y'' + [\lambda - 2h^2 \cos 2t]\, y = 0, \tag{1}$$

where h is a parameter. The above is clearly a Hill's equation with

$$q(t) = 2h^2 \cos 2t,$$

where

$$q(t+\pi) = q(t)$$

$$\int_0^\pi q(t)\,dt = 0$$

$$q(-t) = q(t).$$

Accordingly all results previously derived in this chapter are valid.

Corresponding to every zero of $\Delta(\lambda) - 2$ we must have a solution of period π, which is either even or odd, and similarly for every zero of $\Delta(\lambda) + 2$ there must be a solution of period 2π, which is even or odd. According to the four cases itemized in the last section we obtain respectively

1. for $y'(0) = y'(\pi/2)$

$$ce_{2n}(t) = \sum_{r=0}^{\infty} A_{2n,\,2r} \cos 2rt \tag{2}$$

2. for $y(0) = y(\pi/2) = 0$

$$se_{2n}(t) = \sum_{r=1}^{\infty} B_{2n,\,2r} \sin 2rt \tag{3}$$

3. for $y'(0) = y(\pi/2) = 0$

$$ce_{2n+1}(t) = \sum_{r=0}^{\infty} A_{2n+1,\,2r+1} \cos(2r+1)t \tag{4}$$

4. for $y(0) = y'(\pi/2) = 0$

$$se_{2n+1}(t) = \sum_{r=0}^{\infty} B_{2n+1,\,2r+1} \sin(2r+1)t. \tag{5}$$

Thus we see that with every zero of $\Delta^2(\lambda) - 4$ we can associate at least one of the above solutions of (1). We shall show shortly, that in each of these cases (1) has only one periodic solution. Any other independent solution is necessarily nonperiodic.

By virtue of the simple structure of q much more can be said about the Fourier coefficients in the above series. If we insert (2), for example in (1) use the fact that

$$\cos 2t \cos 2rt = \tfrac{1}{2} \cos 2(r+1)t + \tfrac{1}{2} \cos 2(r-1)t,$$

and rearrange terms we obtain

$$-\sum_{r=0}^{\infty} [4r^2 - \lambda] A_{2n,\,2r} \cos 2rt - h^2 \times \sum_{r=0}^{\infty} A_{2n,\,2r}[\cos(2r-2)t + \cos(2r+2)t] = 0$$

By setting the coefficient of every harmonic equal to zero we obtain the following system of recurrence formulas.

$$\begin{gathered} \lambda A_{2n,\,0} - h^2 A_{2n,\,2} = 0 \\ 2h^2 A_{2n,\,0} + [4 - \lambda] A_{2n,\,2} + h^2 A_{2n,\,4} = 0 \\ h^2 A_{2n,\,2r-2} + [4r^2 - \lambda] A_{2n,\,2r} + h^2 A_{2n,\,2r+2} = 0 \qquad r = 2, 3, \ldots. \end{gathered} \tag{6}$$

The above is an infinite system of linear, homogeneous equations. Its determinant is given by

$$\delta(\lambda) = \begin{vmatrix} \lambda & -h^2 & 0 & 0 & 0 & \cdots \\ \dfrac{2h^2}{4} & 1 - \dfrac{\lambda}{4} & \dfrac{h^2}{4} & 0 & 0 & \cdots \\ 0 & \dfrac{h^2}{16} & 1 - \dfrac{\lambda}{16} & \dfrac{h^2}{16} & 0 & \cdots \\ 0 & 0 & \dfrac{h^2}{36} & 1 - \dfrac{\lambda}{36} & \dfrac{h^2}{36} & \cdots \\ 0 & 0 & 0 & \dfrac{h^2}{64} & 1 - \dfrac{\lambda}{64} & \cdots \\ \vdots & \vdots & \vdots & \vdots & \vdots & \ddots \end{vmatrix}. \tag{7}$$

In order for (6) to have nonvanishing solutions we require that $\delta(\lambda) = 0$. $\delta(\lambda)$ is an entire function and its zeros coincide with those of $\Delta(\lambda) - 2$ for which (1) has even solutions. Similar considerations apply to the other solutions (3), (4) (5). Thus with $se_{2n}(t)$ we associate the recurrence formulas

$$\begin{gathered} (4 - \lambda) B_{2n,\,2} + h^2 B_{2n,\,4} = 0 \\ h^2 B_{2n,\,2r-2} + (4r^2 - \lambda) B_{2n,\,2r} + h^2 B_{2n,\,2r+2} = 0. \qquad r = 2, 3, \ldots \end{gathered} \tag{8}$$

According to the oscillation theorem of 9.3 all zeros of $\Delta^2(\lambda) - 4$ can be arranged in a sequence

$$\lambda_0 < \lambda_1' \leq \lambda_2' < \lambda_3 \leq \lambda_4 < \lambda_3' \leq \lambda_4' < \cdots. \tag{9}$$

In general one cannot tell when the non-strict inequalities in (9) reduce to equalities. When all inequalities in (9) are strict all zeros of $\Delta^2(\lambda) - 4$ are simple and for each of these Hill's equation has one periodic and one non-periodic solution. There is no general technique for deciding when all of the inequalities in (9) are strict. For Mathieu's equation, however, we have the following result.

THEOREM. For $h \neq 0$ in (1), no instability interval vanishes.

Proof. We assume that for some n $\lambda_{2n-1} = \lambda_{2n}$. In that case (2) has as solutions both $ce_{2n}(t)$ and $se_{2n}(t)$. Let

$$u_r = \begin{vmatrix} B_{2n,\,2r} & A_{2n,\,2r} \\ B_{2n,\,2r+2} & A_{2n,\,2r+2} \end{vmatrix}, \qquad r = 1, 2, \cdots.$$

Since the series (2) and (3) are uniformly convergent Fourier series all coefficients must tend to zero. Then

$$\lim_{r\to\infty} u_r = 0 \tag{10}$$

If we replace the second row in u_r by its equivalent from the recurrence formulas (6) and (8) we find

$$u_r = u_{r-1}.$$

By a repetition of this argument we have finally

$$u_r = u_1 = \begin{vmatrix} B_{2n,\,2} & A_{2n,\,2} \\ B_{2n,\,4} & A_{2n,\,4} \end{vmatrix} = \begin{vmatrix} B_{2n,\,2} & A_{2n,\,2} \\ 0 & -2A_{2n,\,0} \end{vmatrix} = -2A_{2n,\,0}\,B_{2n,\,2}.$$

A comparison with (10) shows that either $A_{2n,\,0}$ or $B_{2n,\,2} = 0$. But inspection of (6) and (8) shows then $ce_{2n}(t) \equiv 0$ or $se_{2n}(t) \equiv 0$ respectively. Hence two periodic solutions cannot coexist. For $h = 0$ the last conclusion breaks down, and as we know all zeros of $\Delta^2(\lambda) - 4$, with the exception of λ_0 are double zeros. ∎

8. Energy Bands in Crystals

An interesting application of the theory of Hill's equation can be found in studying the energy states in crystals from a quantum mechanical point of view. The one dimensional Schroedinger equation is given by

$$\frac{h^2}{8\pi^2 m}\frac{\partial^2\Psi}{\partial x^2} - V(x)\Psi = \frac{h}{2\pi i}\frac{\partial\Psi}{\partial t} \tag{1}$$

Where $V(x)$ is the potential field, h Planck's constant, and m the mass of an electron. In a crystal, which is a periodic structure, $V(x)$ is necessarily a periodic function. We denote the electron energy by E and seek solutions of (1) in the form

$$\Psi(x) = u(x)e^{-(2\pi i E/h)t} \tag{2}$$

where $u(x)$ then satisfies

$$u'' + \frac{8\pi^2 m}{h^2}[E - V(x)]u = 0. \tag{3}$$

Equation (3) is clearly a Hill's equation and E must be so selected that u is bounded for all x. We know that this can happen if and only if $\Delta(E)$, the discriminant is bounded between $+2$ and -2. The permissible energy states of (1) must therefore lie on the stability regions of equation (3). According to Floquet's theorem (3) will have solutions of the form

$$u = p(x)e^{2\pi ix/\lambda} \tag{4}$$

Where $p(x)$ is periodic and λ depends on E. To deduce the relationship between λ and E we see that, if R is the period of $V(x)$, then

$$u(x + R) = \rho u(x) = e^{2\pi iR/\lambda}u(x)$$

so that

$$\rho + \frac{1}{\rho} = 2 \cos \frac{2\pi R}{\lambda} = \Delta(E) \tag{5}$$

For λ to be real in (4), clearly, by (5), E has to be in a stability region. In that case, combining (2) and (4) we see that

$$\Psi = p(x)e^{2\pi i(x/\lambda - (E/h)t)}$$

so that Ψ behaves almost like a plane wave. In fact for $V(x)$ constant, $p(x)$ will be constant and Ψ will be a plane wave. λ is the wave length of that wave. We can use the asymptotic structure of $\Delta(E)$ to obtain an asymptotic relationship between E an λ. From (3) we can deduce as in previous sections, that $\Delta(E) \approx 2 \cos (2\pi R/h)\sqrt{2m\ E}$ and using (5) we note that

$$E \approx \frac{1}{2m}\left(\frac{h^2}{\lambda}\right). \tag{6}$$

(6) is only an asymptotic result, and holds only in the stability regions of $\Delta(E)$. In physics the instability regions are referred to as the forbidden energy bands, whereas the stability regions are the allowed energy bands.

Appendix

We shall devote this section to a proof of the theorem used in section 4.

THEOREM. Let H be a Hilbert space and $\{\phi_n\}$ a complete orthonormal set. Let $\{\Psi_n\}$ be an orthonormal set on H such that

$$\sum_{n=0}^{\infty} \|\phi_n - \Psi_n\|^2 < \infty.$$

Then $\{\Psi_n\}$ is also complete.

Proof. Let

$$(\Psi_n, \phi_k) = a_{kn} \qquad k \neq n$$
$$= a_{nn} + 1 \qquad k = n.$$

This defines an infinite matrix A whose i, j element is a_{ij}. Clearly since $\{\phi_n\}$ is complete and orthonormal we have

$$\Psi_n = \phi_n + \sum_{k=0}^{\infty} a_{kn} \phi_k .$$

Note that

$$\|\Psi_n - \phi_n\|^2 = \sum_{k=0}^{\infty} |a_{kn}|^2$$

so that

$$\sum_{n=0}^{\infty} \|\Psi_n - \phi_n\|^2 = \sum_{n=0}^{\infty} \sum_{k=0}^{\infty} |a_{kn}|^2 < \infty.$$

Since the Ψ_n are orthonormal we find

$$(\Psi_n, \Psi_m) = \left(\phi_n + \sum_{k=0}^{\infty} a_{kn} \phi_k , \phi_m + \sum_{k=0}^{\infty} a_{km} \phi_k\right)$$
$$= (\phi_n, \phi_m) + (\bar{a}_{nm} + a_{mn}) + \sum_{k=0}^{\infty} a_{kn} \bar{a}_{km} .$$

In view of the fact that both $\{\phi_n\}$ and $\{\Psi_m\}$ are orthonormal we find that

$$\bar{a}_{nm} + a_{mn} + \sum_{k=0}^{\infty} a_{kn} \bar{a}_{km} = 0. \tag{1}$$

The matrix A defines an operator on the whole space, since it is defined on $\{\phi_n\}$ which is a complete set. With A we associate an adjoint operator A^*, its adjoint. The adjoint is defined by

$$A\phi_n = \sum_{k=0}^{\infty} a_{kn} \phi_k .$$

$$A^*\phi_n = \sum_{k=0}^{\infty} \bar{a}_{nk} \phi_k .$$

The matrix associated with A^* is the hermitean transpose of the matrix A. Symbolically (1) is equivalent to

$$A + A^* + A^*A = 0.$$

We shall now show that

$$\phi_n = \Psi_n + A^*\Psi_n . \tag{2}$$

Assuming the above we see that we can express all $\{\phi_n\}$ in terms of $\{\Psi_n\}$. In that case the latter set must be complete, thus proving the theorem.

To verify (2) we write

$$\begin{aligned}\Psi_n + A^*\Psi_n &= \phi_n + \sum_{k=0}^{\infty} a_{kn}\phi_k + A^*\phi_n + \sum_{l=0}^{\infty} a_{ln} A^*\phi_l \\ &= \phi_n + \sum_{k=0}^{\infty} a_{kn}\phi_k + \sum_{k=0}^{\infty} \bar{a}_{nk}\phi_k + \sum_{l=0}^{\infty} a_{ln} \sum_{k=0}^{\infty} \bar{a}_{lk}\phi_k \\ &= \phi_n + \sum_{k=0}^{\infty}\left(a_{kn} + \bar{a}_{nk} + \sum_{l=0}^{\infty} a_{ln}\bar{a}_{lk}\right)\phi_k = \phi_n\end{aligned}$$

by use of (1). ■

Exercises

1. Show that the equation

$$x^2y'' + xy' + \left(a + bx + \frac{c}{x}\right)y = 0$$

in terms of the new independent variable $t = (1/2i)\log\sqrt{c/b}\,x$ becomes

$$y'' - 4[a + 2\sqrt{bc}\cos 2t]\,y = 0.$$

2. Show that if $q(-t) = q(t)$ in Hill's equation then

$$\begin{aligned} y_1(\pi) + 1 &= 2y_1\left(\frac{\pi}{2}\right)y_2'\left(\frac{\pi}{2}\right) \\ y_1(\pi) - 1 &= 2y_1'\left(\frac{\pi}{2}\right)y_2\left(\frac{\pi}{2}\right) \\ y_1'(\pi) &= 2y_1\left(\frac{\pi}{2}\right)y_1'\left(\frac{\pi}{2}\right) \\ y_2'(\pi) &= 2y_2\left(\frac{\pi}{2}\right)y_2'\left(\frac{\pi}{2}\right). \end{aligned}$$

3. Show that if y satisfies Hill's equation then

$$y(t+\pi) + y(t-\pi) = \Delta(\lambda)y(t).$$

4. Supply detailed proofs for all four parts of the theorem in 9.6.

5. Consider the equation

$$u'' + [\lambda - q(t+\tau)]u = 0.$$

Let $\{\mu_n(\tau)\}$ denote its eigenvalues when subjected to the boundary conditions

$$u(0) = u(\pi) = 0.$$

Similarly let $\{\nu_n(\tau)]$ denote the eigenvalues corresponding to

$$u'(0) = u'(\pi) = 0.$$

Show that

$$\lambda_1' \leq \mu_1(\tau) \leq \lambda_2', \quad \lambda_1 \leq \mu_2(\tau) \leq \lambda_2, \ldots$$

and

$$\nu_0(\tau) \leq \lambda_0, \; \lambda_1' \leq \nu_1(\tau) \leq \lambda_2', \quad \lambda_1 \leq \nu_2(\tau) \leq \lambda_2, \ldots.$$

6. Consider the equation

$$y'' + \lambda p(t) y = 0$$

where

$$\begin{aligned} p(t) &= 1 \qquad 0 \leq t < 1 \\ &= a^2 \qquad 1 \leq t \leq L \end{aligned}$$

and is continued outside $(0, L)$ as an even function of period $2L$.

Show that $\Delta^2(\lambda) - 4$ has double zeros, if and only if, $a(L - 1)$ is a rational number.

Bibliography

The references immediately following pertain to some or all the chapters of this book, and also relate to the many applications arising in mathematical physics.

Bateman, H., *Partial Differential Equations of Mathematical Physics*, Cambridge University Press, London, 1959.

Bateman, H., *The Mathematical Analysis of Electrical and Optical Wave Motion on the Basis of Maxwell's Equations*, Dover Publications, Inc., New York, 1955.

Bateman, H., manuscript project, *see* Erdelyi.

Courant, R., and D. Hilbert, *Methods of Mathematical Physics*, Interscience Publishers, New York, volume 1, 1953, volume 2, 1962.

Erdelyi, A.,W. Magnus, *F. Oberhettinger*, and F. G. Tricomi, *Higher Transcendental Functions*, 3 volumes, McGraw-Hill Book Co., New York, 1953.

Frank, P., and R. Von Mises, *Die Differential und Integralgleichungen der Mechanik und Physik*, 2 volumes, 2nd enlarged edition, Dover Publications, Inc., New York, 1961.

Hochstadt, H., *Special Functions of Mathematical Physics*, Holt Rinehart and Winston, Inc., New York 1961.

Jahnke, E., and F. Emde, *Tables of Higher Functions*, 6th edition, revised by F. Lösch, McGraw-Hill Book Co., New York 1960.

Jeffreys, H., and B. S. Jeffreys, *Methods of Mathematical Physics*, 3rd. edition, Cambridge University Press, London, 1956.

Kratzer, A., and W. Franz, *Transzendente Funktionen*, Akademische Verlags-gesellschaft, Leipzig, 1960.

Lebedev, N. N., *Special Functions and their Applications*, translated from the Russian by R. A. Silverman, Dover Publications, Inc., New York, 1972.

Lense, J., *Reihenentwicklungen in der Mathematischen Physik*, 3rd edition, Walter de Gruyter & Co., Berlin, 1953.

Macrobert, T. M., *Functions of a Complex Variable* 3rd. edition, Macmillan, London, 1947.

Magnus, W., and F. Oberhettinger, *Formulas and Theorems for the Functions of Mathematical Physics*, translated from German by J. Wermer, Chelsea Publishing Co., New York, 1954, 3rd edition, Springer Verlag, Berlin 1966.

Miller, W. J., Jr., *Lie Theory and Special Functions*, Academic Press, New York, 1968.

Morse, P. M., and H. Feshbach, *Methods of Theoretical Physics*, 2 volumes, McGraw Hill Book Co., New York, 1953.

Rainville, E. D., *Special Functions*, The Macmillan Co., New York, 1960.

Schäfke, F. W., *Einfuehrung in die Theorie der Speziellen Funktionen der Mathematischen Physik*, Springer Verlag, Berlin, 1963.

Sneddon, I. N., *Special Functions of Mathematical Physics and Chemistry*, 2nd edition, Oliver and Boyd, London, 1961.

Truesdell, C., *An Essay toward a Unified Theory of special Functions Based upon the Functional Equation* $\partial/\partial z F(z, a) = F(z, a+1)$, Princeton University Press, Princeton, N. J., 1948.

Vilenkin, I. A., *Special Functions and the Theory of Group Representation*, A.M.S., Providence, 1968.

Whittaker, E. T., and G. N. Watson, *A Course of Modern Analysis*, 4th edition, Cambridge University Press, London, 1963.

Chapters 1 and 2

Jackson, D., *Fourier Series and Orthogonal Polynomials*, Carus Mathematical Association of America, State University of New York, Buffalo, New York, 1941.

Kaczmarz, S., and H. Steinhaus, *Theorie der Orthogonalreihen*, Chelsea Publishing Co., New York, 1951.

Sansone, G., *Orthogonal Functions*, revised 2nd edition, translated from the Italian by A. H. Diamond, Interscience Publications, Inc., New York 1959.,

Szego, G., *Orthogonal Polynomials*, revised edition, American Mathematical Society, Providence, R. I., 1959.

Tricomi, F. G., *Vorlesungen über Orthogonalreihen*, Springer Verlag, Berlin, 1955.

Chapter 3

Artin, E., *The Gamma Function*, translated from the German by M. Butler, Holt, Rinehart and Winston, New York, 1964.

Belardinelli, M. G., *Résolution Analytique des Equations Algebraiques Generales*, Mémorial des Sciences Mathematiques, Fascicule 145, Gauthiers-Villars, Paris 1960.

Nielson, M., *Handbuch der Theorie der Gammafunktion*, B. G. Teubner, Leipzig, 1906.

Chapter 4

Bailey, W. N., *Generalized Hypergeometric Series*, Cambridge Tracts in Mathematics and Mathematical Physics, no. 32, Cambridge University Press, London, 1935.

Hochstadt, H., *Differential Equations–A Modern Approach*, Holt, Rinehart and Winston, New York, 1964 (rpt. Dover Publications, Inc., New York, 1975).

Ince, E. L., *Ordinary Differential Equations*, Dover Publications, Inc. New York, 1956.

Kampe de Feriet, J., *La Fonction Hypergéométrique*, Memorial des Sciences Mathématiques, Fascicule 85, Gauthier-Villars, Paris, 1937.

Klein, F., *Vorlesungen uber die Hypergeometrische Funktion*, Springer Verlag, Berlin, 1933.

Nehari, Z., *Conformal Mapping*, McGraw-Hill Book Co., New York, 1952 (rpt. Dover Publications, Inc., New York, 1975).

Forsyth, A. R., *Theory of Functions of a Complex Variable*, 2 Volumes, Dover Publications, Inc., New York, 1965.

Snow, C., *The Hypergeometric and Legendre Functions with Applications to Integral Equations of Potential Theory*, National Bureau of Standards Applied Mathematics Series, no. 19, U.S. Government Printing Office, Washington, D. C., 1956.

Zassenhaus, H., *The Theory of groups*, Chelsea Publishing Co. New York, 1956.

Chapter 5

Heine, E., *Handbuch der Kugelfunktionen*, Physica-Verlag, Würzburg, 1961.

Hobson, E. W., *The Theory of Spherical and Ellipsoidal Harmonics*, Chelsea Publishing Co., New York, 1955.

Lense, J., *Kugelfunktionen*, Akademische Verlags-gesellschaft, Leipzig, 1954.

Macrobert, T. M., *Spherical Harmonics*, Methuen, London, 1928.

Chapter 7

Buchholz, H., *Die Konfluente Hypergeometrische Funktion, mit besonderer Berücksichtigung ihrer Anwendungen*, Springer Verlag, Berlin, 1953.

Tricomi, F. G., *Fonctions Hypergeometriques Confluentes*, Memorial des Sciences Mathematiques, Fas. 140, Gauthiers-Villars, Paris, 1960.

Chapter 8

Gray, A., G. B. Mathews, and T. M. MacRobert, *Bessel Functions*, Macmillan London, 1922.

McLachlan, N. W., *Bessel Functions for Engineers*, Oxford, 1946.

Watson, G. N., *A Treatise on Bessel Functions*, 2nd edition, Cambridge, 1948.

Weyrich, R., *Die Zylinderfunktionen und ihre Anwendungen*, Teubner, 1937.

Chapter 9

Arscott, F. M., *Periodic Differential Equations*, Pergamon, 1964.

McLachlan, N. W., *Theory and Application of Mathieu Functions*, Clarendon Press, Oxford, 1947.

Magnus, W., and S. Winkler, *Hill's Equation*, Interscience Publishers, New York, 1966.

Meixner, J., and F. W. Schäfke, *Mathieusche Funktionen und Sphäroidfunktionen*, Springer Verlag, Berlin, 1954.

Strutt, M. J. O., *Lamésche, Mathieusche und Verwandte Funktionen in Physik und Technik. Ergebnisse der Mathematik und ihrer Grenzgebiete*, volume I., no. 3, Berlin, 1933.

Index

A CATALOG OF SELECTED DOVER BOOKS IN ALL FIELDS OF INTEREST

SMOCKING: TECHNIQUE, PROJECTS, AND DESIGNS, Dianne Durand. Foremost smocking designer provides complete instructions on how to smock. Over 10 projects, over 100 illustrations. 56pp. 8¼ × 11. 23788-5 Pa. $2.00

AUDUBON'S BIRDS IN COLOR FOR DECOUPAGE, edited by Eleanor H. Rawlings. 24 sheets, 37 most decorative birds, full color, on one side of paper. Instructions, including work under glass. 56pp. 8¼ × 11. 23492-4 Pa. $3.95

THE COMPLETE BOOK OF SILK SCREEN PRINTING PRODUCTION, J.I. Biegeleisen. For commercial user, teacher in advanced classes, serious hobbyist. Most modern techniques, materials, equipment for optimal results. 124 illustrations. 253pp. 5⅝ × 8½. 21100-2 Pa. $4.50

A TREASURY OF ART NOUVEAU DESIGN AND ORNAMENT, edited by Carol Belanger Grafton. 577 designs for the practicing artist. Full-page, spots, borders, bookplates by Klimt, Bradley, others. 144pp. 8⅜ × 11¼. 24001-0 Pa. $5.95

ART NOUVEAU TYPOGRAPHIC ORNAMENTS, Dan X. Solo. Over 800 Art Nouveau florals, swirls, women, animals, borders, scrolls, wreaths, spots and dingbats, copyright-free. 100pp. 8⅛ × 11. 24366-4 Pa. $4.00

HAND SHADOWS TO BE THROWN UPON THE WALL, Henry Bursill. Wonderful Victorian novelty tells how to make flying birds, dog, goose, deer, and 14 others, each explained by a full-page illustration. 32pp. 6½ × 9¼. 21779-5 Pa. $1.50

AUDUBON'S BIRDS OF AMERICA COLORING BOOK, John James Audubon. Rendered for coloring by Paul Kennedy. 46 of Audubon's noted illustrations: red-wingèd black-bird, cardinal, etc. Original plates reproduced in full-color on the covers. Captions. 48pp. 8¼ × 11. 23049-X Pa. $2.25

SILK SCREEN TECHNIQUES, J.I. Biegeleisen, M.A. Cohn. Clear, practical, modern, economical. Minimal equipment (self-built), materials, easy methods. For amateur, hobbyist, 1st book. 141 illustrations. 185pp. 6⅛ × 9¼. 20433-2 Pa. $3.95

101 PATCHWORK PATTERNS, Ruby S. McKim. 101 beautiful, immediately useable patterns, full-size, modern and traditional. Also general information, estimating, quilt lore. 140 illustrations. 124pp. 7⅞ × 10¾. 20773-0 Pa. $3.50

READY-TO-USE FLORAL DESIGNS, Ed Sibbett, Jr. Over 100 floral designs (most in three sizes) of popular individual blossoms as well as bouquets, sprays, garlands. 64pp. 8¼ × 11. 23976-4 Pa. $2.95

AMERICAN WILD FLOWERS COLORING BOOK, Paul Kennedy. Planned coverage of 46 most important wildflowers, from Rickett's collection; instructive as well as entertaining. Color versions on covers. Captions. 48pp. 8¼ × 11. 20095-7 Pa. $2.50

CARVING DUCK DECOYS, Harry V. Shourds and Anthony Hillman. Detailed instructions and full-size templates for constructing 16 beautiful, marvelously practical decoys according to time-honored South Jersey method. 70pp. 9¼ × 12¼. 24083-5 Pa. $4.95

TRADITIONAL PATCHWORK PATTERNS, Carol Belanger Grafton. Cardboard cut-out pieces for use as templates to make 12 quilts: Buttercup, Ribbon Border, Tree of Paradise, nine more. Full instructions. 57pp. 8¼ × 11. 23015-5 Pa. $3.50

JAPANESE DESIGN MOTIFS, Matsuya Co. Mon, or heraldic designs. Over 4000 typical, beautiful designs: birds, animals, flowers, swords, fans, geometrics; all beautifully stylized. 213pp. 11⅜ × 8¼. 22874-6 Pa. $7.95

THE TALE OF BENJAMIN BUNNY, Beatrix Potter. Peter Rabbit's cousin coaxes him back into Mr. McGregor's garden for a whole new set of adventures. All 27 full-color illustrations. 59pp. 4¼ × 5½. (Available in U.S. only) 21102-9 Pa. $1.75

THE TALE OF PETER RABBIT AND OTHER FAVORITE STORIES BOXED SET, Beatrix Potter. Seven of Beatrix Potter's best-loved tales including Peter Rabbit in a specially designed, durable boxed set. 4¼ × 5½. Total of 447pp. 158 color illustrations. (Available in U.S. only) 23903-9 Pa. $10.80

PRACTICAL MENTAL MAGIC, Theodore Annemann. Nearly 200 astonishing feats of mental magic revealed in step-by-step detail. Complete advice on staging, patter, etc. Illustrated. 320pp. 5⅜ × 8½. 24426-1 Pa. $5.95

CELEBRATED CASES OF JUDGE DEE (DEE GOONG AN), translated by Robert Van Gulik. Authentic 18th-century Chinese detective novel; Dee and associates solve three interlocked cases. Led to van Gulik's own stories with same characters. Extensive introduction. 9 illustrations. 237pp. 5⅜ × 8½. 23337-5 Pa. $4.50

CUT & FOLD EXTRATERRESTRIAL INVADERS THAT FLY, M. Grater. Stage your own lilliputian space battles. By following the step-by-step instructions and explanatory diagrams you can launch 22 full-color fliers into space. 36pp. 8¼ × 11. 24478-4 Pa. $2.95

CUT & ASSEMBLE VICTORIAN HOUSES, Edmund V. Gillon, Jr. Printed in full color on heavy cardboard stock, 4 authentic Victorian houses in H-O scale: Italian-style Villa, Octagon, Second Empire, Stick Style. 48pp. 9¼ × 12¼. 23849-0 Pa. $3.95

BEST SCIENCE FICTION STORIES OF H.G. WELLS, H.G. Wells. Full novel *The Invisible Man*, plus 17 short stories: "The Crystal Egg," "Aepyornis Island," "The Strange Orchid," etc. 303pp. 5⅜ × 8½. (Available in U.S. only) 21531-8 Pa. $4.95

TRADEMARK DESIGNS OF THE WORLD, Yusaku Kamekura. A lavish collection of nearly 700 trademarks, the work of Wright, Loewy, Klee, Binder, hundreds of others. 160pp. 8¾ × 8. (Available in U.S. only) 24191-2 Pa. $5.95

THE ARTIST'S AND CRAFTSMAN'S GUIDE TO REDUCING, ENLARGING AND TRANSFERRING DESIGNS, Rita Weiss. Discover, reduce, enlarge, transfer designs from any objects to any craft project. 12pp. plus 16 sheets special graph paper. 8¼ × 11. 24142-4 Pa. $3.50

TREASURY OF JAPANESE DESIGNS AND MOTIFS FOR ARTISTS AND CRAFTSMEN, edited by Carol Belanger Grafton. Indispensable collection of 360 traditional Japanese designs and motifs redrawn in clean, crisp black-and-white, copyright-free illustrations. 96pp. 8¼ × 11. 24435-0 Pa. $3.95

KEYBOARD WORKS FOR SOLO INSTRUMENTS, G.F. Handel. 35 neglected works from Handel's vast oeuvre, originally jotted down as improvisations. Includes Eight Great Suites, others. New sequence. 174pp. 9⅜ × 12¼.
24338-9 Pa. $7.50

AMERICAN LEAGUE BASEBALL CARD CLASSICS, Bert Randolph Sugar. 82 stars from 1900s to 60s on facsimile cards. Ruth, Cobb, Mantle, Williams, plus advertising, info, no duplications. Perforated, detachable. 16pp. 8¼ × 11.
24286-2 Pa. $2.95

A TREASURY OF CHARTED DESIGNS FOR NEEDLEWORKERS, Georgia Gorham and Jeanne Warth. 141 charted designs: owl, cat with yarn, tulips, piano, spinning wheel, covered bridge, Victorian house and many others. 48pp. 8¼ × 11.
23558-0 Pa. $1.95

DANISH FLORAL CHARTED DESIGNS, Gerda Bengtsson. Exquisite collection of over 40 different florals: anemone, Iceland poppy, wild fruit, pansies, many others. 45 illustrations. 48pp. 8¼ × 11. 23957-8 Pa. $1.75

OLD PHILADELPHIA IN EARLY PHOTOGRAPHS 1839-1914, Robert F. Looney. 215 photographs: panoramas, street scenes, landmarks, President-elect Lincoln's visit, 1876 Centennial Exposition, much more. 230pp. 8⅜ × 11¾.
23345-6 Pa. $9.95

PRELUDE TO MATHEMATICS, W.W. Sawyer. Noted mathematician's lively, stimulating account of non-Euclidean geometry, matrices, determinants, group theory, other topics. Emphasis on novel, striking aspects. 224pp. 5⅜ × 8½.
24401-6 Pa. $4.50

ADVENTURES WITH A MICROSCOPE, Richard Headstrom. 59 adventures with clothing fibers, protozoa, ferns and lichens, roots and leaves, much more. 142 illustrations. 232pp. 5⅜ × 8½. 23471-1 Pa. $3.95

IDENTIFYING ANIMAL TRACKS: MAMMALS, BIRDS, AND OTHER ANIMALS OF THE EASTERN UNITED STATES, Richard Headstrom. For hunters, naturalists, scouts, nature-lovers. Diagrams of tracks, tips on identification. 128pp. 5⅝ × 8. 24442-3 Pa. $3.50

VICTORIAN FASHIONS AND COSTUMES FROM HARPER'S BAZAR, 1867-1898, edited by Stella Blum. Day costumes, evening wear, sports clothes, shoes, hats, other accessories in over 1,000 detailed engravings. 320pp. 9⅜ × 12¼.
22990-4 Pa. $10.95

EVERYDAY FASHIONS OF THE TWENTIES AS PICTURED IN SEARS AND OTHER CATALOGS, edited by Stella Blum. Actual dress of the Roaring Twenties, with text by Stella Blum. Over 750 illustrations, captions. 156pp. 9 × 12.
24134-3 Pa. $8.50

HALL OF FAME BASEBALL CARDS, edited by Bert Randolph Sugar. Cy Young, Ted Williams, Lou Gehrig, and many other Hall of Fame greats on 92 full-color, detachable reprints of early baseball cards. No duplication of cards with *Classic Baseball Cards*. 16pp. 8¼ × 11. 23624-2 Pa. $3.50

THE ART OF HAND LETTERING, Helm Wotzkow. Course in hand lettering, Roman, Gothic, Italic, Block, Script. Tools, proportions, optical aspects, individual variation. Very quality conscious. Hundreds of specimens. 320pp. 5⅜ × 8½.
21797-3 Pa. $4.95

HOW THE OTHER HALF LIVES, Jacob A. Riis. Journalistic record of filth, degradation, upward drive in New York immigrant slums, shops, around 1900. New edition includes 100 original Riis photos, monuments of early photography. 233pp. 10 × 7⅞. 22012-5 Pa. $7.95

CHINA AND ITS PEOPLE IN EARLY PHOTOGRAPHS, John Thomson. In 200 black-and-white photographs of exceptional quality photographic pioneer Thomson captures the mountains, dwellings, monuments and people of 19th-century China. 272pp. 9⅜ × 12¼. 24393-1 Pa. $12.95

GODEY COSTUME PLATES IN COLOR FOR DECOUPAGE AND FRAMING, edited by Eleanor Hasbrouk Rawlings. 24 full-color engravings depicting 19th-century Parisian haute couture. Printed on one side only. 56pp. 8¼ × 11. 23879-2 Pa. $3.95

ART NOUVEAU STAINED GLASS PATTERN BOOK, Ed Sibbett, Jr. 104 projects using well-known themes of Art Nouveau: swirling forms, florals, peacocks, and sensuous women. 60pp. 8¼ × 11. 23577-7 Pa. $3.50

QUICK AND EASY PATCHWORK ON THE SEWING MACHINE: Susan Aylsworth Murwin and Suzzy Payne. Instructions, diagrams show exactly how to machine sew 12 quilts. 48pp. of templates. 50 figures. 80pp. 8¼ × 11. 23770-2 Pa. $3.50

THE STANDARD BOOK OF QUILT MAKING AND COLLECTING, Marguerite Ickis. Full information, full-sized patterns for making 46 traditional quilts, also 150 other patterns. 483 illustrations. 273pp. 6⅞ × 9⅝. 20582-7 Pa. $5.95

LETTERING AND ALPHABETS, J. Albert Cavanagh. 85 complete alphabets lettered in various styles; instructions for spacing, roughs, brushwork. 121pp. 8¾ × 8. 20053-1 Pa. $3.95

LETTER FORMS: 110 COMPLETE ALPHABETS, Frederick Lambert. 110 sets of capital letters; 16 lower case alphabets; 70 sets of numbers and other symbols. 110pp. 8⅛ × 11. 22872-X Pa. $4.50

ORCHIDS AS HOUSE PLANTS, Rebecca Tyson Northen. Grow cattleyas and many other kinds of orchids—in a window, in a case, or under artificial light. 63 illustrations. 148pp. 5⅜ × 8½. 23261-1 Pa. $2.95

THE MUSHROOM HANDBOOK, Louis C.C. Krieger. Still the best popular handbook. Full descriptions of 259 species, extremely thorough text, poisons, folklore, etc. 32 color plates; 126 other illustrations. 560pp. 5⅜ × 8½. 21861-9 Pa. $8.50

THE DORÉ BIBLE ILLUSTRATIONS, Gustave Doré. All wonderful, detailed plates: Adam and Eve, Flood, Babylon, life of Jesus, etc. Brief King James text with each plate. 241 plates. 241pp. 9 × 12. 23004-X Pa. $8.95

THE BOOK OF KELLS: Selected Plates in Full Color, edited by Blanche Cirker. 32 full-page plates from greatest manuscript-icon of early Middle Ages. Fantastic, mysterious. Publisher's Note. Captions. 32pp. 9¾ × 12¼. 24345-1 Pa. $4.50

THE PERFECT WAGNERITE, George Bernard Shaw. Brilliant criticism of the Ring Cycle, with provocative interpretation of politics, economic theories behind the Ring. 136pp. 5⅜ × 8½. (Available in U.S. only) 21707-8 Pa. $3.00

THE RIME OF THE ANCIENT MARINER, Gustave Doré, S.T. Coleridge. Doré's finest work, 34 plates capture moods, subtleties of poem. Full text. 77pp. 9¼ × 12. 22305-1 Pa. $4.95

SONGS OF INNOCENCE, William Blake. The first and most popular of Blake's famous "Illuminated Books," in a facsimile edition reproducing all 31 brightly colored plates. Additional printed text of each poem. 64pp. 5¼ × 7. 22764-2 Pa. $3.50

AN INTRODUCTION TO INFORMATION THEORY, J.R. Pierce. Second (1980) edition of most impressive non-technical account available. Encoding, entropy, noisy channel, related areas, etc. 320pp. 5⅜ × 8½. 24061-4 Pa. $4.95

THE DIVINE PROPORTION: A STUDY IN MATHEMATICAL BEAUTY, H.E. Huntley. "Divine proportion" or "golden ratio" in poetry, Pascal's triangle, philosophy, psychology, music, mathematical figures, etc. Excellent bridge between science and art. 58 figures. 185pp. 5⅜ × 8½. 22254-3 Pa. $3.95

THE DOVER NEW YORK WALKING GUIDE: From the Battery to Wall Street, Mary J. Shapiro. Superb inexpensive guide to historic buildings and locales in lower Manhattan: Trinity Church, Bowling Green, more. Complete Text; maps. 36 illustrations. 48pp. 3⅞ × 9¼. 24225-0 Pa. $2.50

NEW YORK THEN AND NOW, Edward B. Watson, Edmund V. Gillon, Jr. 83 important Manhattan sites: on facing pages early photographs (1875-1925) and 1976 photos by Gillon. 172 illustrations. 171pp. 9¼ × 10. 23361-8 Pa. $7.95

HISTORIC COSTUME IN PICTURES, Braun & Schneider. Over 1450 costumed figures from dawn of civilization to end of 19th century. English captions. 125 plates. 256pp. 8⅜ × 11¼. 23150-X Pa. $7.50

VICTORIAN AND EDWARDIAN FASHION: A Photographic Survey, Alison Gernsheim. First fashion history completely illustrated by contemporary photographs. Full text plus 235 photos, 1840-1914, in which many celebrities appear. 240pp. 6½ × 9¼. 24205-6 Pa. $6.00

CHARTED CHRISTMAS DESIGNS FOR COUNTED CROSS-STITCH AND OTHER NEEDLECRAFTS, Lindberg Press. Charted designs for 45 beautiful needlecraft projects with many yuletide and wintertime motifs. 48pp. 8¼ × 11. 24356-7 Pa. $2.50

101 FOLK DESIGNS FOR COUNTED CROSS-STITCH AND OTHER NEEDLECRAFTS, Carter Houck. 101 authentic charted folk designs in a wide array of lovely representations with many suggestions for effective use. 48pp. 8¼ × 11. 24369-9 Pa. $2.25

FIVE ACRES AND INDEPENDENCE, Maurice G. Kains. Great back-to-the-land classic explains basics of self-sufficient farming. The one book to get. 95 illustrations. 397pp. 5⅜ × 8½. 20974-1 Pa. $4.95

A MODERN HERBAL, Margaret Grieve. Much the fullest, most exact, most useful compilation of herbal material. Gigantic alphabetical encyclopedia, from aconite to zedoary, gives botanical information, medical properties, folklore, economic uses, and much else. Indispensable to serious reader. 161 illustrations. 888pp. 6½ × 9¼. (Available in U.S. only) 22798-7, 22799-5 Pa., Two-vol. set $16.45

HOW THE OTHER HALF LIVES, Jacob A. Riis. Journalistic record of filth, degradation, upward drive in New York immigrant slums, shops, around 1900. New edition includes 100 original Riis photos, monuments of early photography. 233pp. 10 × 7⅞. 22012-5 Pa. $7.95

CHINA AND ITS PEOPLE IN EARLY PHOTOGRAPHS, John Thomson. In 200 black-and-white photographs of exceptional quality photographic pioneer Thomson captures the mountains, dwellings, monuments and people of 19th-century China. 272pp. 9⅜ × 12¼. 24393-1 Pa. $12.95

GODEY COSTUME PLATES IN COLOR FOR DECOUPAGE AND FRAMING, edited by Eleanor Hasbrouk Rawlings. 24 full-color engravings depicting 19th-century Parisian haute couture. Printed on one side only. 56pp. 8¼ × 11. 23879-2 Pa. $3.95

ART NOUVEAU STAINED GLASS PATTERN BOOK, Ed Sibbett, Jr. 104 projects using well-known themes of Art Nouveau: swirling forms, florals, peacocks, and sensuous women. 60pp. 8¼ × 11. 23577-7 Pa. $3.50

QUICK AND EASY PATCHWORK ON THE SEWING MACHINE: Susan Aylsworth Murwin and Suzzy Payne. Instructions, diagrams show exactly how to machine sew 12 quilts. 48pp. of templates. 50 figures. 80pp. 8¼ × 11. 23770-2 Pa. $3.50

THE STANDARD BOOK OF QUILT MAKING AND COLLECTING, Marguerite Ickis. Full information, full-sized patterns for making 46 traditional quilts, also 150 other patterns. 483 illustrations. 273pp. 6⅞ × 9⅝. 20582-7 Pa. $5.95

LETTERING AND ALPHABETS, J. Albert Cavanagh. 85 complete alphabets lettered in various styles; instructions for spacing, roughs, brushwork. 121pp. 8¾ × 8. 20053-1 Pa. $3.95

LETTER FORMS: 110 COMPLETE ALPHABETS, Frederick Lambert. 110 sets of capital letters; 16 lower case alphabets; 70 sets of numbers and other symbols. 110pp. 8⅛ × 11. 22872-X Pa. $4.50

ORCHIDS AS HOUSE PLANTS, Rebecca Tyson Northen. Grow cattleyas and many other kinds of orchids—in a window, in a case, or under artificial light. 63 illustrations. 148pp. 5⅜ × 8½. 23261-1 Pa. $2.95

THE MUSHROOM HANDBOOK, Louis C.C. Krieger. Still the best popular handbook. Full descriptions of 259 species, extremely thorough text, poisons, folklore, etc. 32 color plates; 126 other illustrations. 560pp. 5⅜ × 8½. 21861-9 Pa. $8.50

THE DORÉ BIBLE ILLUSTRATIONS, Gustave Doré. All wonderful, detailed plates: Adam and Eve, Flood, Babylon, life of Jesus, etc. Brief King James text with each plate. 241 plates. 241pp. 9 × 12. 23004-X Pa. $8.95

THE BOOK OF KELLS: Selected Plates in Full Color, edited by Blanche Cirker. 32 full-page plates from greatest manuscript-icon of early Middle Ages. Fantastic, mysterious. Publisher's Note. Captions. 32pp. 9¾ × 12¼. 24345-1 Pa. $4.50

THE PERFECT WAGNERITE, George Bernard Shaw. Brilliant criticism of the Ring Cycle, with provocative interpretation of politics, economic theories behind the Ring. 136pp. 5⅜ × 8½. (Available in U.S. only) 21707-8 Pa. $3.00

CATALOG OF DOVER BOOKS

SOURCE BOOK OF MEDICAL HISTORY, edited by Logan Clendening, M.D. Original accounts ranging from Ancient Egypt and Greece to discovery of X-rays: Galen, Pasteur, Lavoisier, Harvey, Parkinson, others. 685pp. 5⅜ × 8½.
20621-1 Pa. $10.95

THE ROSE AND THE KEY, J.S. Lefanu. Superb mystery novel from Irish master. Dark doings among an ancient and aristocratic English family. Well-drawn characters; capital suspense. Introduction by N. Donaldson. 448pp. 5⅜ × 8½.
24377-X Pa. $6.95

SOUTH WIND, Norman Douglas. Witty, elegant novel of ideas set on languorous Meditterranean island of Nepenthe. Elegant prose, glittering epigrams, mordant satire. 1917 masterpiece. 416pp. 5⅜ × 8½. (Available in U.S. only)
24361-3 Pa. $5.95

RUSSELL'S CIVIL WAR PHOTOGRAPHS, Capt. A.J. Russell. 116 rare Civil War Photos: Bull Run, Virginia campaigns, bridges, railroads, Richmond, Lincoln's funeral car. Many never seen before. Captions. 128pp. 9⅜ × 12¼.
24283-8 Pa. $6.95

PHOTOGRAPHS BY MAN RAY: 105 Works, 1920-1934. Nudes, still lifes, landscapes, women's faces, celebrity portraits (Dali, Matisse, Picasso, others), rayographs. Reprinted from rare gravure edition. 128pp. 9⅜ × 12¼. (Available in U.S. only) 23842-3 Pa. $7.95

STAR NAMES: THEIR LORE AND MEANING, Richard H. Allen. Star names, the zodiac, constellations: folklore and literature associated with heavens. The basic book of its field, fascinating reading. 563pp. 5⅜ × 8½. 21079-0 Pa. $7.95

BURNHAM'S CELESTIAL HANDBOOK, Robert Burnham, Jr. Thorough guide to the stars beyond our solar system. Exhaustive treatment. Alphabetical by constellation: Andromeda to Cetus in Vol. 1; Chamaeleon to Orion in Vol. 2; and Pavo to Vulpecula in Vol. 3. Hundreds of illustrations. Index in Vol. 3. 2000pp. 6⅛ × 9¼. 23567-X, 23568-8, 23673-0 Pa. Three-vol. set $36.85

THE ART NOUVEAU STYLE BOOK OF ALPHONSE MUCHA, Alphonse Mucha. All 72 plates from *Documents Decoratifs* in original color. Stunning, essential work of Art Nouveau. 80pp. 9⅜ × 12¼. 24044-4 Pa. $7.95

DESIGNS BY ERTE; FASHION DRAWINGS AND ILLUSTRATIONS FROM "HARPER'S BAZAR," Erte. 310 fabulous line drawings and 14 *Harper's Bazar* covers, 8 in full color. Erte's exotic temptresses with tassels, fur muffs, long trains, coifs, more. 129pp. 9⅜ × 12¼. 23397-9 Pa. $6.95

HISTORY OF STRENGTH OF MATERIALS, Stephen P. Timoshenko. Excellent historical survey of the strength of materials with many references to the theories of elasticity and structure. 245 figures. 452pp. 5⅜ × 8½. 61187-6 Pa. $8.95

Prices subject to change without notice.

Available at your book dealer or write for free catalog to Dept. GI, Dover Publications, Inc., 31 East 2nd St. Mineola, N.Y. 11501. Dover publishes more than 175 books each year on science, elementary and advanced mathematics, biology, music, art, literary history, social sciences and other areas.

T
3